The Wadden Sea Ecosystem

Springer
Berlin
Heidelberg
New York
Barcelona
Hong Kong
London
Milan
Paris
Singapore
Tokyo

Sabine Dittmann (Ed.)

The Wadden Sea Ecosystem

Stability Properties and Mechanisms

With 68 Figures and 23 Tables

Springer

Editor:

Dr. Sabine Dittmann
Forschungszentrum Terramare
Schleusenstraße 1
26382 Wilhelmshaven
Germany

ISBN-13: 978-3-642-64256-2 e-ISBN-13: 978-3-642-60097-5
DOI: 10.1007/978-3-642-60097-5

Library of Congress Cataloging-in-Publication Data
The Wadden Sea ecosystem: stability properties and mechanisms / Sabine Dittmann (ed.). p. cm. Includes bibliographical references. ISBN-13: 978-3-642-64256-2 1. Tidal flat ecology – Netherlands – Waddenzee. I. Dittmann, Sabine.
QH159.W24 1999 577.69'9–dc21 99-34031 CIP

Softcover reprint of the hardcover 1st edition 1999

Typesetting: camera-ready by the authors
SPIN 10693392 30/3136-5 4 3 2 1 0

Contents

Contributors

Hauke Bietz
Forschungszentrum Terramare
Schleusenstr. 1
26382 Wilhelmshaven
bfnt@mail.lsw.nrw.de

Dr. Sabine Dittmann
Forschungszentrum Terramare
Schleusenstr. 1
26382 Wilhelmshaven
sabine.dittmann@terramare.de

Dr. Volker Grimm
Umweltforschungszentrum Leipzig-Halle
Sektion Ökosystemanalyse
Postfach 2
04301 Leipzig
vogri@oesa.ufz.de

Dr. Carmen-Pia Günther
Alfred-Wegener-Institut für Polar- und Meeresforschung
Columbusstr.
27568 Bremerhaven
cpguenther@aol.com

Dr. Andreas Hild
ICBM
Universität Oldenburg
Postfach 2503
26111 Oldenburg
A.Hild@geo.icbm.uni-oldenburg.de

Hanno Hildenbrandt
Zentrum für Marine Tropenökologie
Fahrenheitstr.
28359 Bremen

Dr. Verena Niesel
Umweltbundesamt
Postfach 330022
14191 Berlin
verena.niesel@t-online.de

Prof. Dr. Ulrike Schleier
FH Wilhelmshaven
Postfach 1465
D-26354 Wilhelmshaven
schleier@fbe.fh-wilhelmshaven.de

Dr. Marlies Villbrandt
Umweltforschungs- und Technologiezentrum (UFT)
Universität Bremen
Leobener Str.
28359 Bremen
marlies@biotec.uni-bremen.de

1 Introduction

This book covers the synthesis of the joint research project ELAWAT (Elastizität des Ökosystems Wattenmeer = Resilience of the Wadden Sea Ecosystem), which was part of the "Ecosystem Research Wadden Sea" in Germany. The major objective of ELAWAT was the analysis of potential stability properties and relevant processes for regeneration in this intertidal ecosystem. The synthesis of this project summarizes the results of several years of field studies on spatial and temporal patterns of several abiotic and biotic parameters and on the effects of disturbances. Based on this analysis, stability properties and mechanisms were identified. The results contribute to the protection of processes in this coastal ecosystem (Chap. 12). This book on the interdisciplinary project ELAWAT also provides a detailed case study on the ecology of a backbarrier system.

The "Ecosystem Research Wadden Sea" was part of the Ecosystem Research Program of the Federal Republic of Germany and a contribution to the MAB-programme of the UNESCO (MAB-5). Results of this interdisciplinary research scheme were meant to form the scientific background of political decisions for the protection and sustainable use of this coastal ecosystem. Thus the "Ecosystem Research Wadden Sea" followed the guidelines of the precautionary principle and environmental research of the federal government. The research was carried out in association with the administrations of the Wadden Sea National Parks in the states of Schleswig-Holstein and Lower Saxony.

The aim of the "Ecosystem Research Wadden Sea" was to investigate the complex interactions between organisms and their environment, the exchange processes with adjacent media (air, sea and land) and the effects of human activities. This comprehensive analysis was realised by a combination of applied and basic research projects. In the applied projects, impacts of fisheries and eutrophication were investigated, monitoring techniques and the application of remote sensing tested, and socio-economic studies were carried out as well.

ELAWAT was the basic research project of the "Ecosystem Research Wadden Sea" in Lower Saxony. It had the task to analyse fundamental ecological processes, improve the understanding of spatial and temporal patterns and investigate effects of disturbances in the Wadden Sea (Chaps. 3–7). These objectives were based on a concept whereby the natural dynamic was seen as essential for the long-term existence of the Wadden Sea and its ability to respond to disturbances (see Chap. 2). In the course of the project, this turned into a critical debate about the analysis of stability properties in the Wadden Sea ecosystem (Chaps. 8–10).

From late 1991 to early 1993, pilot studies were carried out for ELAWAT. In this phase, ten institutions were involved with fifteen sub-projects to develop methods for a proper analysis of temporal and spatial patterns in the Wadden Sea, the appropriate scales of observation and sampling strategies required. This development took place in close collaboration with statisticians which were involved in the project. The integration of a mathematical/statistical team was continued during the main phase of ELAWAT and ensured that statistical requirements were met for the design of the studies and their data analysis. The pilot studies also provided a survey of the main study area with respect to a number of environmental parameters, benthic communities and bird distributions.

The main phase of ELAWAT commenced in October 1993. It was realized as an interdisciplinary research project with 11 (later on 12) sub-projects carried out by 8 (later 9) institutions (see Chap. 11 and appendix). The Research Centre Terramare in Wilhelmshaven was responsible for the administration and coordination of the project. The collaboration between the various disciplines and teams was facilitated by a number of seminars and working groups (Chap. 11). Thus, a substantial degree of interdisciplinary collaboration could be realized, which is also reflected in this synthesis. The research was carried out by 42 scientists, 17 PhD students, a number of Master (Diploma) students and technical staff. The little over 4 years of the main phase were funded by the Federal Ministry for Education, Science, Research and Technology (BMBF) with a budget of 9 Mill. DM.

Three hearings were carried out to present and discuss research results and the work progress with the funding agency and external reviewers. Progress reports were handed in in April 1995 and 1996. The sub-projects submitted their final reports, which formed the base for this synthesis, in spring 1997. The extensive ornithological studies carried out in ELAWAT are covered by several published and submitted publications and a summarizing article is in preparation, therefore only certain aspects are included in the synthesis here. A special volume of the journal Senckenbergiana maritima (Vol. 26 (3/6)), issued in 1996, comprised results of the pilot phase and early main phase. Furthermore, more than 50 scientific publications of ELAWAT appeared in peer-reviewed journals and results were presented in over 70 oral and 30 poster presentations on various symposia. Two workshops on central topics of ELAWAT were carried out with international participation. The contributions to the workshop "The concept of ecosystems" held in 1995 were published in Senckenbergiana maritima (Vol. 27 (3/6)). A second workshop on "The role of disturbances in coastal marine ecosystems" was carried out in February 1997.

The main study area of the ecosystem research in the Lower Saxonian Wadden Sea was the backbarrier system of the island of Spiekeroog (Fig. 3.1.2). A number of investigations had been carried out in this area prior to the ecosystem research. Access to the tidal flats was relatively easy from land and sea. The investigations of ELAWAT focused on two areas: the Swinnplate, partly covered with mussel beds, and the Gröninger Plate, a typical sandflat. Many studies were only possible with the use of the vessel "Terramare I", the research vessel "Senckenberg" and research vessels of the Forschungsstelle Küste.

This synthesis was carried out by 8 scientists from April 1997 to February 1998. In addition to the scientific report, a brochure and leaflets were prepared for the general public. The synthesis did not only compile the results of the project, but evaluated the approach and its realization, thus giving a critical assessment of joint research projects. These aspects will be picked up by the final synthesis of the "Ecosystem Research Wadden Sea", which is currently under way and will give recommendations for future research in the Wadden Sea. The report was first written in German language and translated into English for the book version presented here. However, the book is not a literal translation, as the chapters were improved in the course of a review process.

In this book, we, the 8 scientists who wrote the synthesis report, function as authors for the respective chapters. But in many chapters, especially the chapters 3 to 7, the honour for the work we review goes to our colleagues who did the actual field work. Many colleagues and students, who did their thesis in the course of the project, supplied unpublished data. They are named in the acknowledgements of the respective chapters and whenever possible we have cited their publications. The credit for the results of ELAWAT we have compiled is theirs.

This book commences with an outline of the theoretical background, the concept and approach for this ecosystem research. The main study area, the backbarrier system of the island of Spiekeroog, is described next, introducing both the physical setting and the major benthic organisms living in the tidal flats. The following chapters, giving detailed accounts of spatial and temporal patterns and their underlying causes for a variety of parameters under natural and disturbance situations, often refer to the sampling sites described here. In general, the chapters of this book are tightly linked to each other, as the common thread leading through all of them is the analysis of stability properties and mechanisms. Thus the reader of this book should bear in mind that the chapters are not alone independent accounts, but all part of the story to understand the processes behind the scene of this fascinating intertidal ecosystem.

Acknowledgements

ELAWAT could have never been realized nor come anywhere near its scientific success without the considerable engagement and enthusiasm of all the scientists, students, technicians, administrative personnel and boat staff participating in it. All of them deserve our sincere gratitude for their work and commitment. In all those years a friendly atmosphere had developed between the scientists of the various sub-projects, which is not always a matter-of-course. We are especially thankful to all colleagues who supplied unpublished data for this report. The initiative to establish an ecosystem research on the Lower Saxonian Wadden Sea originated from Thomas Höpner. The Research Centre Terramare administered the project and Gerd Liebezeit functioned as the main project leader for ELAWAT. All colleagues of the project leadership for the Ecosystem Research Wadden Sea in Lower Saxony accompanied the project with great interest. The field station of the University of Münster in Carolinensiel provided an ideal location for field work. Roswitha Keuker-Rüdiger and Detlef Heyen assisted in further compilation of data during the synthesis. Mr. Abbey corrected the English of several chapters. Brigitte Behrends managed the tedious task of preparing most of the figures for this book. It would have been impossible to realize the book without Christa Dohn. She deserves a special thank you for all her technical help and the lay-out which she handled with great care and incredible patience. The chapters in this book were substantially improved by comments from a number of reviewers and colleagues and we sincerely thank Erik Bonsdorf, Roger Bradbury, Bob Clarke, Norbert

Dankers, Folkert de Jong, Rob Dekker, Richard Haedrich, Peter Herman, Hermann Hötker, Udo Hübner, Harald Marencic, Don Morrisey, James Nebelsick, Karsten Reise, Gregor Scheiffarth, Greg Skilleter, Martin Thiel, Justus van Beusekom, Wim van Raaphorst, Jennifer Verduin, Bob Whitlatch and Wim Wolff. Many friends have supported us in various ways to finish this book. We thank the Federal Ministry for Education, Science, Research and Technology (BMBF) for the financial support of the project (03F0112A and B). Prof. Dr. U. Schöttler from the "Projektträger Meeresforschung" accompanied the project with great expertise and interest and we thank him for inspiring discussions and his engagement for the project.

2 ELAWAT: Goals and Conceptual Framework

Sabine Dittmann & Volker Grimm

The "Ecosystem Research Wadden Sea" consisted of four large joint projects carried out since 1989 on the southern North Sea coast of Germany. The projects had two major goals which were closely related. One was to develop a protection and management strategy for the Wadden Sea to preserve the diversity of its biota and landscape. This aspect also addressed the sustainable use of the Wadden Sea, i.e. a use of resources in such a way that no long-term decline of biological diversity results and the needs and aspirations of present and future generations are maintained. This goal was the task of two applied projects within the "Ecosystem Research Wadden Sea", one in the state of Schleswig-Holstein and the other in Lower Saxony (Stock et al. 1996; Dittmann et al. 1997; Kellermann et al. 1997). The second goal was to achieve a fundamental understanding of this ecosystem. This was the task of two research projects, namely SWAP ("Sylter Wattenmeer Austausch Prozesse" = "Exchange processes of the Sylt Rømø-bight"), which was carried out in Schleswig-Holstein (Gätje & Reise 1998), and ELAWAT ("Elastizität des Ökosystems Wattenmeer" = "Resilience of the Wadden Sea ecosystem"; Dittmann et al. 1997), carried out in Lower Saxony.

Although both basic research projects were directed towards the level of the ecosystem, they reflected two different conceptual approaches to ecosystem research. Ecosystem research can mean an assessment of the flow of energy and matter. SWAP followed in the tradition of this approach, which is based on the insight that all ecological processes ultimately constitute a transformation of matter and energy. However, critics of this approach, which goes back to Odum (1971), doubt that energy and matter are the only valid currencies in ecology (Manson & McGlade 1993). They believe it fails to pay attention to the "elements" of ecological systems, the organisms (Jax 1996). SWAP was aware of this critique and did not neglect the study of species.

An alternative approach to ecosystem research follows the original definition of "ecosystem" by Tansley (1935) more closely, according to which an ecosystem consists of all the biotic and abiotic components and interactions within a certain area or landscape (see also Trepl 1988; Golley 1993). ELAWAT followed this approach. However, this alone would not have been sufficient as a conceptual framework for ecosystem analysis, as merely describing all the components does not provide any insight or understanding of essential processes. Interactions can only be studied in a meaningful way when precise questions are asked about certain mechanisms in the system (cf. Reise 1990).

The development of the conceptual framework for ELAWAT was guided by the observation that, despite of short-term, seasonal, annual and historical changes in the Wadden Sea, a certain persistence remains to be seen (Reise 1985). This is in particular the case regarding the long-term pattern of large-scale zonations of sediment types with their associated biota, which is determined by abiotic factors and biotic interactions. The programme of ELAWAT was to study the mechanisms of this persistence. Understanding how the Wadden Sea has "managed" to maintain its - still to be defined - properties and characteristics over a long time will enable an evaluation of the present state of the Wadden Sea and may also allow forecasts concerning its future development.

Conceptual development in ELAWAT was originally based on the working hypothesis "stability by variability" (ARSU 1989), which was subsequently replaced because of its vagueness by an emphasis on spatial and temporal patterns of biotic and abiotic components of the Wadden Sea as well as natural disturbances related to these patterns (Jax et al. 1993). The distribution of species in the Wadden Sea and their association into defined communities reflect the patchy distribution of resources, local interactions between species and local effects of specific disturbances. These distribution patterns exhibit temporal variation on a short-term, a seasonal, and a long-term scale. What processes are relevant for the development of specific temporal and spatial patterns and what effects disturbances have on the patterns and thus on the entire system were the central topics of ELAWAT. Based on the analysis of regeneration following specific disturbance events, stability properties (resilience, persistence) were to be identified (Table 2.1.1 and see Chap. 9). Therefore the sub-projects of ELAWAT carried out investigations and experiments at several hierarchy levels and in various compartments of the ecosystem Wadden Sea.

The common perspective for analysing the results of the sub-projects was the assessment of stability properties. To our knowledge this is the first attempt to use stability properties, instead of energy flow, for the development of an integrated view of an ecosystem, i.e. of a landscape complete with all the components and processes involved therein. In this sense ELAWAT was a methodological pilot study, which – asides from the defined aims concerning the Wadden Sea – could set a new direction for future ecosystem analyses.

Table 2.1.1 Overview on the most important stability concepts and -properties in ecology. After Grimm et al. 1992 and Grimm 1996

Stability property	Definition
Constancy	Remaining essentially unchanged
Resistance	Remaining essentially unchanged despite the presence of disturbances
Resilience	Returning to the reference state (or dynamic) after a temporary disturbance
Elasticity	Speed of return to the reference state (or dynamic) after a temporary disturbance
Persistence	Persistence over time of an ecological system

As ELAWAT thus entered uncharted territory it is not surprising that the conceptual framework developed during the course of the project. This process is documented in the synthesis of the project presented here, along with the actual results of the synthesis. Although resilience stood in the foreground at the beginning of ELAWAT (as reflected in the title of the project with "Elastizität" being German for resilience), it became increasingly obvious that studying resilience would not be an end in itself but a diagnostic tool to investigate basic processes in the Wadden Sea and to integrate the results of the sub-projects. Thus, stability properties were not taken as a mechanistic model imposed on the Wadden Sea ecosystem (cf. Reise 1994) in which all parts interact like cogs and lead to deterministic, predictable behaviour of the entire system (such as a "balance of nature").

The elements of the conceptual framework of ELWAT are described below. Subsequently, the task for the synthesis of ELAWAT and, in turn, the structure of this book are explained.

What do we mean by the Wadden Sea "ecosystem"?

ELAWAT focused on the original ecosystem definition by Tansley (1935), which reads in a modern version by Likens (1992) as follows: An ecosystem is "a spatially explicit unit of the Earth that includes all of the organisms, along with all components of the abiotic environment within its boundaries" (see also Jax 1994a; Jax et al. 1993). Following this definition, the Wadden Sea can be defined as an ecosystem whose spatially explicit unit is given by the tidal area. As the Wadden Sea is a transition zone between land and sea, like any other tidally dominated shallow sea or estuary, the outer margins – the high- and low water line – are open borders to adjacent terrestrial and marine systems (Costanza et al. 1993). The organisms living in this tidal area are adapted to the living conditions changing with each tidal cycle. Yet the habitat of many organisms extends beyond the tidal area (Reise 1985): pelagic larvae of intertidal benthic organisms, for example, occur in North Sea waters. The Wadden Sea is also a major nursery ground for fish and crustaceans of the North Sea. Migratory birds, who rest in the Wadden Sea on their migration between tropical wintering grounds and arctic breeding grounds, have the largest habitat range of Wadden Sea visitors. Thus the Wadden Sea is just a temporary home for many organisms.

Disturbances and disturbance regime: definitions

The stability property "resilience" (see below) can only be assessed in relation to disturbance-induced changes. Thus, ELAWAT focused on the investigation of mainly natural, small-scale and short-term disturbances. Following Jax et al. (1993), the definition of disturbance by Pickett & White (1985) was used in ELAWAT, whereby "a disturbance is any relatively discrete event in time that disrupts ecosystem, community, or population structure and changes resources, substrate availability, or the physical environment". This definition combines cause and effect. However, forming an objective judgement on disturbance effects is often not easy.

Disturbances can be distinguished by their cause, frequency, intensity and time of occurrence (Probert 1984; Grubb & Hopkins 1986; Skilleter 1995). Certain species or areas are affected depending on the type of disturbance. Natural disturbances, for example, can occur in the course of biotic interactions and enhance small-scale heterogeneity (Johnson 1970, 1972; Thistle 1981). Other natural disturbances can result from climatic events in the Wadden Sea, e.g. from storm and ice winter disturbances resulting in a temporary loss of species or habitats, or increasing the recruitment of some species (Chap. 5.3; Ziegelmeier 1964; Beukema et al. 1988). The ice winter of 1995/96, which occurred during the field studies for ELAWAT, allowed an analysis of such an event (Chaps. 6.4 and 7).

In the opinion of Pickett & White (1985), these natural disturbances are by definition part of a system and play a part in the theories discussed below. According to several other authors, natural abiotic variations should not be seen as disturbances, as they are part of the environmental factors related to the development of the biological components of a system (Waide 1995).

In addition to natural disturbances, anthropogenic disturbances have increased over the past few decades. The oldest anthropogenic disturbance in the Wadden Sea is the construction of dikes, which dates back as far as the 11th century (Homeier 1979; Meier 1994). During the course of coastal engineering, the size of the tidal basins and consequently the hydrography of the Wadden Sea has changed over the centuries (Homeier & Luck 1969). The loss of mudflats described in Chap. 3.4 is ascribed to this development. Fisheries and tourism, as well as the input of nutrients and pollutants, have had substantial effects on the Wadden Sea. However, these anthropogenic disturbances were not dealt with by ELAWAT, but by the applied research projects mentioned above. Only by simulating the increase of macroalgal mats (which were probably caused by eutrophication) in experimental organic enrichments and defaunation experiments, were the effects of these disturbances on the sediment chemistry and regeneration of the biota studied in ELAWAT.

During the synthesis it became clear that concentrating on individual disturbance events can distract from their overall ecological role (see below). To assess the stability properties of individual components of the Wadden Sea, it was essential to consider the entire disturbance regime. A disturbance regime characterizes the spatial and temporal attributes of disturbance events such as storms or ice winters on a longer time scale (50 years or more). What matters is not only the mean frequency and intensity of disturbances, but the precise frequency distribution and intensity, the size of the affected areas and the duration of disturbance-free periods.

The ecological role of disturbances

Disturbances play an important role as events in the ecology of many ecosystems (Dayton 1984; Sousa 1984). They can initiate transitions between phases or cycles of a system and can promote diversity in some cases ("creative disturbances", Connell 1978). Small-scale disturbances (e.g. wave surf on a rocky shore, predation) can create space for recolonization, leading to a mosaic of different successional stages (Paine & Levin 1981). This "patch-dynamic" concept was developed on rocky shores with a dominance of sessile species. Johnson (1970, 1972) and van Blaricom (1982) described similar mechanisms in soft-bottom communities. Yet, the extent to which the patch-dynamic concept is transferable to soft-sediment communities is still under debate (Frid & Townsend 1989; Downes 1990).

A concept related but not identical (Jax et al. 1994b) to the patch-dynamic concept is the mosaic-cycle theory (Remmert 1991). This theory states that a more or less cyclic sequence of communities exists at a certain site (e.g. of the size of a tree-habitat) striving for a climax. The climax exists only for a certain time and then collapses, due to either internal processes (the death of individuals) or external disturbances. Thus a more or less regular cycle occurs on each site, from a pioneer state through intermediate successional stages to a climax state. All these sites or patches of an area form a mosaic of successional stages of differing age, so that a static or cyclic balance appears at the landscape level (Wissel 1992).

Successions in marine benthic communities can be seen in the light of this mosaic-cycle theory (Reise 1991). The application of this theory to the benthic communities of the Wadden Sea is discussed in Chap. 8.

Stability concepts: an approach

"Elasticity" is one of several different stability properties (Orians 1975; Boesch & Rosenberg 1981; Connell & Sousa 1983; Grimm et al. 1992) and defines the return time following a disturbance, i.e. the time before a system returns or comes close to its former state after a temporary disturbance. However, as stated in the project proposal for ELAWAT, elasticity was used here as a synonym for resilience and defined as the property of a system to restore its structure and system properties after a disturbance (this confusion on stability concepts is widespread in the ecological literature and was unavoidable in the beginning of ELAWAT; see also Grimm & Wissel 1997). This definition of resilience referring to the entire system was refined in the course of the project to a definition taking specific ecological situations into account, i.e. precise variables, levels of description, disturbances, reference dynamics as well as spatial and temporal scales (Grimm et al. 1992).

It also became evident that the space and time frame necessary for the investigations of disturbances and the assessment of stability properties (Connell & Sousa 1983) could not be fulfilled by the restricted time frame and defined study area of ELAWAT. Thus the synthesis of stability properties given in Chap. 9 had to exceed the results of ELAWAT by incorporating further knowledge about the Wadden Sea currently available. In Chaps. 10 and 11 we critically discuss how these temporal and spatial restrictions limited the potential of ELAWAT to achieve a fundamental understanding of essential processes in the Wadden Sea and to gain an integrated view of this ecosystem.

Emergent properties: is the whole more than sum of its parts?

Stability properties are emergent properties, i.e. they can only be observed at a higher level of the ecological hierarchy. Resilience, for example, has no direct significance for a single individual, but is relevant to its population. The existence of emergent properties at the level of the ecosystem is doubted by a number of authors (Sommer 1996; Wiegleb & Bröring 1996), and the interpretation of "emergent property" also varies among authors (Müller et al. 1997). This debate became apparent during an international workshop organized by ELAWAT on conceptual aspects of ecosystems (Dittmann et al. 1996). For ELAWAT, the workshop discussions enabled to extend the scope beyond the single stability property addressed by the project title. Thus, ELAWAT turned into a more general search for possible emergent properties of the system investigated.

A distinction must be drawn between "collective" and "emergent" properties (Wiegleb & Bröring 1996). Collective properties are the sum of properties of a system, e.g. the total biomass in a definite area equaling the sum of the biomass of all the organisms in that area. However, the spatial and temporal distributions or variations in biomass cannot be calculated as a sum over all organisms, as population dynamics, interactions between populations and the impact of disturbances come in for this. Emergent properties are best described by the famous sentence "the whole is more than the sum of its parts". This may be the case for many ecosystems, e.g. forests, coral reefs or savannahs, but is it also true for the Wadden Sea? Is this ecosystem a system at all - or is it a more or less loose collection of subsystems (e.g. populations)?

This question is by no means limited to academic relevance, as answering it would have far-reaching methodological consequences: if there were no emergent properties at the level of the entire system, it would be sufficient to investigate subsystems independently, e.g. the decline of mussel beds in the East Frisian Wadden Sea since the 1980s (Zens et al. 1997). However, if there are emergent properties, then the decline of mussel beds cannot be considered detached from the entire system.

Aim of the synthesis

The task for the ELAWAT synthesis was to combine the results of specific system components derived from the sub-projects and thus to deduce possible stability properties of the Wadden Sea (Chap. 9). To do so, the results of the various investigations of all the disciplines involved had to be summarized and evaluated.

Thus the synthesis achieves an overview of the study area and the processes affecting temporal and spatial patterns as well as the regeneration after disturbances (Chaps. 3, 5, 6, 7 and 8). Statistical methods to assess change and, in turn, stability properties are described in Chap. 4. The evaluation of stability properties is given in Chap. 9. Finally, ELAWAT also critically analysed the approach used (Chap. 10) and developed recommendations for future research and the realization of interdisciplinary projects in the Wadden Sea (Chap. 11). Further recommendations address approaches for the conservation of the Wadden Sea (Chap. 12).

Acknowledgements

We thank all our colleagues who have contributed with their comments and discussions to the development of the conceptual framework. The synthesis group and a reviewer gave helpful comments for this chapter. The project was funded by the German Bundesministerium für Bildung, Wissenschaft, Forschung und Technologie (BMBF) under grant number 03F0112 A and B. The responsibility for the contents of the publication rests with the authors.

References

ARSU (1989) Programmkonzeption zur Ökosystemforschung im niedersächsischen Wattenmeer. Umweltbundesamt, Texte 11/1989, Berlin

Beukema JJ, Dörjes J, Essink, K (1988) Latitudinal differences in survival during a severe winter in macrozoobenthic species sensitive to low temperatures. Senckenbergiana marit 20: 19–30.

Boesch DF, Rosenberg R (1981) Response to stress in marine benthic communities. In: Barrett GW, Rosenberg R (Eds) Stress Effects on Natural Ecosystems. John Wiley & Sons Ltd., pp 179–200

Connell JH (1978) Diversity in tropical rain forests and coral reefs. Science 199: 1302–1310

Connell JH, Sousa W (1983) On the evidence needed to judge ecological stability or persistence. Am Nat 121: 789–824

Costanza R, Kemp WM, Boynton WR (1993) Predictability, scale, and biodiversity in coastal and estuarine ecosystems: implications for management. Ambio 22: 88–96

Dayton PK (1984) Patch dynamics and stability of some California kelp communities. Ecol Monogr 54: 253–289

Dittmann S, Kröncke, I, Albers, B, Liebezeit, G (Eds) (1996) The Concept of Ecosystems. Senckenbergiana marit 27: 81–255.

Dittmann S, Marencic H, Roy, M (1997) Ökosystemforschung im Niedersächsischen Wattenmeer. In: Fränzle O, Müller F, Schröder W (Eds) Handbuch der Umweltwissenschaften. Ecomed, Landsberg, V-4.1.2

Downes BJ (1990) Patch dynamics and mobility of fauna in streams and other habitats. Oikos 59: 411–413

Frid CLJ, Townsend CR (1989) An appraisal of the patch dynamics concept in stream and marine benthic communities whose members are highly mobile. Oikos 56: 137–141

Gätje C, Reise K (1998) Ökosystem Wattenmeer. Austausch-, Transport- und Stoffumwandlungsprozesse. Springer, Berlin, Heidelberg, New York

Golley FB (1993) A History of the Ecosystem Concept in Ecology. Yale University Press, New Haven

Grimm V (1996) A down-to-Earth assessment of stability concepts in ecology: dreams, demands and the real problems. Senckenbergiana marit 27: 215–226

Grimm V, Schmidt E, Wissel C (1992) On the application of stability concepts in ecology. Ecol Model 63: 143–161

Grimm V, Wissel C (1997) Babel, or the ecological stability discussions: An inventory and analysis of terminology and a guide for avoiding confusion. Oecologia 109: 323–334

Grubb PJ, Hopkins AJM (1986) Resilience at the level of the plant community. In: Dell B, Hopkins AJM, Lamont BB (Eds). Resilience in Mediterranean-Type Ecosystems. Dr W Junk Publ, Dordrecht, pp 21–38

Homeier H (1979) Die Verlandung der Harlebucht bis 1600 auf der Grundlage neuer Befunde. Forschungsstelle Norderney, Jahresber 1978/30: 105–115

Homeier H, Luck G (1969) Das historische Kartenwerk 1 : 50000 der Niedersächsischen Wasserwirtschaftsverwaltung als Ergebnis historisch-topographischer Untersuchungen und

Grundlage zur kausalen Deutung hydrologisch-morphologischer Gestaltungsvorgänge im Küstengebiet. Veröffent. Niedersächs. Inst. für Landeskunde und Landesentwicklung Universität Göttingen: Reihe A: Forschungs-, Landes- und Volkskunde. Göttingen

Jax K (1994a) Das ökologische Babylon. Bild der Wissenschaft 9: 92–95

Jax K (1994b) Mosaik-Zyklus und Patch-dynamics: Synonyme oder verschiedene Konzepte? Eine Einladung zur Diskussion. Z. Ökologie und Naturschutz 3: 107–112

Jax K, Vareschi E, Zauke G-P (1993) Entwicklung eines theoretischen Konzepts zur Ökosystemforschung Wattenmeer. Umweltbundesamt, Berlin. Texte 47/93

Jax K (1996) Über die Leblosigkeit ökologischer Systeme – Zur Rolle des individuellen Organismus in der Ökologie. In: Ingensiep HW, Hoppe-Sailer R (Eds). NaturStücke. Zur Kulturgeschichte der Natur. Edition Tertium: Ostfildern, pp 209-230

Johnson RG (1970) Variations in diversity within benthic marine communities. Am Nat 104: 285–300

Johnson RG (1972) Conceptual models of benthic marine communities. In: Schopf TJM (Ed). Models in Paleobiology. Freeman Cooper & Co., San Francisco, pp 148-159

Kellermann et al. (1997) Ökosystemforschung im Schleswig-Holsteinischen Wattenmeer. In: Fränzle O, Müller F, Schröder W (Eds) Handbuch der Umweltwissenschaften. Ecomed, Landsberg, V-4.1.1

Likens GE (1992) The Ecosystem Approach: Its Use and Abuse. Excellence in Ecology 3. Kinne O (Ed) Ecology Institute, Oldendorf/Luhe

Månson BÅ, McGlade JM (1993) Ecology, thermodynamics and HT Odum's conjectures. Oecologia 93: 582–596

Meier D (1994) Geschichte der Besiedlung und Bedeichung im Nordseeküstenraum. In: Lozán JL, Rachor E, Reise K, v. Westernhagen H, Lenz W (Eds). Warnsignale aus dem Wattenmeer. Blackwell, Berlin, pp 11–17

Müller F, Breckling B, Bredemeier M, Grimm V, Malchow H, Nielsen, SN, Reiche EW (1997) Emergente Ökosystemeigenschaften. In: Fränzle O, Müller F, Schröder W (Eds). Handbuch der Umweltwissenschaften. Ecomed: Landsberg, III-2.5

Odum EP (1971) Fundamentals of Ecology (3rd edn). Saunders, Philadelphia

Orians GH (1975) Diversity, stability and maturity in natural ecosytems. In: van Dobben WH, Lowe-McConnell RH (Eds). Unifying Concepts in Ecology. Pudoc (Dr W Junk), The Hague. pp 139–150

Paine RT, Levin SA (1981) Intertidal landscapes: disturbance and the dynamics of pattern. Ecol Monogr 51: 145–178

Pickett STA, White PS (1985) The Ecology of Natural Disturbance and Patch Dynamics. Academic Press, Orlando

Probert PK (1984) Disturbance, sediment stability, and trophic structure of soft-bottom communities. J Mar Res 42: 893–921

Reise K (1985) Tidal Flat Ecology. Springer, Berlin Heidelberg New York, Ecological Studies 54

Reise K (1990) Immer erst warum fragen. Plädoyer für ein neues Konzept der Meeres-Ökologie. Waterkant 4/90: 13–16

Reise K (1991) Mosaic cycles in the marine benthos. In: The Mosaic Cylce Concept of Ecosystems. H. Remmert (Ed). Springer, Berlin Heidelberg New York, pp 61–82

Reise K (1994) Ökologische Qualitätsziele für eine ziellose Natur? In: Ökologische Qualitätsziele für das Meer. Schriftenreihe der Schutzgemeinschaft deutsche Nordseeküste e.V., Wilhelmshaven, pp 38–45

Remmert H (1991) The Mosaic Cylce Concept of Ecosystems. Springer, Berlin Heidelberg New York, Ecological Studies 85

Skilleter GA (1995) Environmental disturbance. In: Underwood AJ, Chapman MG (Eds). Coastal Marine Ecology of Temperate Australia. UNSW Press, pp 263–276

Sousa WP (1984) The role of disturbance in natural communities. Ann Rev Ecol Syst 15: 353–391

Sommer U (1996) Can ecosystem properties be optimized by natural selection? Senckenbergiana marit 27: 145–150

Stock M et al. (1996) Ökosystemforschung Wattenmeer -Synthesebericht: Grundlagen für einen Nationalparkplan. Schriftenreihe des Nationalparks Schleswig-Holsteinisches Wattenmeer, Heft 8

Tansley AG (1935) The use and abuse of vegetational concepts and terms. Ecology 16 (3): 284–307

Thistle D (1981) Natural physical disturbances and communities of marine soft bottoms. Mar Ecol Prog Ser 6: 223–228

Trepl L (1988) Gibt es Ökosysteme? Landschaft + Stadt 20: 176–185

van Blaricom GR (1982) Experimental analyses of structural regulation in a marine sand community exposed to oceanic swell. Ecol Monogr 52: 283–305

Waide JB (1995) Ecosystem stability: revision of the resistence-resilience model. In: Patten BC, Jørgensen SE (Eds). Complex Ecology: The Part-Whole Relation in Ecosystems. Prentice Hall, New Jersey, pp 372–396

Wiegleb G, Bröring U (1996) The position of epistemological emergentism in ecology. Senckenbergiana marit 27: 179–193

Wissel C (1992) Modelling the mosaic-cycle of a Middle European beech forest. Ecol Modell 63: 29–43

Zens M, Michaelis H, Herlyn M, Reetz M (1997) Die Miesmuschelbestände der niedersächsischen Watten im Frühjahr 1994. Ber Forsch-Stelle Küste 41, pp 141–155

Ziegelmeier E (1964) Einwirkungen des kalten Winters 1962/63 auf das Makrobenthos im Ostteil der Deutschen Bucht. Helgol wiss Meeresunters 10: 276–282

3 Study Area: The Backbarrier Tidal Flats of Spiekeroog

3.1 Description of the Spiekeroog Backbarrier System

Andreas Hild

The Wadden Sea of the southern North Sea covers an area of about 9.300 km^2 and represents one of the largest intertidal regions on earth (Fig. 3.1.1). Between Den Helder and the Jade Bay (near Wilhelmshaven), the West and East Frisian Wadden Sea forms a classical barrier island chain (Backhaus 1943). The island Spiekeroog is located in the eastern part of this chain. Its backbarrier system, the main research area of ELAWAT, is connected to the North Sea via the Otzumer Balje (Fig. 3.1.2), a tidal inlet which separates Spiekeroog from the neighbouring island Langeoog in the west. Referring to mean tide high water (MThw), the tidal basin of the Otzumer Balje covers an area of about 73.5 km^2 (Walther 1972). The basin is bordered by two watersheds: In the west by the Langeooger Plate, in the east by the Hohe Bank. The distance between the island and the mainland averages 6.5 km and the island has a length of approx. 10 km. The intertidal section of the Spiekeroog backbarrier system covers about 80 % of the area and is comparable to that of other East Frisian tidal basins. The North Frisian and the Danish Wadden Sea have about 70 % tidal flats on average, but the Sylt-Rømø-bight comprises only 30 % intertidal areas (Reise & Riethmüller 1998). With an average tidal range of 2.7 m at the western end of Spiekeroog island and of 2.9 m in the front of the mainland, mesotidal conditions prevail in this backbarrier system (Niemeyer & Kaiser 1994). At standard sea level (NN) the water volume in the catchment area amounts to 0.112 km^3. The volume of water that is exchanged during a tidal cycle via the Otzumer Balje is calculated to be 0.292 km^3 on average (Chap. 3.3).

The whole area is located in the "Wadden Sea National Park of Lower Saxony". This National Park, which has been established in 1986 distinguishes three zones with different degrees of protection. Large parts of Spiekeroog and the backbarrier area fall into the highest protection category zone 1 (quiet zone) and 2 (intermediate zone), where an economic use is not allowed or only with restrictions. The regulations are determined by the National Park decree.

The majority of the investigations were carried out in two sub-areas of the backbarrier system: the Swinn- and the Gröninger Plate (Fig. 3.1.2). The most important sampling points, which are referred to in later chapters, are indicated in Figs. 3.1.3 and 3.1.4 and listed in Table 3.1.1.

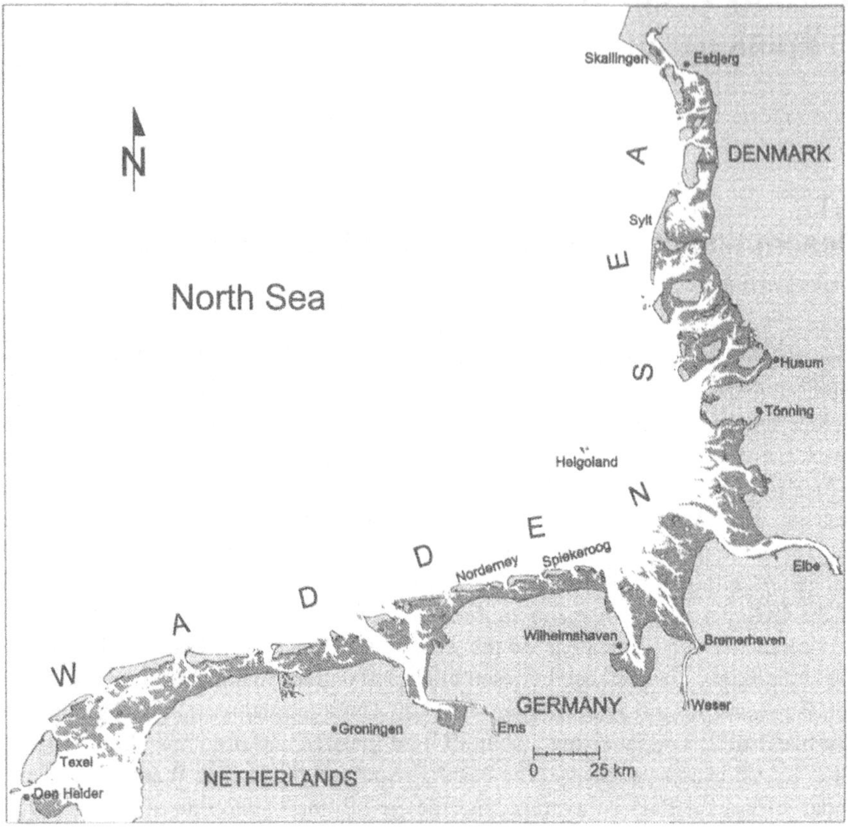

Fig. 3.1.1 The Wadden Sea, extending between the islands and the mainland coast of the southern North Sea. The shaded areas indicate tidal flats

Table 3.1.1 Co-ordinates (Gauß-Krüger) and abbreviations for the sampling sites on the Swinn- and Gröninger Plate, as they are referred to throughout this book

Swinnplate						
sampling site	SP1	SP2	SP3	SP4	SP5	SP6
x-co-ordinate	3417270	3417125	3417080	3416720	3416400	3415880
y-co-ordinate	5958005	5957875	5957935	5957830	5957795	5957670
distance to mussel bed [m]	0	265	355	960	1480	2330
Gröninger Plate						
sampling site	GP1	GP2	GP3	GP4	GP5	GP6
x-co-ordinate	3418025	3418250	3418525	3419020	3419460	3419500
y-co-ordinate	5955025	5955500	5955225	5954970	5954770	5954725

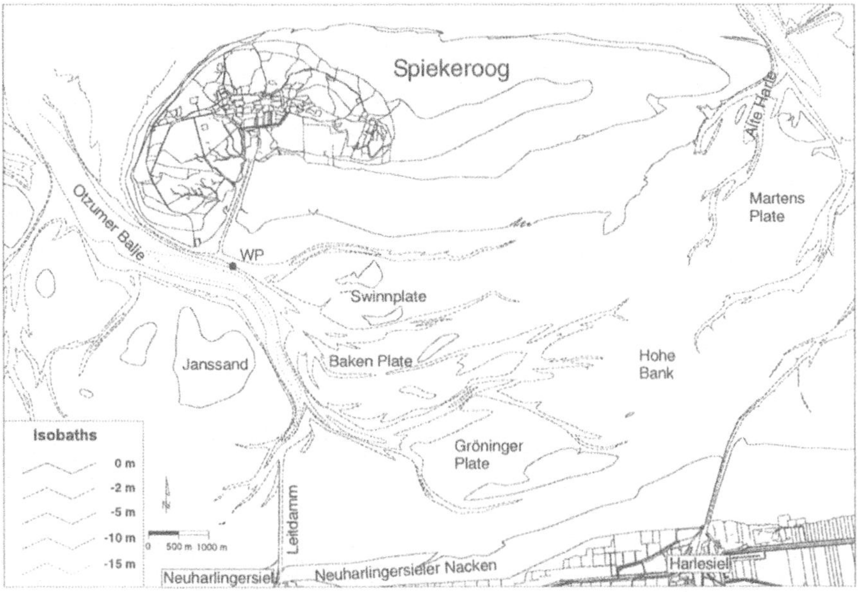

Fig. 3.1.2 Main study area of ELAWAT. WP = phyto- and zooplankton sampling site in the Otzumer Balje

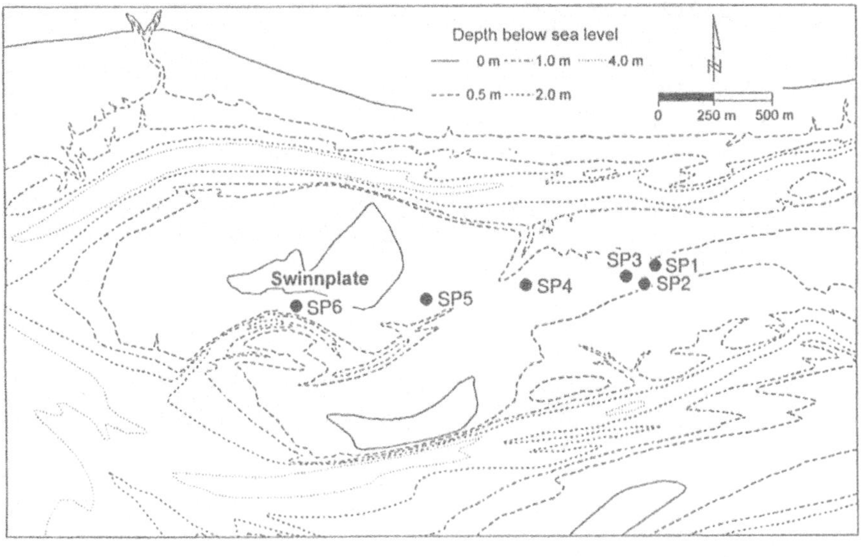

Fig. 3.1.3 Detailed map of the Swinnplate in the Spiekeroog backbarrier system. Sampling sites represent a transect from SP1 (mussel bed) to SP6 (sandflat) following the direction of the ebb tide current

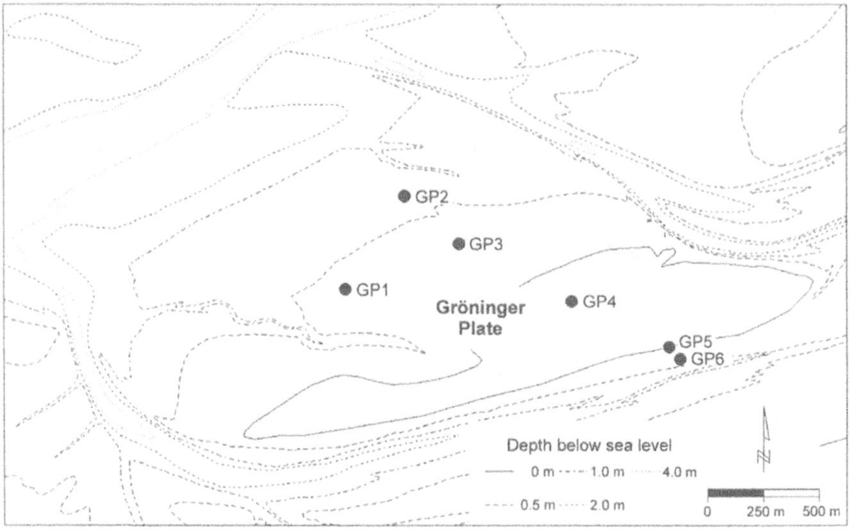

Fig. 3.1.4 Detailed map with the sampling sites on the Gröninger Plate in the Spiekeroog back-barrier system

References

Backhaus H (1943) Die Ostfriesischen Inseln und ihre Entwicklung. Ein Beitrag zu den Problemen der Küstenbildung im südlichen Nordseegebiet. Schriften Wirtschaftswiss. Ges Studiums Niedersachsen A12

Niemeyer HD, Kaiser R (1994) Hydrodynamik im Ökosystem Wattenmeer. Umweltbundesamt, Berlin. Texte 26/94

Reise K, Riethmüller R (1998) Die Sylt-Rømø Wattenmeerbucht: Ein Überblick. In: Gätje C, Reise K (Eds). Ökosystem Wattenmeer. Austausch-, Transport- und Stoffumwandlungsprozesse. Springer, Berlin Heidelberg New York, pp 21–23

Walther F (1972) Zusammenhänge zwischen der Größe der ostfriesischen Seegaten mit ihren Wattgebieten sowie den Gezeiten und Strömungen. Forschungsstelle Küste, Norderney – Jahresbericht 23, 1971, pp 7–32

3.2 Meteorological Conditions in the East Frisian Wadden Sea from 1994–1996

Verena Niesel

Biological and chemical processes were examined in the backbarrier area of Spiekeroog island from 1994 to 1996. Many of these processes are regulated by climatic conditions. Thus, these conditions are relevant for the biotic as well as for the abiotic environment.

Meteorological data were recorded at two sites. One weather station was situated in the salt marshes between Carolinensiel and Harlesiel (Fig. 3.1.2). At this station air temperature and wind speed were measured hourly from March to November. Measurements were not carried out in the winter months. A weather station from the German Weather Service (DWD) is located on the island of Norderney. At this station water temperature, air temperature, wind speed, wind direction and cloud cover were measured during all three years of observation.

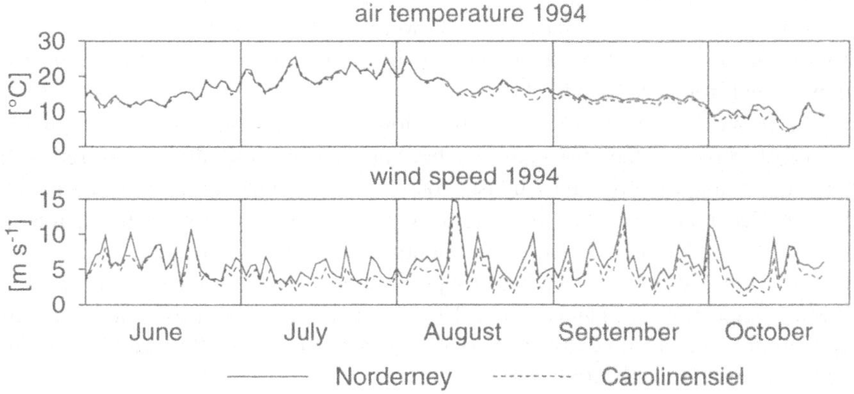

Fig. 3.2.1 Comparison of the air temperature and the wind speed at two different sites of the East Frisian Coast (Norderney and Carolinensiel)

A comparison of the air temperature at the two stations showed only slight differences (Fig. 3.2.1). Therefore the continuous data of the air temperature from Norderney were used for further consideration. A comparison of the wind speed at both stations showed that the fluctuation was similar, but the data from Norderney often reached higher values than those from Carolinensiel. Despite of the higher values, the data from Norderney were used for the comparison of the three years because they were continuously recorded. Table 3.2.1 shows the kind of representation used for the different parameters.

Table 3.2.1 Meteorological data, type of representation and their units

	type of representation	unit
Water temperature	value at 6.00 a.m.	[°C]
Air temperature	24 h - Mean	[°C]
Atmospheric pressure	24 h - Mean	[mbar]
Wind speed	24 h - Mean	[m/s]
Wind direction	24 h - Mean	0° North 90° East 180° South 270° West
Cloud cover	24 h - Mean	Values from 0 - 9 0 = without clouds 9 = complete cloud cover

Air temperature

During the three years of observation air temperatures reached values between - 10 °C and + 28 °C (Fig. 3.2.2). In February 1994 a cold spell of ten days took place with temperatures below 0 °C. Afterwards the air temperature increased continuously. Maximum values of 26 °C were reached in July and August. From August onwards the values decreased slowly. In autumn and winter air temperatures ranged from - 2 °C to 10 °C.

In January 1995 temperatures were below 0 °C for a short period, from the middle of May until the middle of October values over 10 °C were measured with maximum values in July and August around 23 °C. Temperatures dropped below 0 °C in December. The low temperatures lasted, with a few exceptions, for an unusually long duration of nearly four months. In April 1996 the temperatures rose for a short period up to 17 °C, but dropped subsequently below 10 °C. By the end of May temperatures remained above 10 °C. Except for a few record values the air temperatures remained below 20 °C over the summer period. 1996 can be considered as an exceptionally cold year.

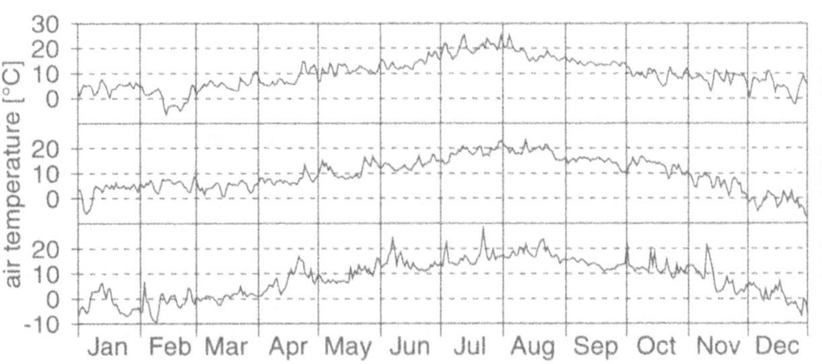

Fig. 3.2.2 Air temperature in the years of observation at the weather station Norderney, data courtesy of the DWD

The winter 1995/1996 was characterized by strong ice formation. Because of the long period of ice cover, it was called "ice winter". However, with 49 days of ice cover the winter was not rated as extreme, but held rank 9 in a 100-year-record (Strübing 1996). After the Ynsen frost index (Ynsen 1991), which is calculated by the number of days with frost (minimum of temperature < 0 °C), the number of days with ice formation (maximum of temperature < 0 °C) and the number of the very cold days (maximum of temperature < -10 °C), the winter 1995/1996 is classified as a strong winter.

Water temperature

With the exception of some anomalies the trend of the water temperature corresponded with the long year mean in all three years and followed the air temperatures (Fig. 3.2.3). Conspicuous were the high temperatures in the summer of 1994 and 1995. The winter 1995/1996 was characterized by long lasting low temperatures, which led to an ice cover in the Wadden Sea. The water temperatures in 1996 were striking by a negative deviation compared to the previous years.

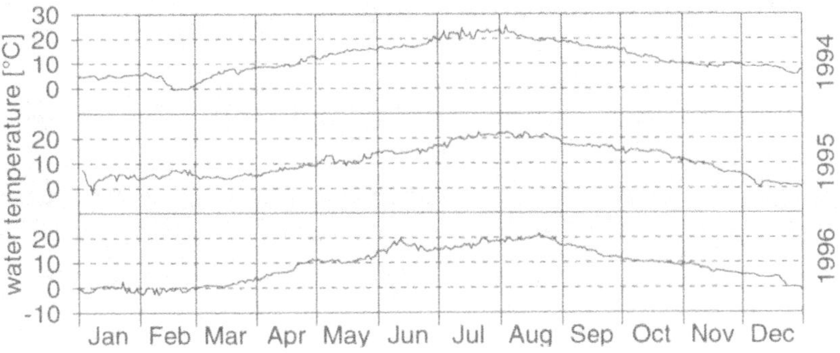

Fig. 3.2.3 Water temperature in the years of observation at the weather station of Norderney, data courtesy of the DWD

Atmospheric pressure

Special features were recorded in the development of the atmospheric pressure in summer 1994 (Fig. 3.2.4). From the end of April to the end of August a general high pressure period persisted. In the beginning of 1995 low pressure prevailed, but in summer a long high pressure period developed similar to the year before. The high pressure periods of the winter 1995/1996 were quite unusual. They were only interrupted for a few days in February 1996. In summer 1996 no typical high pressure period developed. The values varied a lot, indicating frequent weather changes during this time.

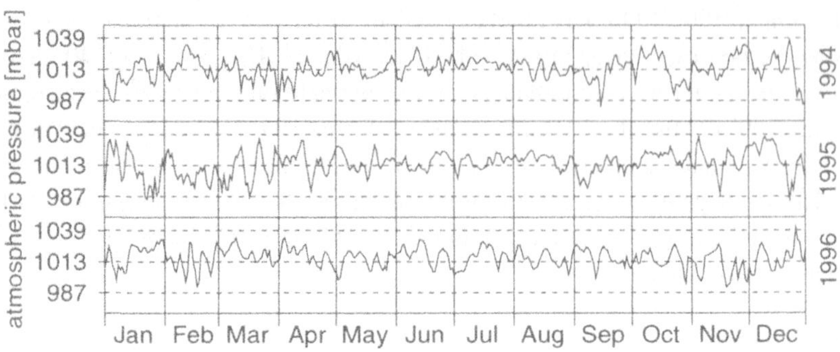

Fig. 3.2.4 Atmospheric pressure in the years of observation at the weather station of Norderney, data courtesy of the DWD

Cloud cover

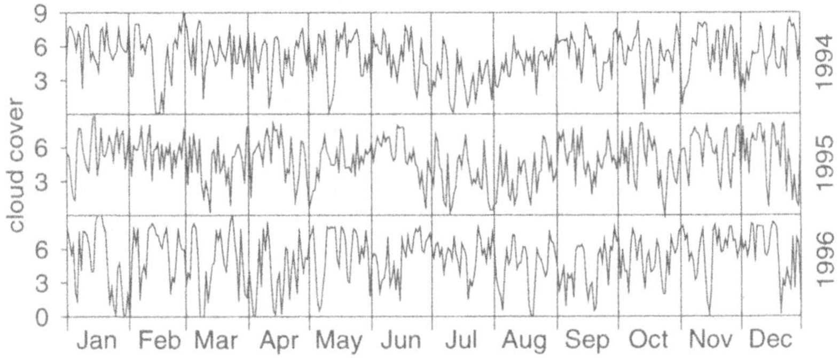

Fig. 3.2.5 Cloud cover in the years of observations at the station of Norderney, data courtesy of the DWD

The first months of 1994 were charcterized by a strong cloud cover, except for some days in February (Fig. 3.2.5). The summer 1994 had little cloud cover. After August, the degree of cloud cover varied strongly. A similar situation was observed in 1995. The year 1996 was different from the two previous years of observation as it had some periods with very low cloud cover in spring, while the cloud cover was unusually dense in summer.

Wind speed

During the period of observation the wind speed at the station of Norderney reached values between 0 and 18 $m*s^{-1}$ (Fig. 3.2.6). Maximum values of daily means correspond to a wind speed of no more than 8, which corresponds to a fresh gale according to the Beaufort scale. A comparison of the weather stations at Norderney and Carolinensiel showed slight differences, so that the wind speed in the observation area of ELAWAT was 1–2 $m*s^{-1}$ lower than those shown here,

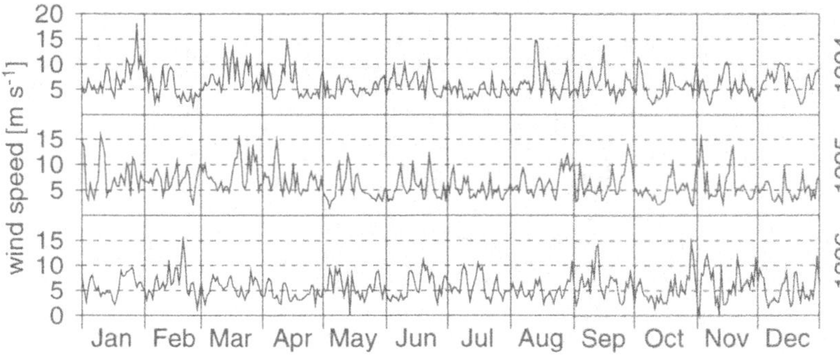

Fig. 3.2.6 Wind speed in the years of observation at the station of Norderney, data courtesy of the DWD

corresponding to a maximal wind speed of 7 (moderate gale). Strong winds were mainly recorded in the first months of the year, especially in January 1994 and 1995. In 1996, the first months were characterized by a low to moderate breeze, which normally occurred during summer months. However, no exceptionally low wind speed was measured in summer 1996.

The frequency of strong wind events, which are defined as events with wind force higher than 6, and the sea level rise, are both considered as indicators for climate change. Looking at these parameters in the German Bight over the last 30 years, a strong increase can be observed in the last 10 years. The number of wind events with a duration of several days increased during this period also (Svendsen 1991; Aagaard et al. 1995). A similar increase was observed in the significant wave-height in the North Atlantic (Bacon & Carter 1991; Bouws et al. 1996). Nevertheless, some authors doubt the increase of strong wind events (Schmidt & von Storch 1993, Kaas et al. 1996). They argue that this apparent increase lies within the range of natural variability and that many of the measurements are based on small time series only.

Wind direction

Frequently changing wind directions are typical for the East Frisian coast (Fig. 3.2.7). In 1994, southwesterly winds were dominant. In May and June the wind turned to NE/NW. Except for some short periods, winds from the SE and SW predominated until the end of the year.

In the beginning of 1995 winds from NW and SW prevailed. In summer the winds often changed direction. In autumn and winter 1995 the wind came from SE and NE. During the first months of 1996 easterly winds occurred for a long period, only interrupted by few days with southwesterly wind. The summer was characterized by NW winds, which later turned to NE.

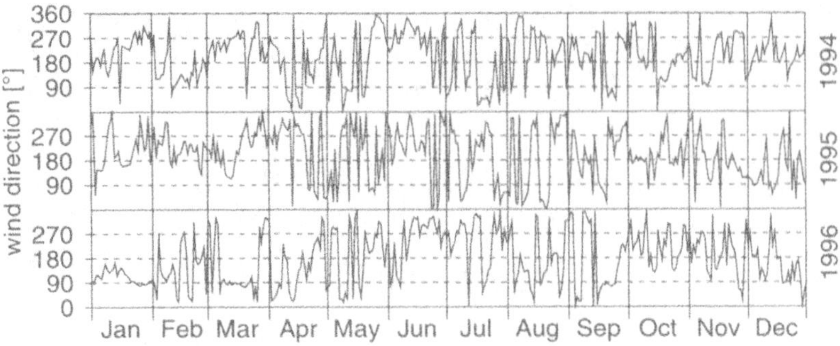

Fig. 3.2.7 Wind direction in the years of observation at the station of Norderney, data courtesy of the DWD

The year 1994

With an average temperature of 10 °C, 1994 was the fourth warmest year since the start of regular meteorological recordings in 1850. The summer was unusually hot with low cloud cover and extreme dryness. From mid April to the beginning of August only gentle breezes occurred. July 1994 was one of the rare months with a high number of sunshine hours (more than 1800 hours, MURSYS 1994). The development of the temperature at the water surface of the North Sea was mainly similar to the years before. Only July showed an anomaly with an increase of the water temperature of over 4.2 °C (MURSYS 1994).

The year 1995

The temperatures in 1995 were higher in comparison to the long-term averages, just as in the year before (long-term averages = monthly average over the last 30 years, MURSYS 1995). The temperatures in February, July, August and October had a positive temperature deviation of up to 3 °C from the mean. In March and April storms occurred with maximum wind force between 6 and 7. During the summer months the wind direction changed often, but the wind speed remained at a low level. Autumn and winter of this year were characterized by a long-lasting high pressure period with varying cloud cover and predominantly gentle breezes from different directions. The air temperature decreased at the end of the year to values below freezing point and the water temperature was around 0 °C.

The year 1996

The year 1996 was 1 °C colder than the 30-year average (1961–1990). It was the fourth coldest year since 1971 (MURSYS 1996). The winter temperatures were not exceptionally low, but stayed low for three months (Strübing 1996). Therefore the longest ice period for 33 years occurred in 1996 (ice winter). The wind situation in the beginning of the year was extremely unusual. Easterly winds dominated and the wind speed was lower than in the previous years. No storms occurred during this time. Water temperatures (0–2 °C) were low until the middle of March, then they quickly rose to 10 °C (within 6 weeks) and remained at this level. Temperatures stayed continuously over 15 °C just in June. Correspondingly, a maximum water temperature was reached late in the year (August). Compared with the previous years, the summer of 1996 was colder with strong cloud cover and unusually high wind speeds for this season. The further meteorological development from August on was inconspicuous.

Acknowledgement

The data from the weather station near Carolinensiel were provided by G. Liebezeit (Terramare). The meteorological data of the station at Norderney were supplied by the DWD (Deutscher Wetterdienst). Thanks for the help in translating to Ute Fuhrhop and Wolfgang Wenzel. The project was funded by the German Bundesministerium für Bildung, Wissenschaft, Forschung und Technologie (BMBF) under grant number 03F0112 A and B. The responsibility for the contents of the publication rests with the author.

References

Aagaard T, Nielsen N, Nielsen J (1995) Skallingen - Origin and evolution of a barrier spit.- Meddel. Skalling-Laboratoriet, XXXV

Bacon S, Carter DJT (1991) Wave climate changes in the North Atlantic and North Sea. Int J Climatol 11: 545–558

Bouws E, Jannink D, Komen GJ (1996) The increasing wave height in the North Atlantic Ocean. Bull Amer Meteor Soc 10: 2275–2277

Kaas E, Li T-S, Schmith T (1996) Statistical hindcast of wind climatology in the North Atlantic and northwestern European region. Climate Research 7: 97–100

Meeresumwelt-Reportsystem (MURSYS) Information aus Nord- und Ostsee Jahresbericht 1994, Bundesamt für Seeschiffahrt und Hydrographie

Meeresumwelt-Reportsystem (MURSYS) Information aus Nord- und Ostsee Jahresbericht 1995, Bundesamt für Seeschiffahrt und Hydrographie

Meeresumwelt-Reportsystem (MURSYS) Information aus Nord- und Ostsee Jahresbericht 1996, Bundesamt für Seeschiffahrt und Hydrographie

Schmidt H, von Storch H (1993) German bight storm analysed. Nature 365: 791

Strübing K (1996) The ice winter of 1995/96 on the German coasts between Ems and Oder, with a survey of the entire Baltic area. Dt Hydrogr Z 48: 73-87

Svendsen E (1991) Climate variability in the North Sea. - Symposium on Hydrobiological Variability in the ICES Area, 1980-1989, Mariehamm, Finnland, Conference Volume, Paper No. 10

Ynsen (1991) Karaktergetallen van de vinters vanaf 1707. Zenit 18: 69–73

3.3
Hydrographic Conditions in the Spiekeroog Backbarrier System

Verena Niesel

Tidal areas are highly dynamic marine systems, which are strongly influenced by tidal cycles. Tidal currents move water masses horizontally and vertically. They can reach high velocities and their forces have consequences on the morphology of the area as well as on the colonization of organisms.

Hydrographical conditions were both measured and simulated within the framework of the Ecosystem Research Wadden Sea (Table 3.3.1). The measurements were carried out by the Research Centre Coast during different measurement campaigns in the backbarrier system of the island of Spiekeroog (Fig. 3.1.2). These measurements comprised: Measurements of the sea level, the velocity and the direction of currents. These measurements were complemented by hydrographical modelling done by the Institute of Marine Science of the University of Hamburg. Modelling was performed with the numeric procedure HAWAM (Hamburger Wattenmeer Modell). This is a non-linear, vertically integrated, two dimensional barotropic model, which was developed for shallow waters with tidal dynamics. Basics about the mathematics and physics and a detailed description of the model can be found in Backhaus (1983). Different hydrographical parameters were simulated in a 200 metres grid and at 10 second intervals, boundary conditions were derived from a North Sea model of the BSH.

The model was validated by comparing the data calculated by the model with data measured in the field and showed a good agreement. Variation of sea levels, exposure and inundation time, directions and velocity of the current were calculated for different time periods. A coupling of the hydrographic model with a dispersion model enabled a calculation of the drift paths of marked tracers. Results of the modelling were published in a current atlas (Hübner & Backhaus 1997).

The tides

The tidal basin of the Otzumer Balje belongs to the zone of activity of the rotated tide in the southern North Sea with an amphidromy at the intersection between Terschelling and Blavands Huk (Defant 1923). Water in the German Bight swirls

Table 3.3.1 Origin of the data used in this chapter

	Measurement	Modelling
Size of the intertidal areas		x
Sea level	x	x
Tidal amplitude	x	-
Water volume	-	x
Currents	x (general)	x (comparison of sub-areas)
Drifting of tracers	-	x
Waves	x	-

counterclockwise, sweeping tides along the coast, i.e. from west to east in Lower Saxony, with a semidiurnal tide of 12 hours and 25 minutes. Near the coast the tidal amplitude and the direction of the tide are increasingly influenced by the interaction of the tides with the morphology of the area.

Tidal range

The mean tidal range is calculated as the difference between the heights of the mean high tide and the mean low tide. With a tidal range of 2.7 m at the western end of Spiekeroog island and of 2.9 m at the mainland the conditions in this area are classified as mesotidal.

Because of the slack tide and the reflection of the tide, the tidal range is higher near the coast than on the seaward margin of the area. An increasing tidal range can also be observed at single sites in the backbarrier system of Spiekeroog. At the water gauge of Spiekeroog the tidal range lies between 2.64 m and 2.88 m, while at the gauge of the Gröninger Plate the tidal range varied between 2.76 m and 3.02 m. These values were taken from measurement campaigns in spring and autumn and include neap as well as spring tides.

Sea level

The sea level in the backbarrier system of Spiekeroog is influenced by the impulse of the North Sea, which in turn is the result of the tidal forces and the meteorological conditions. Southerly and easterly winds cause a decrease of the sea level, whereas north-west winds raise the sea level at the coast (Reineck 1987). During the winter months of the study period, maximal differences of the sea level were measured with ± 60 cm on average. In the remaining months the deviation from the mean was lower, so that the time was characterized by mean sea levels. The highest sea level during the study period was recorded on the 28th of January 1994 with +910 cm on the tide gauge (equals +410 cm above mean sea level). This was the highest record at this tide gauge since it was installed in 1987. A comparison with the tides of earlier storm floods of the years 1906, 1962 and 1976 was not possible.

Water volumes

The transport of the water masses into the catchment area occurs to 70 % over the tidal inlet of the Otzumer Balje. The remaining water masses enter the area over the water sheds. Modelling the conditions during a neap/spring tidal cycle in summer 1994 showed that the water volume during low tide reached values between 30 and 90 million m^3, and during high tide between 240 and 330 million m^3. The difference between the water volume at mean tidal level (equals mean sea level) (110 million m^3) and the maximal volume during high tide showed that the tidal volume can reach 220 million m^3 water. This means that at maximum extension the water volume had trebled.

A comparison of the tidal basin of Spiekeroog with that of the Sylt-Rømø-bight shows that with a mean water volume of 570 million m^3 the basin of Sylt contains substantially more water and that its tidal volume of 559 million m^3 is much higher than that of the backbarrier system of Spiekeroog. But it must be considered that the basin of Spiekeroog only spans an area of 73.5 km^2, whereas the basin of Sylt-Rømø-bight reaches 404 km^2 (Reise & Riethmüller 1998).

Waves

Backbarrier systems have considerably shallower waves than tidal areas without barrier islands (for example Wurster Watt). The decrease of the waves is a result of shelter through the chain of islands. Even with on-shore storms, the wave propagation in the tidal channels decreases so strongly, that the wave height does not exceed the height of wind waves (Niemeyer et al. 1994). The waves and their significant wave heights were measured north (= seaward) of the ebb delta of the Otzumer Balje, in the tidal inlet of the basin (Otzumer Balje), and near the sandflat of the Gröninger Plate. North of the ebb delta the maximum values of the significant wave heights varied between 0.75 m and 2.75 m. In the tidal inlet, the height varied between 0 m and 0.75 m. Similar values were measured on the Gröninger Plate. These levels clearly show that in the backbarrier tidal area of Spiekeroog a strong attenuation of the waves takes place near the island. Inside the backbarrier system the waves were not attenuated, sometimes even higher significant wave heights were measured near the sandflats than in the tidal inlet Otzumer Balje. This can be explained by the influence of a superimposing effect and the morphology in the Wadden Sea.

Drifting of tracers

To examine the drift of passive particles, so-called tracers, the hydrographic model was coupled with a dispersion model to calculate the paths of the tracers using the Monte-Carlo-Procedure (Maier-Reimer 1980). The results showed that the drift of the tracers was influenced by the following parameters: a) the site of start, b) the time of start during the tidal cycle (high tide or low tide) c) and by the duration of the observation.

After one tidal period the tracers were predominantly found near the starting point. Starting at low tide, the tracers drifted until the turning point in the dominant direction of the current and reached the vicinity of the starting site at the end of the tidal period. Tracers which were started during high tide covered much longer distances, but they returned to the starting position after one tide. With longer duration of observation (more than one tidal period) this regularity was no longer observed. Some tracers drifted into other water catchment areas (for example Accumer Ee) and only few were transported back into the backbarrier tidal flat of Spiekeroog after a certain time. Tracers, which were started at high tide in the Otzumer Balje, regularly left the observation area.

Sampling that takes place over a longer period of time (e.g. for more than 5 tidal cycles) does not offer the possibility of comparing the same water body. This has to be taken in consideration when interpreting the data measured in the water column.

Currents

In a backbarrier tidal flat current patterns are mainly characterized by the periodical falling and rising of the water. Areas of extreme currents are the deep channels and the water sheds. In the channel of Otzumer Balje the direction of the dominant current is directed by the topography. With the incoming tide the water flows south-east and with low tide the water streams north-west. At mean wind conditions the velocity of the currents in the deep channel are roughly similar during high tide and low tide. They are highest at half tide when they reach maximal values of 0.7–1.3 m s^{-1}. At the watersheds the direction of the rising and falling waters can not be distinguished that clearly. The direction of the current varies, no clear maximum can be distinguished and the velocities of the current are low with values between 0–0.2 m s^{-1}.

The velocities of the currents are not only influenced by the morphology but also by the spring and neap tide and by the wind. From neap to spring tide the velocity increases. With wind coming from north and west, the velocities intensify. Easterly and southerly winds caused no intensification of the current speed but a reduction depending on the wind strength.

The two studied sandflats, Gröninger Plate and Swinnplate showed a great variability in velocity and direction of the current. Comparing both sandflats, the velocity of the current was higher at the Swinnplate than at the Gröninger Plate. At the Swinnplate a decrease in velocity was found from the edge to the centre. Particularly low values were found in the east of the flat, because this area is sheltered by the higher plateau of the flat. The directions of the ebb and flood tide showed only low variations during a tidal period as well as during neap/spring tidal cycles. In general, the rising water flowed east and the decreasing water flowed off in westerly direction. Because of the higher current velocity during falling water the Swinnplate was designated as ebb tide dominated.

At the Gröninger Plate the current was considerably more variable than at the Swinnplate. The directions of the ebb and flood tide varied in the different areas of the flat. In the central and western areas the incoming waters flowed in south-easterly direction on to the flat and off the flat to the west. In the northern part of the flat the water flowed east-south-east during flood and westward during ebb tide. The higher situated easterly area is flooded from south to east and the ebb water moves in all directions. Because of the higher current velocity during flood, this flat was designated to be flood tide dominated.

Size of the intertidal areas

Intertidal areas (tidal flats) are periodically submerged at high tides. According to the model calculation, the size of the areas emerging at low tide varies, because of the influence of the neap/spring tidal cycle and meteorological conditions. A model calculation for the period from 1st to 10th of August 1994 showed fluctuations in the size of the intertidal areas. Maximum extensions of tidal flats during a neap/spring tidal cycle were calculated to reach 40 and 59 km^2 (48–70 % of the total area) and a minimal size was calculated with values between 10 and 17 km^2 (11–20 % of the total area).

Areas, which fell dry for more than six hours were generally located near the mainland or near the island. Inside the backbarrier tidal flat of Spiekeroog only parts of the Janssand, the Martensplate and the Eversplate emerged for more than six hours. Gröninger Plate and Swinnplate are also intertidal areas, their exposure times varied between one to four hours, depending on neap/spring tide and the wind effect. A minimal extension of the intertidal areas was assumed for 13 minutes after high tide, maximum extension for 26 minutes after low tide.

Duration of inundation and exposure times at the Swinnplate and the Gröninger Plate

The duration of exposure and inundation at the Gröninger Plate and the Swinnplate was determined by geodetic data, meteorological conditions and neap/spring tide. Since the Gröninger Plate lies higher than the Swinnplate, the former is normally flooded longer than the Swinnplate. Because of the varying morphology of the flat some sections are exposed earlier than others.

At the Swinnplate, the central position is characterized by equally long inundation and exposure times (about 6 hours). However, the outer areas are flooded for nine to eleven hours.

The highest point of the Gröninger Plate lies in the east. During flood tide the water comes first from the north and south end of the flat, afterwards it increases from the west, and the east will be flooded last. An exception of this current pattern was observed during the storm flood in January 1994, when the lower situated areas of the flats remained flooded over four days.

Acknowledgement

This chapter is based on the studies in ELAWAT by Jan Backhaus, Udo Hübner, Ralf Kaiser, Hanz Dieter Niemeyer and Susanne Rolinski. Their studies were published as cited in the text, and further unpublished data were kindly provided by them. Udo Hübner and Mrs. Verduin gave helpful comments on the manuscript. Thanks for the help in translating to Ute Fuhrhop and Wolfgang Wenzel. The project was funded by the German Bundesministerium für Bildung, Wissenschaft, Forschung und Technologie (BMBF) under grant number 03F0112 A and B. The responsibility for the contents of the publication rests with the author.

References

Backhaus JO (1983) A semi-implicit scheme for the shallow water equations for application to shelf modelling. Continent Shelf Res 2/4: 243–254

Defant A (1923) Die Gezeiten der Nordsee, Beobachtungen und Theorie. Ann d Hydrogr, LI. Jg., H III

Hübner U, Backhaus, JO (1997) Der küstennahe Gezeitenstrom im Gebiet der östlichen Ostfriesischen Inseln. Forschungszentrum Terramare Berichte 4: 1–65

Maier-Reimer E (1980) On the formation of salt water wedges in estuary. In: Sündermann J, Kolz KP (Eds). Lecture notes on coastal and estuarine studies 1: 91–101

Niemeyer HD, Kaiser R (1994) Hydrodynamik im Ökosystem Wattenmeer. Umweltbundesamt, Berlin. Texte 26/94

Niemeyer HD, Brandt G, Gärtner J, Glaser D, Grüne J, Jensen F, Kaiser, R (1994) Naturuntersuchungen von Wattseegang in der deutschen Nordseeküste. In: Berichte der Forschungsstelle Küste, Band 40, pp 145–186

Reineck H-E (1987) Das Watt - Ablagerungs- und Lebensraum. Kramer, Frankfurt

Reise K, Riethmüller R (1998) Die Sylt-Rømø Wattenmeerbucht: Ein Überblick. In: Gätje C, Reise K (Eds). Ökosystem Wattenmeer. Austausch-, Transport- und Stofffumwandlungsprozesse. Springer, Berlin Heidelberg New York, pp 21–23

3.4 Morphology and Sedimentology of the Spiekeroog Backbarrier System

Andreas Hild

3.4.1 Origin of the Wadden Sea

The origin of the intertidal areas of the southern North Sea is controlled by the influence of tides, ongoing transgressive conditions and significant sediment advection. During the last Ice Age 15 000–18 000 years ago, the sea level was about 110–130 m lower than today (Jelgersma 1979). The melting of the glacial ice-caps in the early phase of the Holocene transgression caused a relatively fast sea level rise. Approximately 9000 years b. p. the sea level reached the northern Doggerbank and was about 45 m below the present sea level. From 8600 to 7100 years b. p. the sea level rose at a rate of approximately 2 cm a^{-1} up to 15 m below mean sea level. The base of the east Frisian islands, approx. 20 m below mean sea level, was reached between 8000 and 7500 years b. p. (Flemming 1990). Afterwards, a reduced rate of sea level rise occurred until 6500 years ago, up to 5–10 m below the present sea level (Streif 1990).

3.4.2 History of the Coastal Evolution

The historical evolution of the east Frisian Wadden Sea is characterized by significant morphological and hydrographical changes (e.g. Ehlers 1988). These processes can be mainly traced back to human activities and have continued, with decreasing tendency, until the 20th century. The analysis of historical maps shows the reduction of the Harle Bay catchment area from 180 km^2 in the year 1362 to only 60 km^2 at present owing to land reclamation (Flemming 1990). This suggests a larger tidal volume in the Wadden Sea before diking took place (Flemming 1990; Flemming & Davis 1994). As a result of the reduction of the catchment area, large amounts of sand were relocated from the former ebb-delta leading to an eastward growth of the island Spiekeroog. Due to the sea level rise, a slow shift of the islands occurs in south-easterly direction following the direction of the mean storm fronts (Nummedal & Penland 1981). This was verified by the results of short drill-cores from the islands that reached down to the Holocene and/or Pleistocene base (Streif 1990).

3.4.3 Morphology and Morphodynamics

The Wadden Sea ecosystem is characterized by the variability of morphology and sediment grain size composition. The morphology of the backbarrier areas is mainly formed by the influence of tidal currents. In the main inlet of the Spiekeroog backbarrier area, the Otzumer Balje, with a depth of up to 17 m, current velocities of 0.7–1.3 m s^{-1} were measured (Bartholomä & Flemming 1993). The velocities above the tidal flats reach values up to 0.3 m s^{-1} (Bartholomä 1993). A further essential factor forming the morphology in the area of the main inlet is swell, storm waves in particular, which enter the Otzumer Balje from NW (Chap. 3.3). In the long-term, the shift and changes in the size of the catchment areas of the tidal basins cause a gradual modification of hydrodynamic conditions. Sedimentation processes are in turn themselves influenced by these variations (Walther 1972; Fitzgerald et al. 1984; Sha & Berg 1993).

In July 1994, March 1995, September 1995, April 1996 and October 1996 a 100 m grid was established on the Swinnplate with a laser theodolite (90 measurement points; Bartholomä, Flemming unpubl. data). The measurements from 1995 and 1996 were converted to mean sea level. For each measurement location, the differences in height between two consecutive measurement campaigns were integrated and summed up for the calculation of volume balances over the area of every measurement unit (10×10^3 m^2).

In October 1994 an area of 1.5 km^2 on the Gröninger Plate was surveyed topographically with a grid distance of 125 m and converted to mean sea level (Bartholomä, Flemming unpubl. data). These data were compared with topographical maps of this area from 1964, 1975 and 1990.

The Swinnplate can be subdivided morphologically into three different areas (Bartholomä, Flemming unpubl. data). The western area changes mainly during wintertime, however, it stabilizes its relief during the summer. The middle, deeper area is covered by small tidal gullies and shows a strong seasonality in topography. The eastern area is a little more elevated and is influenced by biodeposits of the local mussel beds mainly in the summer months. Topographic changes of the Swinnplate, with spatially and temporally variable mussel settlements, show a good spatial agreement with the accumulation of biodeposits in the summer months and their erosion in winter (Sect. 3.4.4).

This seasonal cycle was disturbed by the strong ice winter of 1995/96. After total loss of the mussel beds and locally massive erosion, biodeposition was missing in the early summer and the area was merely subject to hydrodynamic conditions. Nevertheless, the morphology was re-established again approx. within 6 months after the ice winter (Bartholomä, Flemming unpubl. data). This relatively fast reorganization of the morphology after extreme events occurs following the prevailing hydrodynamic conditions and is therefore controlled by the large-scale geomorphology of the tidal basin (Oost & De Boer 1994; Nyandwi 1995; Eitner 1996).

The relief and location of the Gröninger Plate has remained relatively stable over a period of 30 years (Bartholomä, Flemming unpubl. data). The dynamic of this relief which slopes slightly to the north-west, documents a close link to hydrodynamic parameters. The slight decrease in topographical height indicates an increase in strong wind events (Fach 1996). Since the Gröninger Plate, unlike the Swinnplate, is almost solely subject to hydrodynamic influences, it is suitable as an indicator for longer-term modifications of physical water parameters and resulting morphological changes (Bartholomä, Flemming unpubl. data). Changes on the Swinnplate, however, occur in much shorter intervals and vary seasonally.

3.4.4 Sedimentology

Tidal sediments of the southern North Sea are by tradition divided into three sediment types on the basis of their mud content (Sindowski 1973; Gadow & Schäfer 1973): a) muddy sediment (content of the fraction <63 µm of more than 50 %), b) muddy sand (5–50 % of the fraction <63 µm) and c) sandy sediment (less than 5 % of the fraction <63 µm). According to these criteria, there are only small isolated spots of muddy areas in the backbarrier system of Spiekeroog. The by far largest part consists of muddy sand and pure sandflats, which dominante in the central part of the backbarrier system (Flemming & Davis 1994).

The sediment distribution in the Spiekeroog backbarrier system follows an energy gradient from the island to the mainland which is reflected by a particle fining trend in direction towards the mainland (Flemming & Nyandwi 1994). This landward fining trend is the result of a continuous sorting process that is driven by repeated particle resuspension and sedimentation controlled by particle settling velocities (particle size and density) and the local wave energy level (Flemming & Ziegler 1996). This depositional sequence is interrupted by mud patches (mussel beds – biodeposits of *Mytilus edulis* L.) in the central part of the catchment area (Chap. 3.6.1 and see Fig. 3.5.1).

The East- and Westfrisian Wadden Sea is characterized by a pronounced sediment zonation parallel to the coast (Oost 1995). The finest, most landward sediment facies has lately shown a decrease in fine sands accompanied by a widespread lack of mud. This lack in fine grained sediments is a result of elevated energy gradients in the backbarrier area, caused simultaneously by artificial sea dikes and a rising sea level (Flemming & Nyandwi 1994). In the course of this sea level rise the islands are shifted landward. This movement of the whole system is the consequence of a sediment deficit in the backbarrier system resulting from the rising sea level and leads to a further decrease of fine particles in the sediment.

There were only insignificant differences in the grain size distributions between the Swinn- and Gröninger Plate (Bartholomä, Flemming unpubl. data). The Swinnplate has a slightly larger abundance of coarser fine sands (2.0–2.5 phi), whereas the percentage of finer fine sands (2.5–3.0 phi) is somewhat smaller. Both study areas were located outside the zone of physical mud sedimentation, i.e., non-biogenic muds occur only in small and seasonally varying amounts. The Swinnplate is characterized by the accumulation of biogenic muds (Bartholomä, Flemming unpubl. data). These are fine grained biodeposits (faeces and pseudofaeces produced by the mussel *Mytilus edulis* L. (Chap. 3.6.1)) with more than 50 % being contributed by the fraction <63 µm.

Large amounts of this mud are remobilized especially during storm events with high wave energy. The consequences are small- to mid-scale modifications in morphology and sediment composition. During calmer weather conditions, the sediment distribution is reorganized according to the former distribution following the large-scale zonation in accordance with the hydrodynamic equilibrium (Bartholomä, Flemming unpubl. data).

Particle sinking velocities are smaller in the winter months due to higher wave energy and lower water temperatures (increased cinematic viscosity of the water) (i.e., in winter, larger particles can stay longer in suspension; particle sizes are up to 80 µm in winter and approx. 60 µm in summer (Krögel 1997)). This leads to a higher particle load in the water column during these months. This results in a significant decrease in abundance of the finer fractions (in particular the <63 µm fraction) in the sediments during winter.

The sediment distribution in the central part of the Swinnplate clearly reflects the seasonal alternation between a winter situation dominated by hydrodynamical sorting and a summer situation influenced by biodeposition (Bartholomä, Flemming unpubl. data). In the winter, the mud content of the sediments decreases to less than 10 %. The mud is resuspended and exported in suspension. However, the seasonal sediment distribution and/or -zonation remains largely unchanged over longer time intervals.

With regard to the repeated observation of this seasonal alternation, the increased heterogeneity after the strong ice winter of 1995/96 had only a short term effect. The loss of the mussel beds lead to deeper incisions in the topography. After removal of the top sediment layer, the underlying sediments were easily exported. As there was no biodeposition, the effect of the ice winter was even more pronounced on the mud-poor Gröninger Plate (Bartholomä, Flemming unpubl. data). Without biogenic sedimentation, the decrease in mud on the Gröninger Plate during the ice winter will have longer lasting effects than on the Swinnplate (Chap. 3.6.1).

Acknowledgements

This chapter is based on the results and data of A. Bartholomä, B.W. Flemming and their colleagues from the Senckenberg Institute. Their work was published as cited in the text and further unpublished data were kindly provided. A reviewer gave helpful comments on the manuscript. The project was funded by the German Bundesministerium für Bildung, Wissenschaft, Forschung und Technologie (BMBF) under grant number 03F0112 A and B. The responsibility for the contents of the publication rests with the author.

References

Bartholomä A (1993) Zeitliche Variabilität und räumliche Inhomogenität in den Substrateigenschaften und der Zoobenthosbesiedlung im Umfeld von Miesmuschelbänken. D. Hydrodynamik. Senckenberg am Meer Bericht 93/1: 117–123

Bartholomä A, Flemming BW (1993) Zeitliche und räumliche Variabilität in den Sedimentparametern und der Morphologie auf der Gröninger Plate (Spiekerooger Watt). Senckenberg am Meer Bericht 93/1: 5–48

Ehlers J (1988) The morphodynamics of the Wadden Sea. Balkema, Rotterdam

Eitner V (1996) Morphological and sedimentological development of a tidal inlet and its catchment area (Otzumer Balje, southern North Sea). J Coast Res 12: 271–293

Fach B (1996) Die Entwicklung der Windverhältnisse in der Deutschen Bucht seit 1965, als wichtiger Umwelteinfluß auf die Küstenzone. Diplomarbeit, Fachhochschule Wilhelmshaven

Fitzgerald DM, Penland S, Nummedal D (1984) Changes in tidal inlet geometry due to backbarrier filling: East Frisian Islands, West Germany. Shore & Beach 52: 3–8

Flemming BW (1990) Zur holozänen Entwicklung, Morphodynamik und faziellen Gliederung der mesotidalen Düneninsel Spiekeroog (Südliche Nordsee). In: Willems H, Wefer G, Rinski

M, Donner B, Bellman H-J, Eißmann L, Müller A, Flemming BW, Höfle H-C, Merkt J, Streif H, Hertweck G, Kuntze H, Schwaar J, Schäfer W, Schulz M-G, Grube F, Menke B (Eds) Beiträge zur Geologie und Paläontologie Norddeutschlands - Exkursionsführer. Bremen, pp 13–73

Flemming BW, Davis RA (1994) Holocene evolution, morphodynamics and sedimentology of the Spiekeroog barrier island system (Southern North Sea). Senckenbergiana marit 24: 117–155

Flemming BW, Nyandwi N (1994) Land reclamation as a cause of fine grained sediment depletion in backbarrier tidal flats (Southern North Sea). Neth J Aquat Ecol 28: 299–307

Flemming BW, Ziegler K (1996) High-resolution grain size distribution patterns and textural trends in the backbarrier environment of Spiekeroog island (Southern North Sea). Senckenbergiana marit 26: 1–24

Gadow S, Schäfer A (1973) Die Sedimente der Deutschen Bucht: Korngrößen, Tonmineralien und Schwermetalle. Senckenbergiana marit 5: 165–178

Jelgersma S (1979) Sea-level changes in the North Sea basin. In: Oele R, Schüttenhelm RTR, Wiggers AJ (Eds) The Quaternary History of the North Sea. Acta Univ. Upsala, Symp Univ Upsala, Annum Quingentesimum Celebrantis, Upsala, pp 233–248

Krögel F (1997) Einfluß von Viskosität und Dichte des Seewassers auf Transport und Ablagerung von Wattsedimenten (Langeooger Rückseitenwatt, südliche Nordsee). Dissertation, Universität Bremen

Nummedal D, Penland S (1981) Sediment dispersal in Norderneyer Seegat, West Germany. IAS Spec Publ 5: 187–210

Nyandwi N (1995) The nature of the sediment distribution pattern in the Spiekeroog backbarrier area, the East Frisian Islands. Berichte, Fachbereich Geowissenschaften, Universität Bremen, 66

Oost AP (1995) Sedimentological implications of morphodynamic changes in the ebb-tidal delta, the inlet and the drainage basin of the Zoutkamperlaag tidal delta (Dutch Wadden Sea), induced by a sudden decrease in the tidal prism. In: Flemming BW, Bartholomä A (Eds) Tidal Signatures in Modern and Ancient Sediments. Blackwell, Oxford. Int Ass Sed Spec Publ 24: 101–119

Oost AP, de Boer PL (1994) Sedimentology and development of barrier islands, ebb-tidal deltas, inlets and backbarrier areas of the Dutch Wadden Sea. Senckenbergiana marit 24: 65–115

Sha LP, van den Berg JH (1993) Variation in ebb-tidal geometry along the coast of the Netherlands and the German Bight. J Coast Res 9: 730–746

Sindowski K-H (1973) Das ostfriesische Küstengebiet: Inseln, Wattenmeer und Marschen. Sammlung Geologischer Führer Vol 57. Bornträger, Berlin, Stuttgart

Streif H (1990) Das ostfriesische Küstengebiet – Nordsee, Inseln, Watten und Marschen. Sammlung Geologischer Führer Vol. 57. Bornträger, Berlin, Stuttgart

Walther F (1972) Zusammenhänge zwischen der Größe der ostfriesischen Seegaten mit ihren Wattgebieten sowie den Gezeiten und Strömungen. Forschungsstelle Norderney, Jahresbericht 23, (1971) pp 7–32

3.5 Biofacies in the Spiekeroog Backbarrier Tidal Flats

Carmen-Pia Günther

Abstract: Swinnplate and Gröninger Plate are typical sandy tidal flats of the backbarrier area of the island of Spiekeroog. Their biofacies differ mainly in the occurrence, rsp. lack of beds of *Mytilus edulis*. Due to the fact that this species is a biogenic mud collector, sediments of both flats also differ in their mud content (Chap. 3.4). The ice winter 1995/96 resulted in a lack of *Mytilus* beds on both sandflats and of *Lanice conchilega* in large parts of the Swinnplate and nearly entirely on the Gröninger Plate. On the Swinnplate a loss not only of the ecosystem engineer *Mytilus edulis* but also of the bioaccumulated sediment fraction <63 µm was recorded. Directly after the ice winter, *Heteromastus filiformis* was observed in areas where sediment of the fraction <63 µm was newly deposited.

On a large temporal and spatial scale, biofacies provide information on patterns (vertically) and variability (horizontally) of sediment structure and associated macrofauna. The interpretation of sequences of macrofauna associations at one locality as cyclic changes in the sense of the mosaic-cycle theory may be misleading. Vertical sequences of facies may not only result from a temporal succession of macrobenthic associations, but also from a horizontal transport of sediment layers. The vertical sequence of facies may also be affected by erosion due to storms or ice cover leading to a removal of once existing facies. New methods in determining the absolute age of facies, e.g. age determination via the amino acid content in bivalve shells, may provide more precise results in the future.

3.5.1 Introduction

In geology, a facies is defined as a specific character of a sediment with a differentiation between anorganic (lithofacies) and biogenic characteristics (biofacies). The latter may be presented by bodies or skeletons of fossil lifeforms or their traces (Lebensspuren). Actuopaleontology uses recent organisms to interpret biofacies assuming that the same processes lead to the same results independent from the point of time in the earth's history.

Generally, the sediment bodies of the Wadden Sea underlie continuous changes in their morphology and composition (e.g. sediments, biota). Climate, hydrodynamics as well as biological processes interact and produce characteristic facies. Changes of facies indicate changes in sediment composition and associations of biota in vertical and horizontal direction. By estimating the time span in which changes occur, it can be described which macrofauna association characterizes an area, how long an association persisted at one place rsp. whether temporal sequences of associations occurred. Using this biotope history, long-term effects of events in the sediments may be evaluated allowing the consideration of stability properties on a larger temporal scale than the time span of most scientific projects. In the frame of ELAWAT, the evidence based on biofacies in relation to the mosaic-cycle theory was of special importance (compare Chap. 2).

From 1988 to 1993, a large-scale mapping of the backbarrier tidal flats behind the island of Spiekeroog was carried out in order to identify sediment bodies and habitats as well as the Lebensspuren producing animals of the macrozoobenthos (Hertweck 1995). Besides this areal approach, sediment cores of 1 m length were taken at selected stations to document the temporal sequence of facies at these sites. The cores were bisected in the laboratory and hardened with plastic (method described in Reineck 1970). Samples of sediment and shells were collected from each facies in the core. In the frame of diploma theses, sediment structure and biofacies of the Martensplate (Knecht 1991), the Swinnplate (Kurmis 1995; Haberer 1997) and the Gröninger Plate (Engelbrecht 1997) were mapped. Of special interest was the description of the biofacies after the ice winter 1995/96 (Haberer 1997; Engelbrecht 1997).

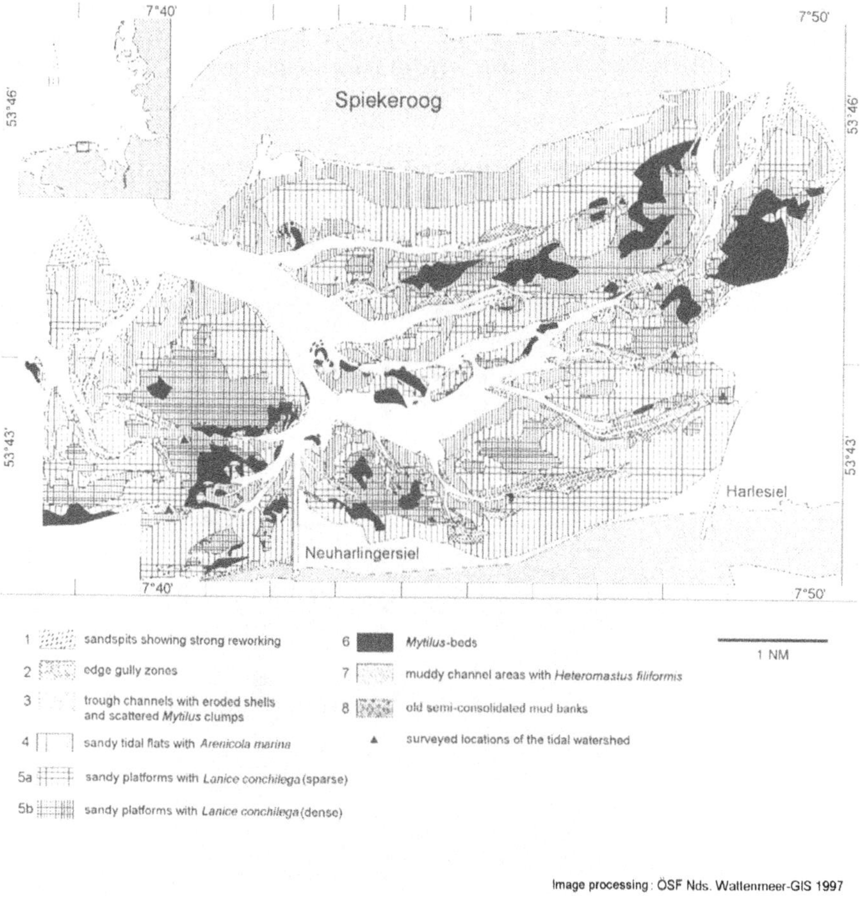

Fig. 3.5.1 Map of the biofacies in the backbarrier system of Spiekeroog Island, surveyed by Hertweck (modified from Hertweck 1995, 1998)

3.5.2
Horizontal and Vertical Aspects of Biofacies

The backbarrier tidal flats behind the island of Spiekeroog were surveyed during the first mapping carried out in 1988 in the area of the tidal water shed between Harle and Otzumer Balje (Hertweck 1995), characterizing the sediment bodies and benthic associations. Hertweck (loc. cit.) discriminated 8 biofacies: (1) sand spits with strong physical reworking, (2) edge gully zones at the border of major tidal channels (description see Reineck, 1995); (3) trough channels with eroded shells and scattered *Mytilus edulis* clumps, (4) sandy tidal flats with *Arenicola marina*, (5) platforms with varying abundances of *Lanice conchilega*, (6) mussel beds (*Mytilus edulis*), (7) muddy channel areas with *Heteromastus filiformis* and *Mya arenaria*, (8) semi-consolidated mudbanks (Fig. 3.5.1). At the time of mapping, the Swinnplate was characterized by areas of sandy high platforms, colonized by *L. conchilega* in varying abundances. In the centre two extended beds of *M. edulis* occurred, which continuously shifted in place during the study period of ELAWAT. Beside the mussel beds, facies of type 1 and 2 were found. Old mud beds and trough channels with shells were of minor importance.

Fig. 3.5.2 Relief cast from a mature mussel bed (*Mytilus edulis*) in the central area of the Swinnplate, backbarrier tidal flat of Spiekeroog Island (photo courtesy of G. Hertweck)

The central part of the Gröninger Plate is a sandy platform, inhabited by *L. conchilega* in varying densities. This facies was surrounded by *A. marina* dominated sandflats. As on the Swinnplate, small-scale areas belonging to facies type 1, 2 and 8 were found. Edge gully zones were found at the border to the northerly and southerly situated tidal channels.

Each facies type also had characteristic features in the vertical direction. At the Swinnplate, six out of the eight characteristic types were sampled with corers. A summary of their characteristics is presented in Table 3.5.1. The stratigraphical time span varied with the situation within the area of investigation. In general, an increase of age from the edge to the centre of the plate could be observed, with the time scale reaching back from a few tides up to several decades or even the order of magnitude of centuries.

Successions can be shown by Lebensspuren of dominant species in different sediment depths. Fig. 3.5.2. shows a relief of sediment in the central part of the Swinnplate, including a sequence of 3 *M. edulis* colonizations at this place. The occurrence of sub-articulated shells in the deeper layer indicates strong and fast sedimentation after the death of the organisms and may be interpreted as proof for locally produced autochthonous shells. In the deepest horizon, tubes of *L. conchilega* from earlier colonization phases can be observed.

Facies sequences of *L. conchilega* and *M. edulis* do not give a clear-cut indication that *Lanice* aggregates and *Mytilus* beds occurred in a temporal sequence and excluded each other. Two possible processes producing such a pattern have to be considered: (1) The patchy distribution of *L. conchilega* and *M. edulis* in the central part of tidal flats, as observed by Hertweck (1995, 1998) includes free areas between the *Mytilus* aggregates inhabited by *Lanice*. Horizontal shifts e.g. of the mussels within the area may lead to a vertical succession of both facies in a sediment core although both species occurred contemporaneously at that site. (2) Settlement of *Mytilus* juveniles in dense *Lanice* aggregates may result in a qualitative shift from one facies to the other (Hertweck 1995). The facies sequence observed is the result of the settling behaviour of *M. edulis* which is dependent on hard substrate (in this case *Lanice*-tubes) and points to a causal biotic interaction.

3.5.3 Case Study *Mytilus* Bed

For the biofacies "*Mytilus* bed" the method of the relief cast was combined with geochemical analyses (C_{org}, N_{tot}, P_{tot}, pH, Eh) of the biogenic mud (Hertweck & Liebezeit 1996). The longer a *Mytilus* bed exists at one place, the deeper the mud layer beyond the mussels and the higher the content of particles < 63 µm in the mud (loc. cit.). A comparison of the mature (with large number of adult specimen) *Mytilus* bed at the Swinnplate with an older receding bed in the area of the Hullplate showed that the latter one had a 15 cm thick persistent layer of mud. The C/N ratio indicated that here, besides erosion and bioturbation, a clear decomposition of the organic material occurred, while in the area of the active *Mytilus* bed a minor decomposition was recorded.

Table 3.5.1 Characteristics of the 8 typical biofacies of the Spiekeroog backbarrier tidal flats (Fig. 3.5.1). The sequence results from the stratigraphical time span from the sediment surface down to ca. 30 cm depth

type	sediment	biogenic structure	situation	stratigraphical time span	characteristics
1	sand	-	sand spits of tidal flats and point bar of large tidal channels,	tides	laminar sediment layers, strong currents
2	sand	-	edge gully zones at the stoss side of large tidal channels	3-5 years	
3	sand	shells of bivalves, *Mytilus*-aggregates	marginal tidal flat position at main channels		shallow trough
4	sand	burrows of *A. marina*, shells of *C. edule*, burrows of *S. armiger*	borders of tidal flats		bioturbation and physical reworking
5	sand	tubes of *L. conchilega*	central parts of tidal flats		single tubes up to dense tube mats of *L. conchilega*
6	mixed sediments	*Mytilus* shells	central parts of tidal flats, at borders of tidal channels fields of shells eroded	> 50 years	often sub-articulated *Mytilus* shells
7	mixed sediments mud	burrows of *H. filiformis*, *Mya*-shells in life position	watershed of backbarrier systems		area of sedimentation
8	semi-consolidated mud		border of tidal channels with erosive character	order of magnitude of several centuries	historical water sheds

From the sedimentologists' point of view, *Mytilus* beds present highly dynamic systems. The age of such systems is difficult to estimate due to the continuous change of sediment import and erosion. Hertweck (1998) estimated the age of the *Mytilus* beds at the Swinnplate to be in the range of 21–50 years. Behrends et al.

(1997) made a first attempt to determine the age of sediment layers from the D/L-ratio of asparagin acid in the shells of bivalves (*Cerastoderma edule*, *Mya arenaria*). The shells of these two species were analysed from two 1 m long sediment cores sampled in the *Mytilus* bed. The maximum age was determined for a 2000 year old *C. edule*, the youngest specimen was estimated to be a living *M. arenaria* belonging to the O-group. From the relationship between age of the shell and sediment depth the authors estimated the mean sedimentation rate to be 34 cm per century. According to this, the sediment depth of 35 cm analysed by Hertweck would represent a century, while he himself estimated the age of successive *Mytilus* facies to be a maximum of 50 years. This points to the need for further development of methods, such as the amino acid method, for dating sediment layers in highly dynamic environments.

3.5.4 Effects of Ice Winters

Beside the mapping of sediment composition and biofacies in the Spiekeroog backbarrier tidal flats (Hertweck 1995), a more detailed study was carried out at the Swinnplate in 1988 (Kurmis 1995). The results are in good agreement with those from Hertweck (1995). After the ice winter of 1995/96, the biofacies distribution was modified. In the western part of the Swinnplate no *L. conchilega* were found and *A. marina* became the dominating species (Haberer 1997). At two stations in the southern part, the polychaete *H. filiformis*, who prefers muddy habitats, was observed together with a local increase of sediments <63 μm up to 5–10 % of the total sediment (surrounding stations 0–5 %). *Mytilus* beds, dominating the centre of the Swinnplate were destroyed by ice-scour. The mud deposited within the mussel beds was removed. At the eastern part of the Swinnplate a sparse to dense colonization of *L. conchilega* was described. It could not be distinguished if these represented either empty tubes of animals living there in preceding years, or juveniles (date of sampling: March/April 1996; Haberer 1997).

As opposed to the Swinnplate, *L. conchilega* was only observed at three sampling points at the Gröninger Plate (sampling dates April–June 1996; Engelbrecht 1997). On the large-scale, the area was dominated by *A. marina* with some patches of *Pygospio elegans*. In the area of the tidal channel protruding from the southwest into the Gröninger Plate, *H. filiformis* was found after the ice winter. This differs from observations made in 1992 when the polychaete did not occur at this area (Flemming & Bartholomä, unpublished; Müller & Widdel, unpublished). As on the Swinnplate, this finding co-occurred with an increase of sediments <63 μm at these stations which could be due to ice-scour.

These changes in the spatial distribution of biofacies and sediments due to a natural disturbance event help with the interpretation of facies sequences in depth profiles.

Acknowledgements

This chapter is based on the studies and data of G. Hertweck and his colleagues and students from the Senckenberg Institute in Wilhelmshaven. I gratefully acknowledge the help from G. Hertweck, who provided me with information, discussion and (hopefully) insight into the field of actuopalaeontology. A reviewer gave helpful comments on the manuscript. The project was funded by the German Bundesministerium für Bildung, Wissenschaft, Forschung und Technologie (BMBF) under grant number 03F0112 A and B. The responsibility for the contents of the publication rests with the author.

References

Behrends B, Goodfriend GA, Liebezeit G (1997) Dating of intertidal flat sedimentation by amino acid racemisation - first results. Forschungszentrum Terramare Berichte 2: 23–26

Engelbrecht I (1997) Geologische Detailkartierung der Gröninger Plate im Spiekerooger Rückseitenwatt im Frühjahr 1996. Diplomarbeit, Universität Bremen

Haberer T (1997) Geologische Detailkartierung der Swinnplate im Spiekerooger Rückseitenwatt im Frühjahr 1996. Diplomarbeit, Universität Bremen

Hertweck G (1993) Zeitliche Variabilität und räumliche Inhomogenität in den Substrateigenschaften und der Zoobenthosbesiedlung im Umfeld von Miesmuschelbänken. A. Biofazies und Aktuopaläontologie. Senckenberg am Meer Bericht 93/1: 49–63

Hertweck G (1995) Verteilung charakteristischer Sedimentkörper und Benthosbesiedlungen im Rückseitenwatt der Insel Spiekeroog, südliche Nordsee. I. Ergebnisse der Wattkartierung 1988–92. Senckenbergiana marit 26: 81–94

Hertweck G (1998) Facies characteristics of backbarrier tidal flats of the East Frisian island of Spiekeroog, southern North Sea. In: Alexander C, Davis jr. RA, Henry J (Eds) Tidalities, Processes and Products. SEPM Special Publ 61: 23–30

Hertweck G, Liebezeit G (1996) Biogenic and geochemical properties of intertidal biosedimentary deposits related to *Mytilus*-beds. P.S.Z.N.I. Mar Ecol 17: 131–144

Knecht S (1991) Morphologische Entwicklung, Sedimentverteilung und Faziesdifferenzierung der Martensplate (östliches Spiekerooger Watt). Diplomarbeit (Teil 1), Justus-Liebig-Universität Gießen

Kurmis R (1995) Geologische Detailkartierung der Swinnplate im Spiekerooger Rückseitenwatt. Sedimentologie und Biofazies. Diplomarbeit, Universität Bremen

Reineck H-E (1970) Reliefguß und projizierbarer Dickschliff. Senckenbergiana marit 2: 61–66

Reineck H-E (1995) Randpriele. Senckenbergiana marit 26: 37–43

3.6 Ecosystem Engineers: *Mytilus edulis* and *Lanice conchilega*

Andreas Hild & Carmen-Pia Günther

The two main study areas of the ecosystem project ELAWAT, the Swinnplate and the Gröninger Plate, were characterized by different dominant species, which were the focus of scientific interest. The Swinnplate (Fig. 3.1.3) represented an area partly covered with mussel beds of different age (*Mytilus edulis* L.), while the Gröninger Plate (Fig. 3.1.4) was covered with the tube-worm *Lanice conchilega* in patches of varying density. Both species influence sediment and biota in their vicinity, but in different ways. The blue mussel affects the surrounding area predominantly by the biogenic deposition of mud in the form of faeces and pseudofaeces, which are deposited in and around the mussel beds. Additionally, the roughness of these beds reduces current velocities which results in the physical settling of particles within the mussel beds. Currents were also reduced in patches with *L. conchilega*. In this case, the tubes affected the current velocities. As a consequence, the surrounding benthic population was modified. Therefore, both species influence the local hydrographic regime by their presence, while *M. edulis* additionally accumulates material.

3.6.1 *Mytilus edulis*

Mussel beds significantly influence the sedimentation of fine grained material in the Wadden Sea. The mussels filter suspended matter (inorganic particles, refractory/labile detritus and living cells) out of the water column, incorporate TOC, bacteria and phytoplankton as food (Wright et al. 1982; Fréchette & Bourget 1985) and excrete non-digested material compressed as faeces and pseudofaeces (faecal pellets) (Verwey 1952; Haven & Morales-Alamo 1966). Faeces have passed the intestine of the mussels while pseudofaeces have not and are rejected before. *Mytilus edulis* L. filters material in the size range of < 100 µm (Bayne et al. 1976) to > 4 µm (sometimes material >1 µm) out of the water column (Kautsky & Evans 1987). Above particle concentrations of ± 5 mg l^{-1}, some of the filtered material is rejected and excreted as pseudofaeces without passing the intestine tract (Widdows et al. 1979). The rate of filtration amounts to 10–20 l per day per mussel (Asmus 1994). In addition to the active deposition, flocculated fine grained material settles passively due to the roughness of the mussel beds (e.g. ten Brinke et al. 1995). In this way, large amounts of biodeposits (faeces and pseudofaeces) as well as shells and fine grained material are deposited within mussel beds and in the surrounding areas. Therefore, mussel beds may be termed biosedimentary systems. Old mussel beds in sheltered parts of the Wadden Sea may be elevated (mussel mudmounds) compared to the surrounding tidal flats (Reineck 1994).

Sinking and/or transportation behaviour of the faecal pellets corresponds to that of quartz particles in the silt/sand size-class (Oost 1996). Therefore, resuspended biodeposits are transported and deposited together with sand particles and/or incorporated into sandy sediments (Hüttel et al. 1996). This can lead to sandflats with a relatively high mud content in the vicinity of mussel beds, resulting from the export of fine grained mussel deposits. Without their activity, the fine grained material deposited by the mussels could only accumulate in sheltered areas of the Wadden Sea (e.g. mudflats near the coast), or would be exported from the backbarrier system towards the North Sea. Biodeposits are muddy sediments deposited independently from the hydrodynamic energy gradients (Chap. 3.4; Flemming 1990).

The resistance of the biodeposits against erosion is governed by the ratio of faecal pellets deposited by filtration and mud particle aggregates settling out of the water column (ten Brinke et al. 1995). These authors showed that the biogenically deposited particles in the form of faecal pellets represent only 30 % of the entire mud content in the upper 5–10 cm of the sediments. Consequently, the existence and resistance against erosion of the biodeposits depends mainly on the physical state of the muds and the local hydrodynamic conditions.

The amount of faeces and pseudofaeces produced varies strongly over the seasons depending on the filtration activity of the mussels. The lowest amounts are produced in winter, while from spring to early summer more and more material is accumulated (Oost 1996). This is in good agreement with the results of Kautsky & Evans (1987) and Jørgensen (1990), which showed that the activity of the mussels in the northern hemisphere is highest from April to September.

The accumulation and erosion of fine grained material in the vicinity of mussel beds in the Dutch Wadden Sea is described by Oost (1996) as an annual cycle which is controlled by variations in physical and biological conditions. Because of storms and ice drift, mud is resuspended and/or eroded. Physical factors like wind, temperature and tides result in a higher suspended sediment load in the water column during winter months (Dronkers 1984; De Wilde & Beukema 1984; Hickel 1984; Krögel 1997) (Chap. 3.4). De Haas & Eisma (1993) give suspended sediment concentrations during storms of 0.8 g/l, Oost (1996) mentions maximum values of up to 50 g l^{-1}. In the context of ELAWAT, maximum suspended sediment concentrations of 0.7 g l^{-1} were determined. Resuspended sediments are exported towards the open North Sea or settle again at the end of the stormy intervals.

The effects of the strong ice winter of 1995/96 showed the sediment stabilizing function of mussel beds. After the removal of the mussel beds, deep incisions (up to 1 m) into the topography developed. The lack of protection against erosion and of sediment accumulation by the mussels, induced an increased erosion of the former underlying sediment layers.

Although the mussel covered areas amount only to approx. 1 % of the total Wadden Sea area, the biomass of the mussels comprises between 20 % and 70 % of the total benthic biomass, depending on the area under consideration (Reise 1991). Differences exist between the East Frisian and North Frisian tidal flats in shape and location of the mussel beds. In the backbarrier systems behind the East Frisian island chain, the mussel beds are frequently found close to the tidal watersheds, whereas in the North Frisian tidal flats they are situated almost exclusively in the lower intertidal zone. Both, the mean population density of the mussels and their biomass per m^2 were at least twice as high in the Königshafen of Sylt compared to the East Frisian tidal flats (Dittmann 1987; Asmus 1987; Michaelis et al. 1995). The distribution pattern of the mussel beds in individual tidal areas remains relatively constant, but the size of the mussel population fluctuates strongly from year to year (Dankers & Koelemaij 1989; Nehls & Thiel 1993). Following calculations of Dankers & Koelemaij (1989), the water volume of the Dutch Wadden Sea is filtered by the mussel population every 8–9 days, in years with large amounts of mussels. This takes about 1 month in years with a small mussel population. In the last years a severe decrease in the mussel population has been observed both for the Dutch (Beukema 1993; Dankers 1993) and the East Frisian Wadden Sea (Michaelis et al. 1995; Herlyn 1996). Possible causes are: increased predation, algal blooms, fishing, pollution and parasite infestation. The continuous existence of a mussel bed is also subject to the influence of mechanical disturbances like storm events, ice drift and the seasonal cycle of physical water parameters as well (Verwey 1952). Nehls & Thiel (1993) discuss the influence of storm events on the distribution and frequency of mussel beds. They distinguish between highly dynamic, exposed mussel beds and mussel beds in wind sheltered areas that can exist for a long time period.

In sediments containing biodeposits a fauna prevails that is adapted to oxygen deficiency as well as high TOC- and H_2S-concentrations. The sediment below mussel beds represents a habitat with a fauna consisting of a small number of species with many individuals. By contrast the epifauna that only exists on mussel beds in the Wadden Sea, exhibits a high diversity (Dittmann 1987, 1990).

Remineralization processes are intensified in the deposits of mussel beds compared to the surrounding sandflats (Asmus et al. 1990). The cycle of production and consumption of organic material is accelerated (Dankers & Zuidema 1995). *In situ* measurements have shown the possibility of high release rates of inorganic nutrients from mussel beds into the water column (Dame & Dankers 1988; Dame et al. 1989; Asmus et al. 1990; Asmus 1994; Prins & Smaal 1994; Zurburg et al. 1994). In this case, the mussels are only responsible for a part of the whole remineralization. The major part is controlled via the activity of micro-organisms and meiofauna living in the faeces and the pseudofaeces (Dankers et al. 1989; Kosfeld 1989). Biodeposits reveal a high bacterial activity (Dahlbäck & Gunnarsson 1981; Grenz et al. 1990). The nutrients which are released during mineralization are available for primary production again. In this way mussel beds stimulate the growth of phytoplankton (Asmus & Asmus 1991; Prins & Smaal 1994). A comprehensive study of the role of mussel beds in deposition and/or consumption of organic matter and the release of inorganic nutrients was published by Smaal & Prins (1993).

Due to the habitat heterogeneity of the mussel bed, the biodiversity is higher compared to mudflats. Since mussel beds influence and modify their surrounding environment due to their presence and activity, they can be regarded as ecosystem engineers in the sense of Jones et al. (1994). In conjunction with the increasing impoverishment of fine grained material in the backbarrier systems in the course of diking and land reclamation (Flemming & Nyandwi 1994), the areas influenced by biogenic sedimentation gain a special significance in the Wadden Sea (potential substitute for mudflats). However, these areas represent no precise analogue to mudflats, since they are subject to a strong temporal and spatial variability. In addition, the benthic communities in mussel beds and tidal flats enriched with biodeposits are different from natural mudflats (Kröncke 1996).

3.6.2
Lanice conchilega

In sandy tidal flats of the Wadden Sea and in the ambient subtidal areas of the North Sea, the tube building polychaete *Lanice conchilega* may reach a high abundance combined with high biomass values. In the East Frisian Wadden Sea the highest abundance of 2800 ind m^{-2} was recorded from the Dornumer Nacken. In the Spiekerooger backbarrier tidal flats, on the Neuharlingersieler Nacken, a maximum abundance of 1600 ind m^{-2} was observed, and at the Gröninger Plate 150–1800 ind m^{-2}. The highest abundance ever reported in the literature was more than 10 000 ind m^{-2} in a subtidal estuarine population (Buhr 1979).

In the intertidal area, *L. conchilega* preferentially occurs in sandy sediments and feeds on suspended algae as well as detritus (Hartmann-Schröder 1996). In the Wadden Sea, this species may reach an age of 1–2 years (Beukema et al. 1978).

As for other tube building polychaetes, it may be assumed that the tubes of *L. conchilega*, protruding 2–3 cm from the sediment surface, strongly affect the hydrodynamical regime in the benthic boundary layer and thus the distribution of co-occurring biota (e.g. Eckman et al. 1981; Eckman 1985; Carey 1983, 1987). During the pilot phase of ELAWAT, studies revealed that the aggregation of a harpacticoid copepod was significantly correlated with the occurrence of *L. conchilega* tubes (Pfeifer et al. 1996). More interactions with other taxa such as meiofauna (nematodes) and macrofauna were investigated during the main phase of ELAWAT (Chap. 5.5).

In summary, this polychaete species is an "ecosystem engineer" (sensu Jones et al. 1994) and as such a central key to the understanding of the community dynamics in *Lanice* flats. ELAWAT results proved that *L. conchilega* does not build up a characteristic association of its own, but enriches an *Arenicola*-dominated sandflat association in abundance and species numbers due to a reduction of current velocities by its tubes (Zühlke et al. 1998). Its effect on the ambient area varies with the density of tubes, which itself is determined by recruitment success of *L. conchilega* (reproduction and initial settlement) (Chaps. 5.3 and 5.4). In the long-term, the population-dynamics of this cold sensitive species are determined by the frequency of cold winters in the Wadden Sea. Directly after cold winters, *L. conchilega* can no longer be observed in the intertidal. Onset of recovery of the intertidal parts of the population will occur the next summer due to recruitment processes (Chap. 5.3; Ziegelmeier 1964; Beukema et al. 1978). High standing stocks of *L. conchilega* in the intertidal develop in periods of several successive mild winters (Beukema et al. 1978; Beukema 1992).

Acknowledgements

The data used in this chapter from ELAWAT are based on the studies by W.-E. Arntz, Alexander Bartholomä, Brigitte Behrends, Carlo van Bernem, Dietrich Blome, Monique Delafontaine, Burghard Flemming, Carmen-Pia Günther, Günther Hertweck, Jens Heuers, Sandra Jaklin, Ingrid Kröncke, Gerd Liebezeit, Jörn Reichert, Günther Sach and Ruth Zühlke. We thank the reviewer for helpful comments on the manuscript. The project was funded by the German Bundesministerium für Bildung, Wissenschaft, Forschung und Technologie (BMBF) under grant number 03F0112 A and B. The responsibility for the contents of the publication rests with the authors.

References

Asmus H (1987) Secondary production of an intertidal mussel bed community related to its storage and turnover compartments. Mar Ecol Prog Ser 39: 251–266

Asmus H (1994) Bedeutung der Muscheln und Austern für das Ökosystem Wattenmeer. In: Lozán JL, Rachor E, Reise K, v. Westernhagen H, Lenz W (Eds) Warnsignale aus dem Wattenmeer. Blackwell Wissenschafts-Verlag, Berlin, pp 127–132

Asmus RM, Asmus H (1991) Mussel beds: limiting or promoting phytoplankton? J Exp Mar Biol Ecol 148: 215–232

Asmus H, Asmus RM, Reise K (1990) Exchange processes in an intertidal mussel bed: a Sylt-flume study in the Wadden Sea. Ber Biol Anst Helgoland 6: 1–79

Bayne BL, Thompson RJ, Widdows J (1976) Feeding and digestion. In: Bayne BL (Ed) Marine Mussels: Their Ecology and Physiology. Int Biol Progr 10, Cambridge Univ Press, Cambridge, pp 121–158

Beukema JJ (1992) Expected changes in the Wadden Sea benthos in a warmer world: lessons from periods with mild winters. Neth J Sea Res 30: 73–79

Beukema JJ (1993) Increased mortality in alternative bivalve prey during a period when the tidal flats of the Dutch Wadden Sea were devoid of mussels. Neth J Sea Res 31: 395–406

Beukema JJ, De Bruin W, Jansen JJM (1978) Biomass and species richness of the macrobenthic animals living on the tidal flats of the Dutch Wadden Sea: Long-term changes during a period of mild winters. Neth J Sea Res 12: 58–77

Buhr K-U (1979) Eine Massensiedlung von *Lanice conchilega* (Polychaeta: Terebellidae) im Weser-Ästuar. Veröff Inst Meersforsch Bremerh 17: 101–149

Carey DA (1983) Particle resuspension in the benthic boundary layer induced by flow around polychaete tubes. Can J Fish Aquat Sci 40 (Suppl. 1): 301–308

Carey DA (1987) Sedimentological effects and palaeoecological implications of the tube-building polychaete *Lanice conchilega* Pallas. Sedimentology 34: 49–66

Dahlbäck B, Gunnarsson LAH (1981) Sedimentation and sulfate reduction under a mussel culture. Mar Biol 63: 269–275

Dame RF, Dankers N (1988) Uptake and release of materials by a Wadden Sea mussel bed. J Exp Mar Biol Ecol 118: 207–216

Dame RF, Spurrier JD, Wolaver TG (1989) Carbon, nitrogen and phosphorus processing by an intertidal oyster reef. Mar Ecol Prog Ser 54: 249–256

Dankers N (1993) Integrated estuarine management - obtaining a sustainable yield of bivalve resources while maintaining environmental quality. In: Dame RF (Ed) Bivalve Filter Feeders in Estuarine and Coastal Ecosystem Processes. Springer, Berlin Heidelberg New York, pp 479–511

Dankers N, Dame R, Kersting K (1989) The oxygen consumption of mussel beds in the Dutch Wadden Sea. In: Ros JD (Ed) Topics in Marine Biology. Scient Mar, pp 473-476

Dankers N, Koelemaij K (1989) Variations in the mussel population of the Dutch Wadden Sea in relation to monitoring of other ecological parameters. Helgoländer Meeresunters 43: 529–535

Dankers N, Zuidema DR (1995) The role of the mussel (*Mytilus edulis* L.) and mussel culture in the Dutch Wadden Sea. Estuaries 18: 71–80

De Haas H, Eisma D (1993) Suspended sediment transport in the Dollart estuary. Neth J Sea Res 31: 37–42

De Wilde PAWJ, Beukema JJ (1984) The role of the zoobenthos in the consumption of organic matter in the Dutch Wadden Sea. Neth Inst Sea Res 10: 145–148

Dittmann S (1987) Die Bedeutung der Biodeposite für die Benthosgemeinschaft der Wattsedimente. Unter besonderer Berücksichtigung der Miesmuschel *Mytilus edulis* L. Dissertation, Georg August Universität Göttingen

Dittmann S (1990) Mussel beds - amensalism or amelioration for intertidal fauna? Helgoländer Meeresunters 44: 335–352

Dronkers J (1984) Import of fine marine sediment in tidal basins. Neth Inst Sea Res 10: 83–105

Eckman JE (1985) Flow perturbation by a protruding animal tube affects sediment bacterial recolonization. J Mar Res 43: 419–435

Eckman JE, Nowell ARM, Jumars PA (1981) Sediment destabilization of animals tubes. J Mar Res 39: 361–374

Flemming BW (1990) Zur holozänen Entwicklung, Morphodynamik und faziellen Gliederung der mesotidalen Düneninsel Spiekeroog (Südliche Nordsee). In: Willems H, Wefer G, Rinski M, Donner B, Bellman H-J, Eißmann L, Müller A, Flemming BW, Höfle H-C, Merkt J, Streif H, Hertweck G, Kuntze H, Schwaar J, Schäfer W, Schulz M-G, Grube F, Menke B (Eds) Beiträge zur Geologie und Paläontologie Norddeutschlands - Exkursionsführer. Bremen, pp 13–73

Flemming BW, Nyandwi N (1994) Land reclamation as a cause of fine grained sediment depletion in backbarrier tidal flats (Southern North Sea). Neth J Aquat Ecol 28: 299–307

Fréchette M, Bourget E (1985) Food-limited growth of *Mytilus edulis* L. in relation to the benthic boundary layer. Can J Fish Aquat Sci 42: 1166–1170

Grenz C, Hermin M-N, Baudinet D, Daumas R (1990) In situ biochemical and bacterial variation of sediments enriched with mussel biodeposits. Hydrobiol 207: 153–160

Hartmann-Schröder G (1996) Die Tierwelt Deutschlands. 58. Teil. Annelida, Borstenwürmer, Polychaeta. 2. Auflage. G. Fischer Verlag, Jena, Stuttgart, Lübeck, Ulm

Haven DS, Morales-Alamo R (1966) Aspects of biodeposition by oysters and other invertebrate filter feeders. Limnol Oceanogr 11: 487–498

Herlyn M (1996) Zur Bestandssituation der Miesmuschelbänke des niedersächsischen Wattenmeeres. Mitt. NNA 1/1996: 56–61

Hickel W (1984) Seston in the Wadden Sea of Sylt (German Bight, North Sea). Neth J Sea Res 10: 113–131

Hüttel M, Ziebis W, Forster S (1996) Flow-induced uptake of particulate matter in permeable sediments. Limnol Oceanogr 41: 309–322

Jones CG, Lawton JH, Shachak M (1994) Organisms as ecosystem engineers. Oikos 69: 373–386

Jørgensen CB (1990) Bivalve Filter Feeding: Hydrodynamics, Bioenergetics, Physiology and Ecology. Olsen & Olsen, Fredensborg

Kautsky N, Evans S (1987) Role of biodeposition by *Mytilus edulis* in the circulation of matter and nutrients in a Baltic coastal ecosystem. Mar Ecol Prog Ser 38: 201–212

Kosfeld C (1989) Mikrobieller Abbau von Faeces der Miesmuschel (*Mytilus edulis* L.). Dissertation, Christian Albrechts Universität Kiel

Krögel F (1997) Einfluß von Viskosität und Dichte des Seewassers auf Transport und Ablagerung von Wattsedimenten (Langeooger Rückseitenwatt, südliche Nordsee). Dissertation, Universität Bremen

Kröncke I (1996) Impact of biodeposition on macrofaunal communities in intertidal sandflats. P.S.Z.N.I. Mar Ecol 17: 159–174

Michaelis H, Obert B, Schultenkötter I, Böcker L (1995) Die Miesmuschelbänke der niedersächsischen Watten, 1989 - 1991. Ber Forsch-Stelle Küste 40: 55–70

Nehls G, Thiel M (1993) Large-scale distribution patterns of the mussel *Mytilus edulis* in the Wadden Sea of Schleswig-Holstein: do storms structure the ecosystem? Neth J Sea Res 31: 181–187

Oost AP (1996) Dynamics and sedimentary development of the Dutch Wadden Sea with emphasis on the Frisian Inlet: a study of the barrier islands, ebb-tidal deltas, inlets and drainage basins. Geologica Ultraiectina 126

Pfeifer D, Bäumer H-P, Ortleb H, Sach G, Schleier U (1996) Modelling spatial distributional patterns of benthic meiofauna species by Thomas and related processes. Ecol Model 87: 285–294

Prins TC, Smaal AC (1994) The role of the blue mussel *Mytilus edulis* in the cycling of nutrients in the Oosterschelde estuary (The Netherlands). Hydrobiologia 282/283: 413–429

Reineck H-E (1994) Landschaftsgeschichte und Geologie Ostfrieslands: Ein Exkursionsführer. von Loga, Köln

Reise K (1991) Dauerbeobachtungen und historische Vergleiche zu Veränderungen in der Bodenfauna des Wattenmeeres. Laufener Seminarbeiträge 7: 55–60

Smaal AC, Prins TC (1993) The uptake of organic matter and the release of inorganic nutrients by bivalve suspension feeder beds. In: Dame RF (Ed) Bivalve Filter Feeders in Estuarine and Coastal Ecosystem Processes. NATO ASI Series G, 33, Springer, Berlin Heidelberg New York, pp 271–298

ten Brinke WBM, Augustinus PGEF, Berger GW (1995) Fine grained sediment deposition on mussel beds in the Oosterschelde (The Netherlands), determined from echosoundings, radioisotopes and biodeposition field experiments. Estuar Coast Shelf Sci 40: 195–217

Verwey J (1952) On the ecology and distribution of cockle and mussel in the Dutch Wadden Sea, their role in sedimentation and the source of their food supply. Arch Neerl Zool 10: 171–239

Widdows J, Fieth P, Worrall CM (1979) Relationships between seston, available food and feeding activity in the common mussel *Mytilus edulis*. Mar Biol 56: 195–267

Wright RT, Coffin RB, Ersing CP, Pearson D (1982) Field and laboratory measurements of bivalve filtration of natural marine bacterioplankton. Limnol Oceanogr 27: 91–98

Ziegelmeier E (1964) Einwirkungen des kalten Winters 1962/63 auf das Makrobenthos im Ostteil der Deutschen Bucht. Helgoländer wiss Meeresunters 10: 276–282

Zühlke R, Blome D, van Bernem K-H, Dittmann S (1998) Effects of the tube building polychaete *Lanice conchilega* (Pallas) on benthic macrofauna and nematodes in an intertidal sandflat. Senckenbergiana marit 29: 131–138

Zurburg W, Smaal AC, Héral M, Dankers N (1994) *In situ* estimations of uptake and release of material by bivalve filter feeders in the bay of Marennes-Oléron (France). In: Dyer KR, Orth RJ (Eds) Changes in Fluxes in Estuaries. ECSA22/ERF Proc. Olsen and Olsen, Fredensborg, pp 239–242

4 Statistical Models and Techniques for the Analysis of Conditions of an Ecological System

Ulrike Schleier

Abstract: This chapter treats the contribution of statistical research and applied statistics to ELAWAT. In the introduction the role of applied statistics in scientific research is described. Definitions of statistical terms and expressions are given to make the subsequent statements definite. The main part of the chapter introduces statistical models and methods. They are arranged according to the ecologically relevant terms in ELAWAT: pattern, process, relation, and distinction between states of a system. Each paragraph starts with an ecological question and treats statistical models and methods useful for finding an answer. Case studies with data from ELAWAT demonstrate some of the methods for data analysis and interpretation. The final paragraph summarizes the methods applied in ELAWAT.

4.1 Introduction

During the interdisciplinary research project of ELAWAT difficulties in understanding each other occurred among scientists from different disciplines. Sometimes, especially in the beginning, this was merely a problem of terminology and lack of information, but often it revealed fundamental differences in conception of sciences and in the way of approaching a research question. This applies particularly to the discipline of statistics. For this reason, in this paragraph an attempt is made to describe the role and view of applied statistics in ecosystem research. Readers who are mainly interested in statistical techniques may omit the paragraph at first reading.

Natural sciences are constantly moving in a circular course from questions, investigations (observations or experiments), evaluations, answers to new questions. Besides common paradigms (e.g. the one that natural phenomena are in principle perceptible to mankind and accessible through measurements and logical thinking) there are also paradigms that constitute basic assumptions to only some disciplines, or even to the scientific work of only some institutions or work groups, where they are often no longer recognized as such. For example: "Environmental studies are mostly done by ship" (Bohle-Carbonell 1992, p. 89). Lamprecht (1992) emphasizes that every scientific work starts with a question, which will always be embedded in the respective paradigm. It is part of the scientific formulation of a question to disclose these paradigms. For the choice of quantitative methods (e.g. statistical indices) they are often of great importance.

A question can derive from theoretical considerations, intuition (e.g. after the personal observation of a phenomenon), tasks coming from outside science (e.g. the demand for a conservation concept) or from the results of former quantitative studies. Also a question's origin, its "quantitative level", plays a role in choosing the adequate mathematical methods to answer it. In the following, conceptual research, which is often situated between answers and new questions, is excluded from the considerations, because it does not require mathematical methods.

To find an answer which is based on scientific reasoning (i.e. objectively found), the question must be in a form that allows quantitative research, i.e. it must be operationalized. The advantage will be that other scientists can follow the argument leading to a certain answer. The disadvantage is that only measurable phenomena can be treated.

A research programme must be planned in such a way that the measurements allow the question to be answered (at least come closer to be answered), logical conclusions must be possible and alternative answers must be excluded. To that extent, quantitative work enforces a sufficiently precise formulation of questions. This includes that the planning of the experimental design can only be started when it is clear in which way the collected data are to be analysed, i.e. in which way they can be used to find an answer. Thus, the choice of a mathematical method is closely connected to the formulation of the question (in its most precise form).

The method of ecological modelling, as used in theoretical ecology, is described by Wissel (1989) as "the pursuit of reasoning in the language of formal logic". Its special aim is to gain understanding of an ecological system. Models are taken as a first step towards a general theory, achieved for example by reproducing mechanisms of a system in the model and studying their plausibility. In the circular course of research from questions over answers to new questions, ecological modelling comes directly after the question. For modelling nothing else is required than a precise question and some primary information about the system. This also allows the inclusion of qualitative knowledge, experiences or suppositions, rules and mechanisms. Answers are often of the form "if ..., then ...". Ecological models show the conclusions that are to be drawn if certain assumptions are made about the system.

Mathematical modelling refers to the abstract description of a system in relation to a certain question (Murthy et al. 1990). This requires a formulation of the system and the question in such a precise form that a translation into the "language of mathematics" becomes possible. In this context models are representations of a system by symbols and formulas whose meaning and rules are defined by mathematics. These models can be deterministic and/or stochastic.

The term simulation model describes a mapping of an ecological or mathematical model into a computer programme. The use of a simulation model is indispensable, if the system cannot be determined analytically, i.e. by mathematical rules only (Example: the model about *Lanice conchilega* in Chap.5.4). A simulation model can also be used to demonstrate the spatial and temporal behaviour of a system, e.g. by graphical representation on the screen (Example: Bäumer 1994).

In contrast to the approaches mentioned so far, models in applied statistics serve primarily for data analysis and are not restricted to a specific system. Here, models are assumptions about the (mathematical) properties of the data and/or form a mental frame for data analysis. They are necessary for assessing the statistical properties of the results. Statistical techniques serve as tools for planning and analysing experiments. In the circular course of research from questions over answers to new questions, applied statistics is directly linked to the investigation (observation or experiment). The aim of applying statistics in ecology is understanding phenomena by interpreting data.

An attempt to bring mathematical modelling and statistical data analysis closer together was made by Richter & Söndgerath (1990). Statistical techniques can be applied to analyse simulated data from ecological models (Samietz & Berger 1997). In general, however, mathematical modelling and statistics are treated separately in theory and application.

Statistical techniques can be classified under different aspects. If the question is the result of a quantitative pilot study and much is known already about the variable under consideration (e.g. its statistical distribution), then the question can be formulated as a statistical null hypothesis with alternative hypothesis. This is part of statistical inference. One important principle in statistical inference is the formulation of the null hypothesis before sampling. On the contrary, in descriptive statistics, the description of data is most important. Exploratory statistics provides tools for finding unknown relations, structures and particularities in the data with the aim of generating models and hypotheses (Bock 1980).

Ecosystem research is characterised by the search for and the analysis of relations between different parts of a system. Regarding data analysis in this context means that it is rather the exception than the rule that a single variable is to be analysed (univariate statistics). Thus, multivariate techniques are of special importance.

Also, the spatial and temporal scales play an important role in ecosystem research. Observations or measurements are often not independent, as required by many statistical techniques. Rather, it is their temporal or spatial relation which is to be investigated.

These different approaches show that there cannot be an overall rule for the application of statistical techniques. The choice of an adequate technique depends rather on the question at hand and of the kind of data sampled. Underwood (1994) gives examples showing in which respect ecological data do not fulfil the requirements of traditional statistical techniques.

4.2 Terminology

A variety of statistical techniques are currently used in ecology. Depending on question and research focus, different models and techniques are preferred. There is an extensive, sometimes confusing vocabulary, which has not been standardized. Thus, depending on the context, there may be different names for the same statistical technique. The other way round, the same name may be used for different techniques, depending on the context. In different disciplines, terms may have different meanings, which may complicate the understanding. In this section an attempt is made to set out statistical paradigms and terms in order to prevent misunderstandings as much as possible.

As part of mathematics the field of mathematical statistics is based on the axioms of probability, it rests on the laws of chance. Therefore it does not matter whether biological phenomena only seem to function randomly (possibly because of many deterministic influences working simultaneously) or whether they really do function randomly. For example, the mean values from two samples will be considered significantly different if the probability of observing or measuring two values with this difference is very low, given that the samples have the same underlying mean. Statistics gives no evidence whether a difference, which is considered significant in the statistical sense, is an ecologically relevant difference. In the field of applied statistics, data are analysed by means of graphical, mathematical and statistical techniques. Often, the technique is based on a probability concept. It is assumed that the measurements or observations are random samples from a real or imaginary sampling universe. (In statistical textbooks the term population is used synonymously with sampling universe. In the following the word population will be reserved for populations in the biological sense.) Results are to be referred to that sampling universe, which cannot be examined as a whole for reasons of time and money or in principle. Only for a random sample is it possible to draw inference from the sample to the sampling universe by means of probability theory. Otherwise, the principle of randomisation can be used to make probability arguments about the sample (for details see Edgington 1995). The techniques belonging to descriptive and exploratory statistics are permitted in any case. However, generalization of results to a larger set than the one observed is rarely allowed.

In the following, measurements and observations are termed "variables".

Statistical hypothesis testing means falsification of the null model. A statistically significant conclusion is only permitted if the observed data would have occurred under a valid null hypothesis (null model) with a very small probability only. Thus, in contrast to the models mentioned above, the null model serves as a reference. A formal requirement is the knowledge of the statistical distribution of the test statistic under the null hypothesis.

A stochastic process is a random function. For statistical modelling it does not really matter whether this is a function of time or of space. In ecology, generally, the word process denotes a phenomenon, which is characterized by a temporal change (e.g. biochemical transformation processes in a mussel bed). In the following, the word process will be used in this general sense. If a process is meant in the mathematical sense, it will be specified by using the term stochastic process.

A pattern is regarded as a structure of objects, like e.g. their spatial arrangement. In Sect. 4.3.1 the term is defined more precisely depending on the context. In ecology the terms pattern and spatial and temporal distribution are often used as synonyms.

The statistical distribution (probability distribution, distribution function, probability function) of a random variable specifies the probabilities of the possible values (e.g. for a fair dice the statistical distribution is: 1-$^{1}/_{6}$, 2-$^{1}/_{6}$, ..., 6-$^{1}/_{6}$). The empirical distribution of an observed variable specifies the cumulative frequencies of the observed values. In statistics, the term parameter denotes a value characterizing a statistical distribution. The mean is a parameter, for example. In other fields of science, parameter is used synonymously with variable and distribution denotes for example the spatial arrangement in an area.

In everyday language and in natural sciences as well, the term correlation is used synonymously with relation. In statistics its meaning is much more restricted. The correlation coefficients of Bravais-Pearson and of Spearman (see e.g. Sachs 1992) solely measure the linear relationship between two variables and their ranks, respectively. Relationships following different functional forms (e.g. quadratic) or more general relations are not covered by this term. There are more measures of correlation (Kendall 1990), each of which measures a special kind of relation.

Measurements made in a temporal sequence often show autocorrelation, i.e. every two or more succeeding values show high correlation (in the sense of the Bravais-Pearson correlation coefficient). A series of measurements in temporal order is called time series or longitudinal data in statistics.

In statistics, the terms variability, spread, noise are used synonymously with the term variance. In ecology they are also used to describe differences in repeated measurements no matter from which source (Lozán 1992, p 40) or as changes of structure and parameter values in both spatial and temporal respect.

It is a task for mathematics to provide adequate similarity measures for the assessment of system structures. First of all, however, it is an ecological question, what is meant by similarity in a specific context. For a comparison of biological communities, for example, only the presence or absence of species might be of interest without regard to abundance (qualitative sampling). The ecological requirements form the background for the choice of an adequate similarity measure to be found by mathematical reasoning. Pfeifer et al. (1998) show the properties of some common similarity measures by way of example. The sampling properties are only known for few similarity measures (Dixon 1993). Therefore, it is mostly not possible to extend similarity-based results from samples to the sampling universe.

4.3 Statistical Methods for the Description and Measurement of Stability Properties

In this paragraph statistical methods are described, which are relevant for the ecological question of stability properties. They are arranged according to the terms pattern, process, relation, and distinction between states of a system. In addition to the presentation of methods, examples are given in the form of case studies. The aim is to demonstrate the statistical method starting with the ecological question, explaining the technique as succinctly as possible, and ending up with an indicated interpretation. It is hoped that the exemplary character of the case studies is strong enough to make them interesting also for reader outside ELAWAT.

4.3.1 Description of Patterns and Processes

As a first step it is necessary to define the scale on which a pattern or process shall be investigated, because a pattern or process is not completely defined without its scale. It depends on the question to be studied which phenomenon may be called a pattern, thus there are no generally applicable recipes for how to analyse patterns statistically. If constancy (see Chap. 9) of a pattern or process is to be used as a stability property, then it is necessary to define the pattern or process in such a way that it is possible to decide whether a situation has or has not changed between two different points in time. In ELAWAT the following patterns and processes were investigated among others.

4.3.1.1
Small-Scale Distribution Patterns

Questions: In what way are individuals arranged in an area? Are they aggregated? Are there repeated structures?

Sampling was done with a multicorer, with single corers arranged in a quadrate without spaces in-between.

In the language of statistics, the arrangement of individuals is a stochastic process in two-dimensional space. The Poisson Process serves as a null model. It is clearly defined by the statement that the probability for a certain number of individuals in an area only depends upon the size of the area, but not upon its location within a larger area or other influences. Such a pattern is sometimes called purely random arrangement. To test the hypothesis, whether an observed arrangement is purely random, aggregated or regular, the index of dispersion may be used. It distinguishes itself from other measures of aggregation by the fact that its statistical distribution under the null model is known, so that a statistical test is possible (Pfeifer et al. 1994).

For a description of the spatial pattern of the arrangement with respect to abundance, autocorrelogram- and variogram-analysis may be applied (Matheron 1971; Jongman et al. 1995). For an example see Blome et al. (1999). Computationally, autocorrelogram and variogram are equivalent, mathematically they are not. A discussion on the use of autocorrelogram and variogram can be found in Journel & Huijbregts (1991).

4.3.1.2
Spatial and Temporal Patterns

Questions: How variable are patterns in aggregations of *L. conchilega*? In which way can the dynamics and the spatial aggregation of wading birds be modelled?

A criterion for the decision between equality and inequality of patterns derives from the way in which the pattern has been described. If the model for the pattern is a Poisson Process with parameter λ, then change in pattern will be change of the parameter λ and/or change in the statistical distribution. If the pattern has been described by means of the variogram, then this function is the criterion for constancy in the pattern. If more sophisticated models are used, specially designed techniques will be necessary (see e.g. Ver Hoef & Cressie 1993). Borovkov et al. (1996) show in which way stochastic networks can be used to model spatial and temporal patterns.

4.3.1.3
Multivariate Spatial and Temporal Patterns

Question: Which patterns emerge in chemical, biological and sedimentological variables along a transect from a mussel bed to a sandflat in the course of one year?

For the spatial and temporal pattern of chemical, biological and sedimentological variables, there is *a priori* no statistical reference model. This makes the problem different from the one concerning the distribution pattern of *L. conchilega*. Moreover, the multivariate pattern is to be investigated, i.e. the pattern of several variables simultaneously. Therefore, techniques from multivariate exploratory statistics are recommended.

A flexible technique, which allows the interpretation of the term "pattern" in different ways depending on the context, is Multidimensional Scaling (MDS). The technique was applied to benthic communities by Warwick & Clarke (1991) and has frequently been used in benthic ecology.

Starting point of the technique is the transition from the data matrix (values of the variables in the samples) to a matrix of dissimilarities between pairs of samples. The choice of the measure of dissimilarity reflects the ecological question. The result of the procedure is a graphical representation of all the samples as points in a low-dimensional Euclidean system of co-ordinates (mostly two-dimensional) and is called the configuration. The algorithm, which transforms the matrix of dissimilarities into the configuration, arranges the samples in such a way that very similar samples will be placed close to each other and very dissimilar samples will be placed far away from each other. This requirement cannot be met in all cases. Whether a configuration found by the algorithm is an adequate representation of the matrix of dissimilarities should be inspected by use of the Shepard diagram and the stress value. The Shepard diagram should be monotonic and the stress value should not be beyond 0.15. The technique is explained and discussed in detail e.g. by Clarke (1993) and Cox & Cox (1994).

4.3.1.4
Case Study: Spatial and Temporal Patterns in Geochemical Variables

Data provided by A. Hild.

Question: Which of the geochemical variables are similar to each other with respect to their spatial and temporal pattern?

Sampling: At 6 stations along a transect one unreplicated sample was taken per sampling date. For this case study, five time points were chosen and 7 variables included. The variables had also been measured in the different grain size fractions of the sediment.

Statistical technique: The Spearman rank-correlation coefficient was used to measure the similarity between two spatial and temporal patterns. It assumes the maximum value of 1 if the measurements of two variables have the same order along the transect and in the course of time. Thus, pattern is defined here as arrangement of the measured values. For a graphical representation of the similarity structure MDS is used (compare Sect. 4.3.1.3).

Results: For the data from the clay fraction the configuration in two dimensions had a stress value of 0.09. The result is represented in Fig. 4.3.1. TOC, C_{min} and Al_2O_3 form an equilateral triangle. From the dissimilarity matrix it can be seen that its side length corresponds to a negative correlation of about -0.3. K_2O and Al_2O_3 are placed close to each other, i.e. the ordering of their ranks is nearly identical, while TiO_2 is weakly correlated with Al_2O_3 and K_2O. While P_2O_5 is weakly correlated with TOC, all of the other variables are not or negatively correlated with TOC and C_{min}.

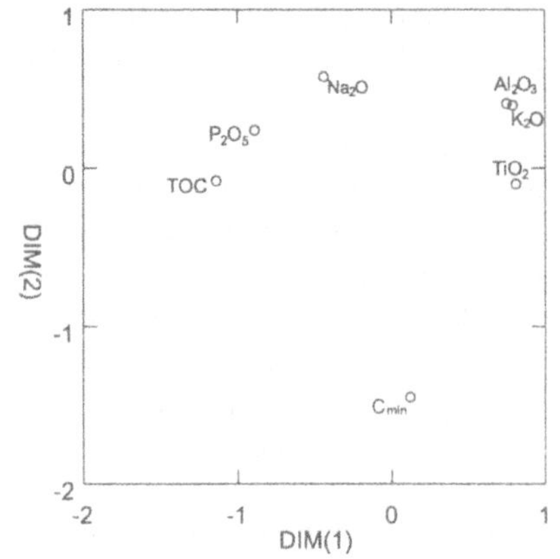

Fig. 4.3.1 Two-dimensional MDS configuration representing distances in the spatial and temporal patterns of seven geochemical variables in the clay fraction based on the Spearman correlation coefficient

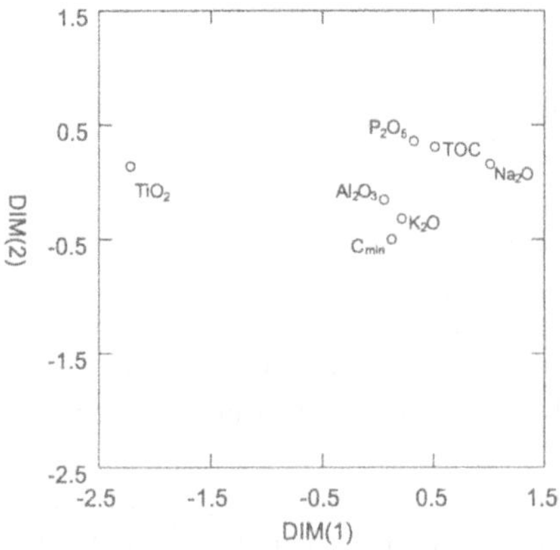

Fig. 4.3.2 Two-dimensional MDS configuration representing distances in the spatial and temporal patterns of seven geochemical variables based on the Spearman correlation coefficient calculated from the same samples as in Fig. 4.3.1 but related to the total sediment fraction

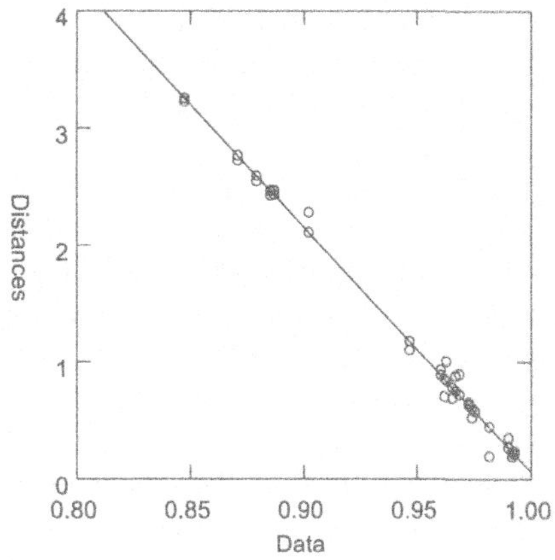

Fig. 4.3.3 Shepard diagram of the distances in the MDS configuration of Fig. 4.3.2

For the data from the total sediment fraction (Fig. 4.3.2) the picture is different (stress=0.07).

TiO_2 is placed far away from all the others. From the corresponding Shepard diagram (Fig. 4.3.3) one can see that the reason is not negative correlation. The axis "Data" shows that all the calculated rank correlations are high or very high (between 0.84 and 0.99). The interpretation of the MDS plot in this case is the following: All of the variables are highly correlated, i.e. show a nearly identical pattern.

4.3.1.5
Variability in Distribution Patterns

Question: Assessment of the background noise in the number of individuals of macrozoobenthic species with the objective to describe a "reference situation" in order to recognize changes caused by hydrography.

In the univariate case the background noise in the values is identified with their variance. There is a high number of statistical parameters to characterize variability (see e.g. Lozán 1992; Sokal & Rohlf 1981), whose application is dependent upon the question and the scale of the data. A "reference situation" may be described by an interval around a location parameter (in general mean or median). The interval is chosen in such a way that values outside the interval will be fairly rare. Such an interval can be represented graphically for example as a "Box-and-Whisker-Plot" (median and quartiles) or as a "variance-interval" around the mean (arithmetic mean and double standard deviation). Values inside the interval can be termed "within normal fluctuation", the interval marks the variability of the empirical distribution. In a similar way a graphical test can be carried out. In this case, the length of the interval must be equal to the confidence region of the respective parameter. Thus, the notches in "notched box plots" mark the 95 % confidence interval for the median, and an interval of the length of four times the standard error is at least a 75 % confidence interval for the mean (by the Tschebyscheff inequality). If the intervals belonging to two samples intersect, a respective statistical test will not reject the null hypothesis of equality of the two location parameters.

In the multivariate case, the variables should be represented simultaneously in a figure, i.e. from the representation one should be able to see which values of several variables were assumed at the same sample. As in the univariate case, a variance interval of the distribution and a confidence interval for the common location parameter can be computed. Both require knowledge about the two-dimensional distribution. In common statistical software packages, a multivariate normal distribution is generally assumed. The confidence region is a (multidimensional) ellipse, whose length and width are determined by the standard deviation of the variables and its orientation is determined by the covariance of the two variables. In the graphical representation of two variables, the two axes must be used for the variables, the temporal pattern is lost. With a few measurements already, the graphical representation will look rather confusing (Fig. 4.3.4). For more than two variables, a confidence-ellipse can be calculated, but a clear graphical representation is no longer possible.

The generalisation of the variance from the univariate to the multivariate case is the covariance matrix, which comprises all the variances and pairwise covariances, which are included in the calculation of the confidence ellipse. The use of this matrix implies that Euclidean distance is used as a measure of dissimilarity and Bravais-Pearson correlation as a measure of relation. If this is not desirable for ecological reasons, a technique should be chosen that allows a choice of the dissimilarity measure, e.g. MDS (see Sect. 4.3.1.3) or Cluster Analysis. The multivariate measure of variability then depends upon the measure of dissimilarity chosen.

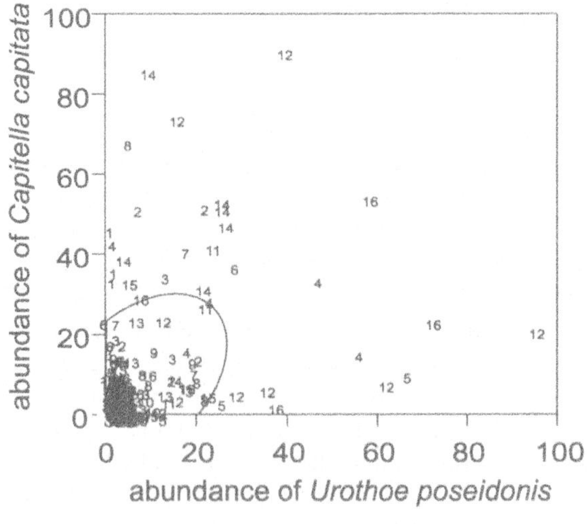

Fig. 4.3.4 Graphical representation of the two-dimensional distribution with 95 % confidence region for abundances of two species in 196 samples. One symbol represents one sample, numbers refer to the 16 consecutive sampling days from May, 4th to May, 19th in 1994

Graphically, variability can be described as "spread of the cloud". Two such "clouds" will be considered different, if their overlap is small compared to their size. This is the main idea underlying the multivariate tests described in Sect. 4.3.3.

If two consecutive sampling times were relatively close to each other, it will be necessary to check the data for temporal correlation (see the sections of this chapter on processes: 4.3.1.7, 4.3.1.9, 4.3.1.10, 4.3.1.12).

4.3.1.6
Case Study: Variability in Benthos Data

Data provided by J. Steuwer.

Question: How strongly do abundances of macrofauna organisms vary on consecutive days in spring? Species with highest abundance were selected: *Capitella capitata* and *Urothoe poseidonis*.

Sampling: From 4.5.94-19.5.94 sixteen replicate benthos samples were taken daily.

Statistical technique: Notched box plot, confidence plot, variance and confidence ellipses, variability described by chord distance.

Results: Both the Notched Box Plots (95 %-intervals) and the confidence regions for the mean (75 %-intervals) (Figs. 4.3.5, 4.3.6) show that there are few significant differences between the average abundances of *Capitella capitata*. Abundance was significantly lower on 13.5.94 than on many other days and significantly higher on 17.5.94 than on some other days.

The ellipse in Fig. 4.3.4 shows the region, where for a normal distribution with mean, variance and covariance as for the sample considered here, 95 % of all pairs of abundance of *C. capitata* and *Urothoe poseidonis* are expected. The numbers denote the day of sampling.

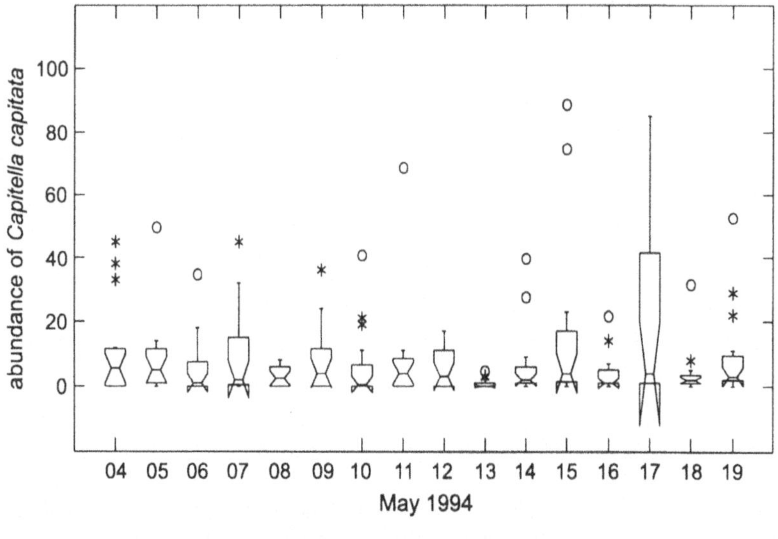

Fig. 4.3.5 Notched Box Plots showing the empirical distribution of abundances of *Capitella capitata* in 16 samples per sampling day in 1994

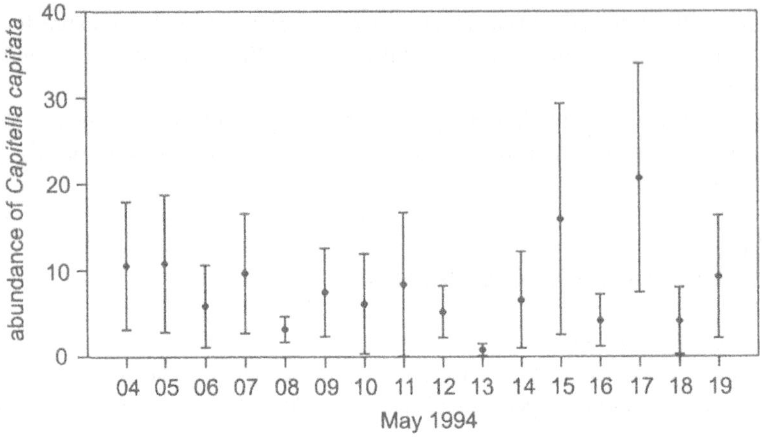

Fig. 4.3.6 Average abundances and confidence intervals for the mean abundances of *Capitella capitata* calculated from 16 samples per sampling day in 1994

Observations outside the ellipse can be interpreted as extreme values. Some replicates of days 2, 4, 12, 14 and 16 are remarkable. Some observations from the 17.5.94 (day number 14) show extremely high abundances of both species. Testing this day against the day before, two confidence ellipses are generated, which overlap (Fig. 4.3.7). The two-dimensional means are thus not significantly different.

Frequently, abundance data are log transformed to reduce skewness and the way the variance is a function of the mean. The statistical techniques outlined above are applicable to transformed data, but interpretation will relate to transformed abundances, not abundances.

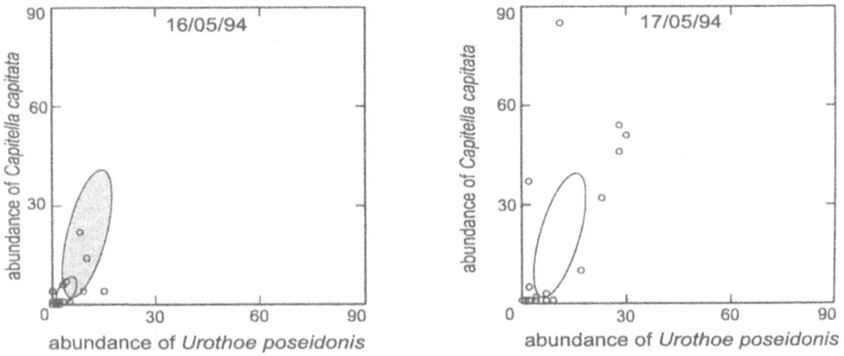

Fig. 4.3.7 Confidence ellipses for simultaneous mean abundances of *Capitella capitata* and *Urothoe poseidonis* calculated from 16 samples per sampling day in 1994

For the representation of variability within one sample and between two samples from two different days, and considering all species simultaneously, Chord-distance was used here as an example. It is a measure of similarity of two samples with respect to species' composition. The distance is maximum if the samples have no species in common, it is minimum if all the species appear in both samples with identical relative frequencies. Variability between days is graphically represented by MDS (see Sects. 4.3.1.3, 4.3.1.5) (Fig. 4.3.8).

4.3.1.7
Univariate Processes

The distinctive feature of data sampled in the course of time is the fact that consecutive measurements are often not independent but correlated (in the statistical sense). Standard techniques for statistical evaluation are summarized under the heading of "time series analysis" (Schlittgen & Streitberg 1994). It consists of fitting a function to a series of measurements or modelling periodic fluctuation or investigating the correlation of consecutive measurements. These techniques are not applicable to data from ELAWAT, because the time series observed there were too short. An alternative is non-parametric time series analysis, e.g. for the estimation of trend (Bortz et al. 1990; Büning & Trenkler 1994).

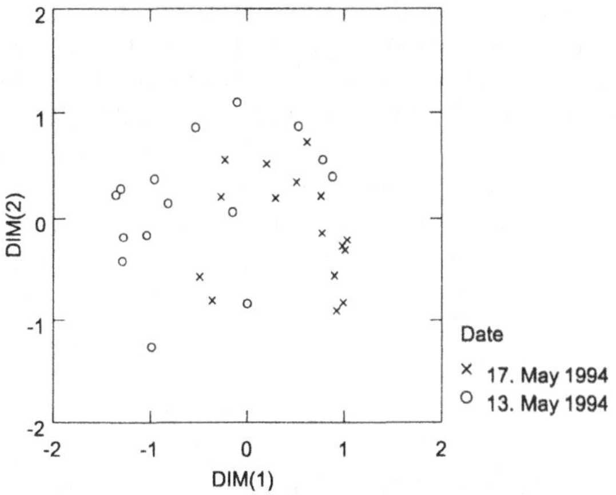

Fig. 4.3.8 Two-dimensional MDS configuration of 16 samples from two days based on Chord-distances, representing variation within and between samples from two days

4.3.1.8
Case Study: Composition of Suspended Matter

Data provided by A. Hild.

Hypothesis: Variation in Mn-content of suspended matter in winter is caused by different percentages of fine material (and thus by Al_2O_3-content), whereas this link does not exist in summer. Hence, the quotient of Mn and Al_2O_3 differs over the year.

Question: The hypothesis can be specified by two different questions. If it is important to examine the way the values develop during the year, then the question will be: is there an increase in the Mn to Al_2O_3-relation from the beginning of the year until midsummer and a decrease from midsummer to the end of the year (reverse U-shape)? If only the difference between summer and winter is important, then the question will be: are, on average, the summer values higher than the winter values?

Sampling: In the months of February and March, May to September, November and December samples were taken on two consecutive days at several time points.

Statistical technique: As sampling was done within short time intervals, it has to be assumed that measurements of one or of two consecutive sampling dates are not independent. For each month the median was calculated from the sub-samples. It is assumed that the monthly median values are correlated. Because a statistical hypothesis has to be generated before sampling and not - as was the case here - after investigation of the data, the test result in the current case study is merely a hint to a phenomenon, not its statistical proof.

To answer the question about annual development, the time series of medians (Fig. 4.3.9) was tested for monotonicity in its first and second part (before the month of July and after the month of July, respectively) by means of a modification of the test of first differences by Moore & Wallis (1943) (compare Sect. 4.3.1.13). The test statistic is the number of those differences between consecutive values whose signs are not in agreement with the null hypothesis. For a comparison of winter- and summer values, the U-Test was carried out.

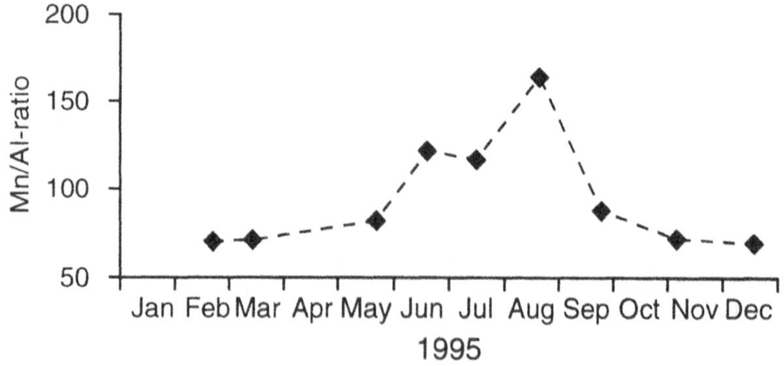

Fig. 4.3.9 Seaonal development in the ratio of the monthly medians in two geochemical variables (Mn and Al) measured in 1995

Results: There is one situation in the first part and one situation in the second part of the series in which the series does not increase or decrease, respectively, as is assumed under the null hypothesis. This number of situations is the test statistic for the Moore-Wallis test of first differences and it corresponds to a p-value of 0.042 (Bortz et al. 1990, Table 40).

The U-Test of the median values from summer against winter results in a p-value of 0.0275 (Bortz et al. 1990, Table 6).

Both test results indicate that the quotient of Mn and Al_2O_3 differs over the year.

4.3.1.9 Multivariate Temporal Processes

Questions: What is the temporal development of sediment chemistry after a disturbance? What is the temporal development of occurrence and abundance of zooplankton during a year? In what way does recolonization by microphythobenthos, nematodes and macrofauna develop after disturbance?

Multivariate measurements cannot, in general, be graphically represented as points in two dimensions without loss of information. Multivariate series of measurements are difficult to represent graphically, because usually it will not be possible to find a ranking. A solution is to change over to dissimilarities between pairs of measurements followed by MDS (compare Sect. 4.3.1.3).

If there is a valid configuration in one dimension, it will be possible to rank the measurements and represent them as usual in a time diagram. If there is a valid configuration in two dimensions, it will be possible to represent time linking the points by arrows.

Another possibility is the choice of a reference point in time, to which all the others are compared. In the time diagram, dissimilarity between the respective measurement and the reference measurement is shown.

4.3.1.10 Multivariate Temporal Processes Along a Transect

Questions: In what way do for example chemical, sedimentological or biological variables develop along a transect from a mussel bed to a sandflat during one year? Frequently, for the data analysis of such a question, temporal and spatial development are separated from each other and analysed one by one. This results in statements concerning the development along the transect, averaged over time, and statements concerning the temporal development, averaged over the transect stations. Simultaneous statements about the spatial and temporal development are achieved by a joint analysis of temporal and spatial multivariate measurements.

4.3.1.11 Case Study: Geochemical Variables Along a Transect During one Year

Data provided by A. Hild.

Question: Which factor is more important for the structure of geochemical variables along a transect: seasonal difference or spatial distance?

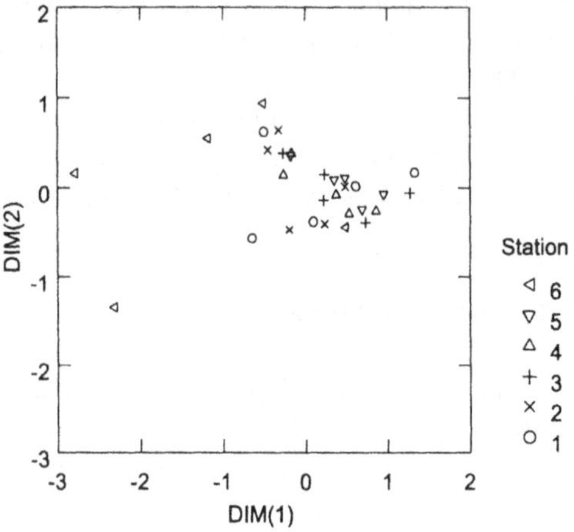

Fig. 4.3.10 Two-dimensional MDS configuration of 30 samples taken at 6 stations in 5 months in 1995 based on Euclidean distance

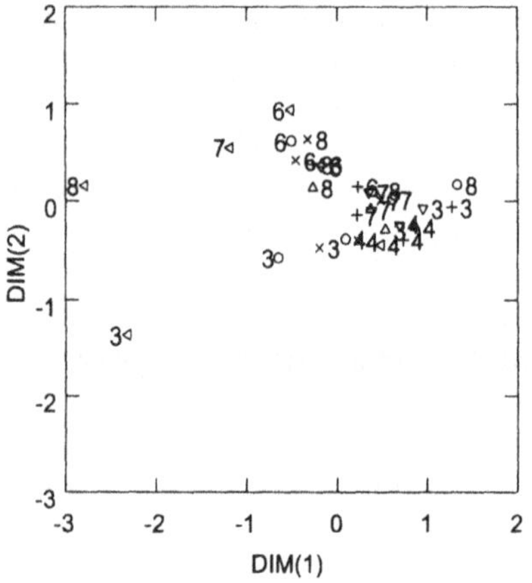

Fig. 4.3.11 Two-dimensional MDS configuration as in Fig. 4.3.10 with numbers representing months in 1995

Sampling (compare Sect. 4.3.1.4): At 6 stations of a transect one unreplicated sample was taken per sampling date. For this case study five dates were chosen from 1995 (March, April, June, July, August) and measurements of seven variables in the clay fraction (K_2O, TiO_2, Al_2O_3, Na_2O, P_2O_5, C_{min}, TOC) included.

Statistical technique: Dissimilarity between two samples was measured by Euclidean distance. Each sample is characterised by space and time, indicated here by number of the transect station (1 to 6) and month of sampling (3, 4, 6, 7, 8). The

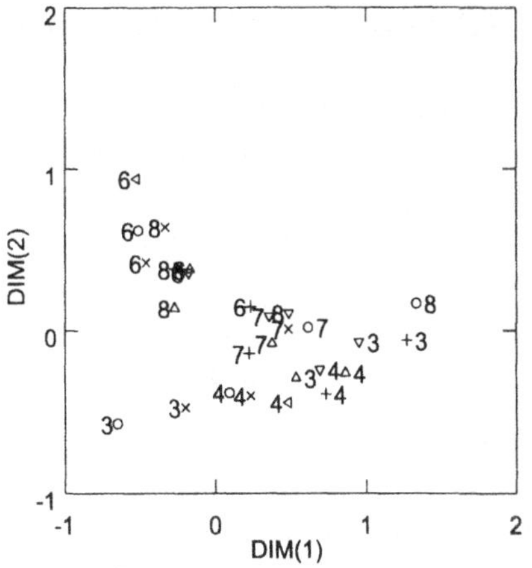

Fig. 4.3.12 Two-dimensional MDS configuration as in Fig. 4.3.10 with numbers representing months in 1995 and enlargement of part of the figure

dissimilarity structure of all the samples with respect to the seven variables is graphically represented by MDS (compare Sect. 4.3.1.3).

Result: Fig. 4.3.10 shows that variation of the measurements is fairly high at station 1 and 6, whereas the measurements belonging to the other stations are closer to each other (stress=0.07). In Fig. 4.3.11, the number of the respective months was inscribed additionally, so that temporal and spatial variation are visible simultaneously.

The months of March and August are responsible for the great variation at stations 1 and 6.

Enlarging part of the Fig. (Fig. 4.3.12), it becomes obvious that season is of greater influence than the transect. The samples belonging to the same month are relatively close to each other, while the samples from the same station are placed relatively far apart from each other. Essentially the same result is achieved here, when the MDS is repeated without the outlying samples. This need not be the case in every analysis, because, without the outliers, the MDS will have more flexibility to get the relationships correctly displayed amongst the remaining samples.

4.3.1.12
Comparison of Processes

Analysis of difference or parallelism between temporal developments can be done using techniques of non-parametric time series analysis (Bortz et al. 1990; Büning & Trenkler 1994; Lienert 1978). Use of the Bravais-Pearson or Spearman correlation coefficients should be avoided in this context, because tests based on them rest on the assumption of independent measurements (i.e. without temporal relation). It has been known for long (Chatfield 1975) that even two uncorrelated time series may result in a high correlation coefficient.

4.3.1.13
Case Study: Temporal Development of Nutrients over a Tidal Cycle

Data provided by G. Liebezeit and V. Niesel.

Question: Do phosphate and silicate content develop in a parallel way during one tide? (Fig. 4.3.13)

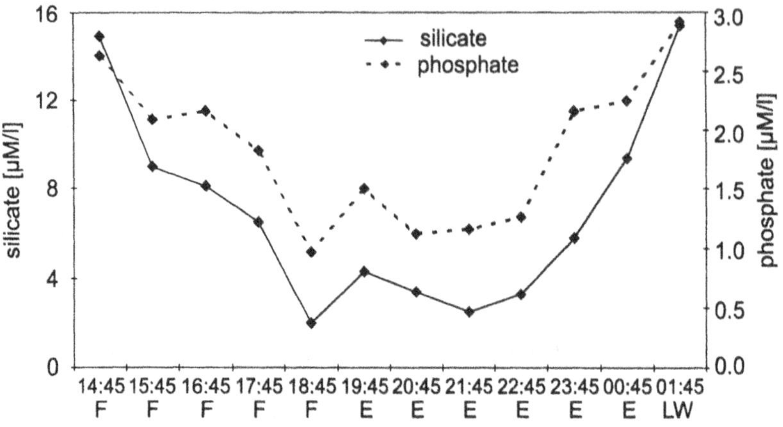

Fig. 4.3.13 Hourly changes in silicate and phosphate concentrations during one tidal cycle. F: flood tide, E: ebb tide, LW: low water

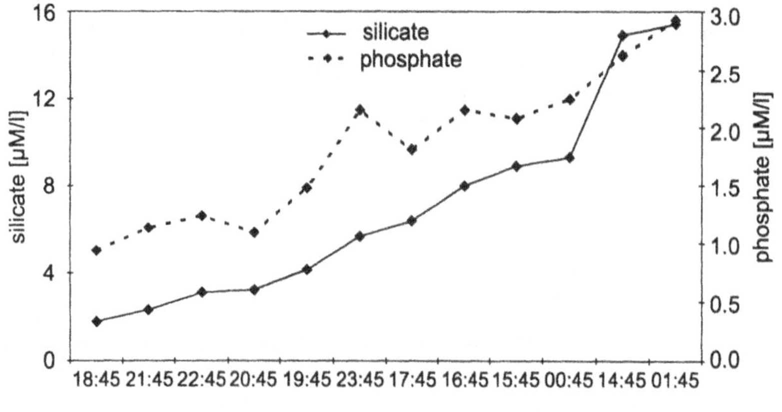

Fig. 4.3.14 Hourly changes in silicate and phosphate concentrations during one tidal cycle as in Fig. 4.3.13, sorted by values of silicate

Technique: One of the series is used as an "anchoring series" (Ofenheimer 1971). The time points are arranged in such a way that the anchoring series (silicate in this case) is monotonic. Then, the other series is rearranged according to the new ordering (Fig. 4.3.14). The null hypothesis is: the order of the values of the newly sorted phosphate series is purely random. According to the principle of randomization the argument proceeds as follows. Under the null hypothesis, a strictly or nearly monotonous ordering is so improbable, that the null hypothesis is rejected. To what extent the series may depart from monotonicity without rejection of the null hypothesis depends upon the significance level and the test procedure. One test of monotonicity is the Moore-Wallis test of first differences (compare 4.3.1.8).

Result: There are three situations in the newly arranged phosphate series, where the series does not increase but decreases. This is the value of the Moore-Wallis test statistic, which corresponds to a p-value of 0.022 (Bortz et al. 1990, Table 40). The test result indicates a parallel development of silicate and phosphate during one tide.

4.3.2 Analysis of Relations

In ELAWAT, processes were investigated with the objective of linking them to patterns.

From the mathematical point of view, the simplest situation is the one of independent replicates with two variables per object. In statistical text books (see e.g. Kendall 1990) there are many measures of dependence, especially of correlation. Which measure is suitable depends mostly on the question and also whether a test is required. The sample properties are known for only some of the measures.

The standard techniques are no longer applicable if the replicates are not independent, but show spatial or temporal dependence, as is the case by definition with patterns and processes. For those cases specially elaborated models are required (Smith et al. 1993; Cressie 1993; Mardia & Goodall 1993). A modification of known tests exists for binary data (Cerioli 1997).

In ELAWAT a special technique was developed for the investigation of the relation between the abundances of two species (the tubeworm *L. conchilega* and the copepod *Harpacticus obscurus*), where abundances of *H. obscurus* in neighbouring samples were not independent from each other (Pfeifer et al. 1996b).

If the replicates are independent, but several variables were measured per object, multivariate techniques will have to be applied. An introduction to many techniques of multivariate statistics with examples from terrestrial ecology is given e.g. by Jongman et al. (1995) and Manly (1986).

If the objective is to investigate the relation of one variable, which is viewed as "target variable" (e.g. abundance of one species), and several other variables, which are viewed as "influential variables", then multiple regression has to be considered as an adequate technique. Its solution consists of the estimation of statistical parameters, which quantify a possible dependence of the "target variable" on the "influential variable". An example of the application of multiple regression to the analysis of a relation between the distribution of birds in an area and the abiotic conditions there is given by Scheiffarth et al. (1996). Regression analysis is well developed in theory and practice, but its assumptions are often not met in ecology. If the "target variable" has only few possible values (e.g. season, habitat), then Cluster Analysis can be applied on the basis of the "influential variables" to see whether objects similar with respect to the "influential variables" are also similar with respect to the "target variable". However, the most frequently applied form of Cluster Analysis, Hierarchical Cluster Analysis, is only applicable to measurements with an underlying hierarchical structure (in the mathematical sense of hierarchical). This assumption is often not considered when the technique is chosen.

If the objective is to investigate the relation between two (or more) groups of variables (e.g. biological and chemical) the following techniques should be taken into account.

The graphical result of Canonical Correspondence Analysis is an arrangement of species (biological variables) and stations as points in a co-ordinate system. Influential variables, e.g. sedimentological variables, are represented as arrows. They point into the direction of those species to which they are highly correlated. Their lengths give information about their importance for the biological variables. The basis is a Multiple Regression Analysis, in which the co-ordinates of the stations are the "target variables". Canonical Correspondence Analysis rests on the assumption that the stations represent a gradient, along which abundances of species develop unimodally. Furthermore, all the assumptions of Multivariate Regression Analysis have to be met (e.g. about correlation among the influential variables). Canonical Correspondence Analysis has been successfully applied in benthic ecology (e.g. Kröncke et al. 1996). Some further multivariate standard techniques are for example Canonical Correlation Analysis (Van der Meer 1991), Principal Component Analysis and Redundancy Analysis, which received different responses in ecological applications (Gittins 1985; Green 1993; James & McCulloch 1990; Minchin 1987). A survey on properties and model assumptions is given e.g. by Jongman et al. (1995, p. 154).

4.3.3 Tests for a Statistical Distinction Between States of a System

A statistical test is a decision procedure stating in which way a test statistic is to be calculated from the sampled data, and for which values of the test statistic the null hypothesis is to be rejected. For parametric tests the statistical distribution of the test statistic is known, because a parametric model is assumed for the data. If no parametric model is or can be assumed, non-parametric (distribution-free) techniques must be applied. Lienert (1969) defines them as techniques "which are not bound to the normal distribution and may be applied even if the distribution of the sampling universe departs from normality or is unknown."

In ecology, multivariate techniques are required, because the state of a system cannot usually be measured by one single variable. The classical parametric multivariate test is Multivariate Analysis of Variance, which implicitly uses Euclidean metric and assumes, besides the normal distribution, that both sampling universes have the same covariance structure, which will usually not be the case with ecological data. Therefore, non-parametric multivariate test procedures are required.

Two trends can be noticed. On the one hand, there are tests suggested from a mathematical point of view (Weiss 1960; Bickel 1969; Friedman & Rafsky 1979; Henze 1988), which have not found many applications yet. On the other hand, there are test procedures, which are applied for example in biology (e.g. Smith et al. 1990; Clarke 1993; Schleier & van Bernem 1996), without precise knowledge of their mathematical properties. In textbooks on the subject (e.g. Büning & Trenkler 1994; Bortz et al. 1990) these tests are not mentioned. Thus, further research is required on this topic.

4.3.3.1
Non-Parametric Multivariate Tests

All the multivariate tests mentioned above follow the principle of randomization, which has already been referred to in Sect. 4.2. As they are based on the pairwise dissimilarities of the sampled objects, their application is strongly connected with a decision about standardization and transformation of the data and with the choice of a measure of dissimilarity. A discussion on this topic is found e.g. in Clarke (1993), Kaufmann & Rousseeuw (1990), Ludwig & Reynolds (1988), Schleier & van Bernem (1996) and Sokal & Rohlf (1981).

To graphically illustrate a result found with the help of non-parametric multivariate tests, MDS (compare Sect. 4.3.1.3) may be used, as it is also related to the pairwise dissimilarities of objects.

4.3.3.2
Case Study: Communities After Experimental Defaunation

Data provided by A. Tecklenborg (compare Chap. 6).

Question: Are there differences between the communities on the control and on the experimental sites?

Sampling: Five replicates were taken per sampling day from the control and from each of the experimental sites.

Statistical technique: Calculation of the pairwise dissimilarities between all pairs of replicates, Minimum Spanning Tree Test, graphical representation by MDS.

Results: After the experimental defaunation, no individuals were left on the experimental sites, while on the control sites the number of individuals was "normal". In this case, a test is not necessary.

On 07.07.94, i.e. 21 days after colonization commenced, there were still clearly less species and individuals on the experimental than on the control sites (Table 4.3.1).

The Minimum Spanning Tree (Schleier & van Bernem 1996) calculated with Chord-distance is not unique. A number of 100 Minimum Spanning Trees was calculated, which gave 11 subtrees on average. Under the null hypothesis of all replicates coming from the same sampling universe, 16 subtrees are expected (p-value 0.028). That means replicates from the experimental and from the control sites represent two fairly strictly separated groups. Thus, the communities still differ at that time.

The graphical representation of pairwise Chord-distances (Fig. 4.3.15) by MDS shows that there are some experimental replicates that do not differ very much from the control replicates. On the whole, however, there is a strict separation between the replicates from the experimental (E) and from the control sites (C).

Table 4.3.1. Species and abundances (7. 6. 1994)

Species	abundance in experimental sites	abundance in control sites
Cerastoderma edule	1	0
Macoma balthica	2	0
Eteone longa	7	1
Nephtys hombergii	5	0
Scoloplos armiger	14	8
Pygospio elegans	104	2
Capitella capitata	6	0
Lanice conchilega	2	0
Crangon crangon	3	1
Ostracoda	28	17

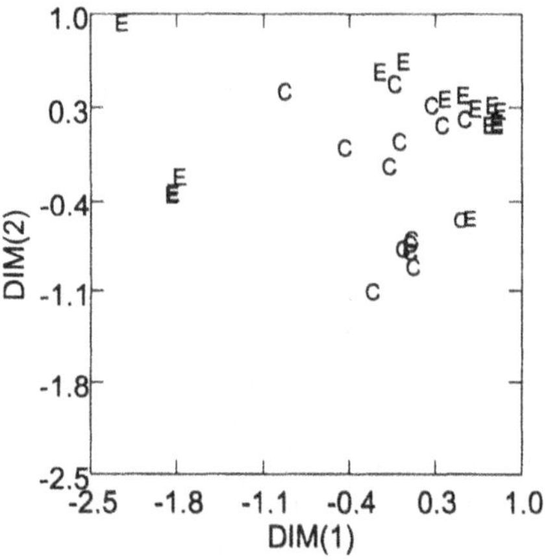

Fig. 4.3.15 Two-dimensional MDS configuration of samples taken on 7.6.1994 from the control (C) and experimental (E) sites based on Chord-distance

4.4 Statistics in ELAWAT

The contribution of statistics to ecosystem research consists primarily in making ecological terms measurable, because what cannot be measured cannot be treated statistically. Thereby, Applied Statistics represents a bridge between system properties on a conceptual level and the measurements on the level of investigation. Stability properties of ecological systems in the sense of Grimm (1996) refer to ecological variables (which are not necessarily measurable in the sense of the present chapter) and their change related to disturbances. The objective is a comparison of states or developments after a disturbance to a reference state or reference dynamic, respectively. From the attempt to make terms measurable, the following questions arose in ELAWAT: Which components of the system, which ecological variables, can be quantified? Which measurable variables characterize their state? In which case are states to be regarded similar? Which distance between states lies within a reference dynamic? In which way can dissimilarity of states be proven statistically? After the statistical analysis and the interpretation of the results another bridge must be constructed. Statistical techniques help understanding ecological phenomena, they do not replace the elaboration of ecological results. Statistical analysis can, at best, support but not replace ecologists' expertise and creative reasoning.

For application in ecology a great number of, partly numerically demanding, statistical techniques are offered. During ELAWAT mainly exploratory statistical techniques were applied (e.g. Box-and-Whisker-Plots, MDS, variograms). Especially MDS, as a very flexible and evident technique, was generally accepted and therefore frequently mentioned and applied in the case studies of the present chapter. Also the other multivariate techniques presented in this chapter are of great potential for the understanding of relations in ecosystems. But they were less accepted. When test procedures were applied, these were mostly non-parametric tests (e.g. U-Test, multivariate randomization tests).

Statistical techniques are sometimes selected with regard to the statistical software package available, but not with regard to statistical reasoning. This mechanism will often be inevitable for the scientist. Nevertheless, one should be aware of it and consider this critically for the assessment of results. For ecologists this requires a good overview of statistical methods beyond those implemented in the statistical software packages at hand. For statisticians it requires a fast transition from research results into available software, as was done in ELAWAT.

A successful application of statistical models and techniques demands from ecologists a basic knowledge in statistics and openness for new techniques. From statisticians it demands the ability of translating statistical subjects into the language of ecology. It also requires the chance for personal exchange so that, as was done in ELAWAT, ecological questions and statistical techniques can be explained and discussed in common.

Statistics is only one of many ways of advancing scientific progress. Its potential ends when no measurements were made or data do not have the required quality. All the statistical techniques explicitly or implicitly assume a model. The part of reality, which is to be described or explained by the model, might be more complex or of a different kind than covered by the statistical model.

The positive co-operation of statisticians and other natural scientists realized in ELAWAT suggested further common developments. Thus, there is a demand for analysis techniques applicable if variables are not spatially or temporally independent. Also, the statistical properties of dissimilarity measures and multivariate non-parametric tests are mostly unknown. Therefore, a combination of ecological questions and appropriate statistical techniques, including software development, is necessary.

Acknowledgements

This chapter owes much to the work of many people in ecosystem research. I especially acknowledge the contribution of the statistical working group: D. Pfeifer, H.-P. Bäumer and H. Ortleb. I am grateful to A. Hild, G. Liebezeit, V. Niesel, J. Steuwer and A. Tecklenborg for permission to use their data for the case studies. I thank H.-P. Bäumer, a reviewer and S. Dittmann for many helpful comments and suggestions on the manuscript. The project was funded by the German Bundesministerium für Bildung, Wissenschaft, Forschung und Technologie (BMBF) under grant number 03F0112 A and B. The responsibility for the contents of the publication rests with the author.

References

Bäumer H-P (1994) Stochastic simulation of dynamic point patterns. In: Faulbaum F (Ed) SoftStat '93: Advances in Statistical Software 4. The 7th Conference on the Scientific Use of Statistical Software. Heidelberg, March, 14-18, 1993. G. Fischer, Stuttgart, pp 145–152

Bickel PJ (1969) A distribution free version of the Smirnov two-sample test in the multivariate case. Ann Math Statist 40: 1–23

Blome D, Schleier U, van Bernem K-H (1999) Analysis of the small-scale spatial patterns of free-living marine nematodes from tidal flats in the East Frisian Wadden Sea. Marine Biology 133: 717–726

Bock HH (1980) Explorative Datenanalyse. In: Victor N, Lehmacher W & van Eimeren W (Eds). Explorative Datenanalyse. Springer, Berlin Heidelberg New York, pp 6–37

Bohle-Carbonell M (1992) Empirische Variabilitätsabschätzung und Methodik der Datenerhebung. Abschlußbericht. Bundesamt für Seeschiffahrt und Hydrographie, Hamburg

Borovkov K, Pfeifer D, Bäumer H-P (1996) Modeling dynamics and spatial aggregation of biological populations by stochastic networks. Senckenbergiana marit 27: 129–136

Bortz J, Lienert GA, Boehnke K (1990) Verteilungsfreie Methoden in der Biostatistik. Springer, Berlin Heidelberg New York

Büning H, Trenkler G (1994) Nichtparametrische statistische Methoden. de Gruyter, Berlin

Cerioli A (1997) Modified tests of independence in 2 x 2 tables with spatial data. Biometrics 53: 619–628

Chatfield C (1975) The Analysis of Time Series: Theory and Practice. Chapman & Hall, London

Clarke K R (1993) Non-parametric multivariate analyses of changes in community structure. Austr J Ecol 18: 117–143

Cox TF, Cox MAA (1994) Multidimensional Scaling. Chapman & Hall, London

Cressie N (1993) Spatial prediction in a multivariate setting. In: Patil GP, Rao CR (Eds) Multivariate Environmental Statistics. Elsevier, Amsterdam, pp 99–107

Dixon PM (1993) The bootstrap and the jackknife: Describing the precision of ecological indices. In: Scheiner SM, Gurevitch J (Eds) Design and Analysis of Ecological Experiments. Chapman & Hall, New York. pp 290–318

Edgington ES (1995) Randomization Tests. Marcel Dekker, New York

Friedman JH, Rafsky LC (1979) Multivariate generalizations of the Wald-Wolfowitz and Smirnov two-sample tests. Ann Statist 7: 697–717

Gittins R (1985) Canonical Analysis. A Review with Applications in Ecology. Springer, Berlin Heidelberg New York

Green RH (1993) Relating two sets of variables in environmental studies. In: Patil GO, Rao CR (Eds) Multivariate Environmental Statistics. Elsevier, Amsterdam, pp 149–163

Grimm V (1996) A down-to-earth assessment of stability concepts in ecology: Dreams, demands, and the real problems. Senckenbergiana marit 27: 215–226

Henze N (1988) A multivariate two-sample test based on the number of nearest neighbor type coincidences. Ann Statist 16: 772–783

James FC, McCulloch CE (1990) Multivariate analysis in ecology and systematics: Panacea or Pandora's box? Ann Rev Ecol Syst 21: 129–166

Jongman RHG, ter Braak, CJF, van Tongeren OFR (1995) Data Analysis in Community and Landscape Ecology. Cambridge University Press, Cambridge

Journel AG, Huijbregts CJ (1991) Mining Geostatistics. Academic Press, London

Kaufmann L, Rousseeuw PJ (1990) Finding Groups in Data. Wiley, New York

Kendall MG, Gibbons JD (1990) Rank Correlation Methods. Edward Arnold, London

Kröncke I, Zeiß B, Dahms S (1996) Makrofauna-Langzeitreihe im Inselvorfeld von Norderney. Bericht des Umweltbundesamtes. UFOPLAN-Nr. 10204270

Lamprecht J 1992. Biologische Forschung: Von der Planung bis zur Publikation. Parey, Berlin

Lienert GA (1969) Testaufbau und Testanalyse. Beltz Verlag, Weinheim

Lienert GA (1978) Verteilungsfreie Methoden in der Biostatistik. Anton Hain, Meisenheim

Lozán JL (1992) Angewandte Statistik für Naturwissenschaftler. Parey, Berlin

Ludwig JA, Reynolds JF (1988) Statistical Ecology. Wiley, New York

Manly BFJ (1986) Multivariate Statistical Methods - A Primer. Chapman & Hall, London

Mardia KV, Goodall CR (1993) Spatial-temporal analysis of multivariate environmental monitoring data. In: Patil GP, Rao CR (Eds) Multivariate Environmental Statistics. Elsevier, Amsterdam, pp 347–386.

Matheron G (1971) The Theory of Regionalized Variables and its Applications. Les Cahiers du Centre de Morphologie Mathematique de Fontainebleau, Nr. 5. Ecole Nationale Superieure des Mines de Paris.

Minchin PR (1987) An evaluation of the relative robustness of techniques for ecological ordination. Vegetatio 67: 1167–1179

Moore GH, Wallis WA (1943) Time series significance tests based on sign of difference. JASA 38: 153–164

Murthy DNP, Page NW, Rodin EY (1990) Mathematical Modelling. A Tool for Problem Solving in Engineering, Physical, Biological and Social Sciences. Pergamon Press, Oxford

Ofenheimer M (1971) Ein Kendall-Test gegen U-förmigen Trend. Biometrische Zeitschrift 13: 416–420

Pfeifer D, Bäumer H-P, Schleier-Langer U (1994) The analysis of spatial data from marine ecosystems. In: H.-H. Bock, H-H, Lenski W, Richter MM (Eds) Information Systems and Data Analysis. Springer, Berlin Heidelberg New York. pp 340–349

Pfeifer D, Bäumer H-P, Dekker R, Schleier U (1998) Statistical tools for monitoring benthic communities. Senckenbergiana marit 29: 63–79

Pfeifer D, Bäumer H-P, Ortleb H, Sach G, Schleier U (1996b) Modeling spatial distributional patterns of benthic meiofauna species by Thomas and related processes. Ecol Model 87: 285–294

Richter O, Söndgerath D (1990) Parameter Estimation in Ecology. The Link between Data and Models. VCH, Weinheim

Sachs L (1992) Angewandte Statistik. Springer, Berlin Heidelberg New York

Samietz J, Berger U (1997) Evaluation of movement parameters in insects – bias and robustness with regard to resight numbers. Oecologia 110: 40–49

Scheiffarth G, Nehls G, Austen I (1996) Modelling distribution of shorebirds on tidal flats in the Wadden Sea and visualisation of results with the GIS IDRISI. In: Lorup E, Strobl J (Eds) Salzburger Geografische Materialien, Heft 25. Selbstverlag des Instituts für Geographie der Universität Salzburg

Schleier U, van Bernem K-H (1996) A method to compare samples of soft bottom communities. Senckenbergiana marit 26: 135–144

Schlittgen R, Streitberg BHJ (1994) Zeitreihenanalyse. Oldenbourg, München

Smith EP, Pontasch KW, Cairns J (1990) Community similarity and the analysis of multispecies environmental data: a unified statistical approach. Water Res 24: 507–514

Smith EP, Rheem S, Holtzman GI (1993) Multivariate assessment of trend in environmental variables. In: Patil GP, Rao CR (Eds) Multivariate Environmental Statistics. Elsevier, Amsterdam, pp 489–507

Sokal RR, Rohlf FJ (1981) Biometry. Freeman, New York

Underwood AJ (1994) Things environmental scientists (and statisticians) need to know to receive (and give) better statistical advice. In: Fletcher DJ, Manly BFJ (Eds) Statistics in Environmental Monitoring. University of Otago Press, Otago, pp 33–61

van der Meer J (1991) Exploring macrobenthos-environment relationship by canonical correlation analysis. J Exp Mar Biol Ecol 148: 105–120

ver Hoef JM, Cressie N (1993) Spatial Statistics: Analysis of Field Experiments. In: Scheiner SM, Gurevitch J (Eds) Design and Analysis of Ecological Experiments. Chapman & Hall, New York. pp 319–425

Warwick RM, Clarke KR (1991) A comparison of some methods for analysing change in benthic community structure. J mar biol Ass. UK 71: 225–244

Weiss L (1960) Two-sample tests for multivariate distributions. Ann Math Statist 31: 159–164

Wissel C (1989) Theoretische Ökologie. Springer, Berlin Heidelberg New York

5 Spatial and Temporal Distribution Patterns and Their Underlying Causes

5.1 Distribution of Nutrients, Algae and Zooplankton in the Spiekeroog Backbarrier System

Verena Niesel & Carmen-Pia Günther

Abstract: Temporal developments of dissolved nutrients, phytoplankton and zooplankton in the water column were investigated on different time scales from 1994 to 1996. Measurements of single tidal cycles allowed to assess the fate of some species in the Wadden Sea. Dissolved nutrients showed some regularly recurring patterns, which could be explained by the tidal mixing of water bodies with different nutrient concentrations. Seasonal cycles of the nutrients, phytoplankton and zooplankton showed a large interannual variability. Timing and extension of algal blooms influenced the occurrence of many meroplanktonic larvae. Most of the algal species in the backbarrier system of Spiekeroog were part of the phytobenthos and the phytoplankton.

5.1.1 Introduction

The seasonal cycles of nutrients, phytoplankton and zooplankton in the Wadden Sea are closely tied. The amount of dissolved nutrients (phosphate, silicate, ammonia, nitrite, nitrate) together with other factors influences the strength of algal blooms and the composition and change of phytoplankton communities (Cadée & Hegemann 1991). The spring bloom is normally built up by diatoms and usually ends due to silicate depletion. If enough nitrogen and phosphate are available, a flagellate bloom can follow, which is often dominated by *Phaeocystis* spp. Algae are food for many of the zooplankton and macrozoobenthic organisms. As timing and extension of the phytoplankton bloom influences the occurrence of the meroplanktonic larvae, some authors called the algal blooms a trigger for the spawning of several macrofauna species (Starr et al. 1991; Himmelman 1975). Therefore, the seasonal dynamic of the planktonic primary producers is of importance for the qualitative and quantitative composition of the zooplankton.

The Wadden Sea is a highly dynamic and turbulent system, where exchange processes between the water volume and the sediment are important (Postma 1984). The consequences of this coupling on the algae species in the benthos and water column were of special interests. Microphytobenthic algae are temporarily suspended into the water column by currents and waves (Baillie & Welsh 1980; Drebes 1974; de Jonge & von Beusekom, 1995). The flexibility of diatoms in the Wadden Sea to live either in the benthic or pelagic environment will be reported in this study.

Temporal and spatial distributions of nutrients, algae and zooplankton were analysed on different time scales in several sub-projects of ELAWAT. On the small-scale, single tidal cycles were investigated. These results can give information about the source of the species and their fate in the Wadden Sea. On the large scale, the seasonal developments of the parameter mentioned above were analysed.

5.1.2 Material and Methods

The studies were carried out in the backbarrier system of the island of Spiekeroog, southern North Sea (Chap. 3.1). Sampling sites and temporal resolutions were chosen by the sub-projects according to their specific research questions (Table 5.1.1). Detailed descriptions on the analysis of nutrients are found in Grasshoff et al. (1983) and Niesel (1997). Phytoplankton analysis was done after Utermöhl (1958), details are given in Niesel (1997). The analysis of zooplankton is described in Günther et al. (1998). Microphytobenthos was sampled regularly once a month, in 1994 and 1995 from April to September, in 1996 from February to September, at different sample sites in the backbarrier tidal flats of Spiekeroog. Dead and living cells were analysed for the qualitative comparison between algae in the sediment and in the water body.

Table 5.1.1 Measured parameter with time and site of the sampling. The precise sampling sites are given in Chap. 3.1 and in Gehm (1996), Günther (1998) and Niesel (1997). WP = sampling site in the Otzumer Balje, see Fig. 3.1.2.

Record	Parameter	Sampling time	Sampling site
tidal cycle	soluble nutrients	hourly during a tidal cycle	Gröninger Plate, Swinnplate, Otzumer Balje (WP)
	phytoplankton	hourly during a tidal cycle	Gröninger Plate, Swinnplate, Otzumer Balje (WP)
	zooplankton	ebb and flood samples	Otzumer Balje (WP)
seasonality	soluble nutrients	at high tide	Otzumer Balje (WP)
	phytoplankton	one hour before high tide	"
	zooplankton	"	"
benthic-pelagic coupling	microphytobenthos		Gröninger Plate, Swinnplate, Breakwater
	phytoplankton		Gröninger Plate. Swinnplate, Breakwater, Otzumer Balje (WP)

5.1.3
Tidal Variations

5.1.3.1
Dissolved Nutrients

The backbarrier system of Spiekeroog is strongly influenced by the North Sea in contrast to the Dutch Wadden Sea (de Jonge et al. 1993). The fluvial input is only of minor importance along the East Frisian coast. Consequently the nutrient concentrations at high tide reflect the situation of the North Sea coastal zone. The nutrient concentration in the Wadden Sea showed a large variability: The magnitude of the values was comparable to those of other Wadden Sea areas, but often about one order of magnitude higher than in the adjacent coastal zone (Martens & Elbrächter 1998). Especially for phosphate and silicate strong gradients were observed. Maximum values were measured mostly during low tide (Table 5.1.2). During every tide nutrient poor coastal waters mixes with nutrient rich Wadden Sea water and results in a nutrient export to the coastal zone.

Dissolved nutrients like silicate, phosphate and nitrogen were measured in spring and summer hourly at different sites in the backbarrier system (Otzumer Balje, breakwater). Dissolved phosphate and silicate showed a regular pattern: At high tide the concentrations of nutrients were low. At low tide and during the beginning flood, concentrations of nutrients increased (Fig. 5.1.1).

Differences in the nutrient concentrations were caused by a variety of processes. Chemical and microbial conversion takes place in the sediment and in the water column. Currents near the sediment boundary can lead to an exchange of pore water (Shum 1993; Hüttel et al. 1996), which is enhanced by the bioturbation of several worms and mussels (Asmus & Asmus 1991; Dame et al. 1991). Furthermore, the nutrient concentrations are increased by heterotrophic processes (Asmus & Asmus 1998). This can explain the higher nutrient concentration in the Wadden Sea as compared to the coastal zone.

The above regular pattern was not found for the dissolved nitrogen. The concentrations of nitrate, nitrite and ammonia varied irregularly during a tidal cycle. The reason for this variability is the high reactivity of nitrogen. Under anoxic conditions, which occur close to the sediment surface microbial denitrification takes place. In this process nitrate is conversed to molecular nitrogen.

Table 5.1.2 Maximum concentrations of dissolved nutrients measured in the Wadden Sea, during flood tide at the breakwater in July 1995. Minimum concentrations of all nutrients reached their detection limits and were always measured in spring. (Data from Niesel 1997)

	Phosphate	Silicate	Nitrate	Nitrite	Ammonia
[µmol / l]	6	100	60	1	65

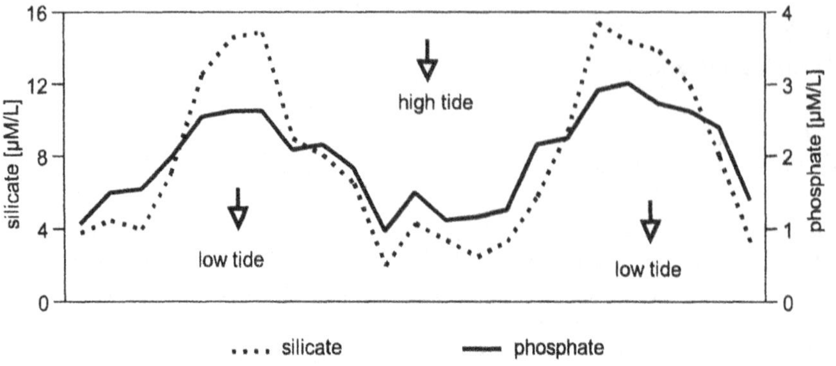

Fig. 5.1.1 Dissolved nutrients, silicate and phosphate, during two tidal cycles in the Otzumer Balje in August 1994. (Data from Niesel 1997)

5.1.3.2
Phytoplankton

Phytoplankton was analysed during 19 tidal cycles in spring and summer at two different sites (Otzumer Balje and breakwater, Fig. 3.1.2). The results of the tidal studies showed that the different algal populations reacted differently to the transport into the Wadden Sea. The species were divided into three groups (Table 5.1.3, Niesel 1997).

1. Species that are transported from the North Sea into the Wadden Sea and cannot grow there because of the unfavourable light situation and strong predation pressure.
2. Species that are transported from the North Sea into the Wadden Sea and which can grow there under these conditions.
3. Species that prefer the Wadden Sea habitat. This group comprises predominantly benthic living algae. During rough weather they are resuspended into the water column and at calm conditions they settle down on the sediment.

Table 5.1.3 Classification of the algal species according to their reproduction potential in the Wadden Sea. The enlisted algal species were dominant during the sampling programme, but they represent only a small part of all species which occur in the Wadden Sea. See text for group distinction. (Data from Niesel 1997)

Group 1	Group 2	Group 3
Thalassiosira rotula	*Asterionellopsis glacialis*	*Paralia sulcata*
Guinardia delicatula	*Chaetoceros* sp.	*Navicula* sp.
Leptocylindrus danicus		*Cylindrotheca closterium*
Leptocylindrus minimus		
Phaeocystis pouchettii		
Phaeocystis globosa	*Coscinodiscus concinnus*	

The diatom *Coscinodiscus concinnus* was assigned to group two, because it grew in the Wadden Sea under the special conditions in spring 1996 (Chaps. 3.2 and 7.4). This centric diatom is a cold adapted, large species (radius of the valve 110–450 μm), which occurred regularly with low numbers in the German bight (Baars 1979; Drebes & Elbrächter 1976). Until 1996 the growth of this diatom species was not observed in the Wadden Sea. However, *C. concinnus* blooms have occurred before in the North Sea (Grøntved 1952; Roskam 1970; Bauerfeind et al. 1990). Grøntved reported a lipid film on the water surface of the central North Sea, which formed after a bloom of *C. concinnus*. Along the Dutch coast the lipids produced by a *C. concinnus* bloom caused a severe oxygen deficiency in 1964 (Roskam 1970). In May 1996 a biogenic lipid film of 640 km^2 was observed in the southern German Bight, which could be related to the bloom of *C. concinnus*. Part of the lipid film, which was transported into the Wadden Sea, as well as the end products of the bloom were regarded as one cause for the development of the anoxic areas behind the East Frisian Islands in June 1996 (Höpner 1996; Niesel 1997; Chap. 7.4).

The haptophyceae *Phaeocystis* spp. was found regularly in the Wadden Sea. But its chance of growing under the conditions in the Wadden Sea was low, so it was placed in group 1 (Niesel 1997). Losses of that species in the Wadden Sea were caused by filtration by mussels (Smaal & Twisk 1997; Petri & Vareschi 1997).

Species like *Brockmaniella brockmannii* or *Biddulphia alternans* are considered to be typical algae of tidal areas (Drebes 1974). These species were found regularly in the backbarrier system of Spiekeroog, but the massive growth described for the Sylt-Rømø-bight was not observed (Schneider et al. 1998). This can be caused by different exchange rates of water in the two areas. Unpublished data from a hydrographic model showed that the average rate of water exchange in the Sylt-Rømø-bight was lower than in the backbarrier system of Spiekeroog. Therefore the growth conditions for the algae are more variable in the backbarrier system. Furthermore, the higher exchange rates lead to quick transport of algae in and out of the Wadden Sea. Pelagic algae, which need constant growth conditions for a certain amount of time, have only a low chance to establish here. Thus, the phytoplankton composition in the backbarrier system of Spiekeroog was characterized by species imported from the North Sea and species resuspended by the currents, wind and wave action (see de Jonge & von Beusekom 1995; Sect. 5.1.8).

5.1.3.3 Zooplankton

Zooplankton was analysed monthly in the tidal inlet of the Otzumer Balje (WP, see Fig. 3.1.2) during low and high tide. Additionally, the variations in the zooplankton were measured once during every hour of a tidal cycle. The variability within one tidal cycle was so large that for most of the zooplankton species no tidal difference could be assessed. Only the larvae of the polychaetes *Lanice conchilega* and *Magelona papillicornis* occurred in higher numbers in flood than in ebb water. Total amounts of zooplankton individuals were mostly higher during flood than during ebb tide. Only in the summer months an opposite situation occurred due to a higher production of larvae of species living in the Wadden Sea (*Hydrobia ulvae*, spionids, cirripedes).

5.1.4 Seasonal Variability

Dissolved nutrients, phytoplankton and zooplankton were analysed in ELAWAT under different questions. Only zooplankton was studied explicitly for its seasonal development, but nutrients and phytoplankton were sampled according to criteria, which were not relevant for recording seasonal development. However, an attempt is made here to describe all three parameters over the 3 study years in spite of their varying sampling schemes.

5.1.4.1 Dissolved Nutrients

Dissolved nutrients were measured irregularly in 1994, and a description will not be given here. In 1995, the seasonal variation of the nutrient concentration was continuously recorded (Fig. 5.1.2). During the first months of the year concentrations of all of the measured dissolved nutrients were high. With the growth of the algae in spring, nutrient concentrations decreased, and in some cases they reached the detection limit. Except for phosphate all nutrient concentrations began to rise at the end of August. The concentrations of phosphate already began to rise much earlier in June. This increase, which was also observed in the Dutch Wadden Sea, is due to the reduction of iron(oxy) hydroxides and the subsequent release of adsorbed phosphates (Jensen et al. 1995).

In 1996 the nutrient concentrations developed in a similar seasonal pattern as in the previous year. With the beginning of the diatom bloom the nutrient concentrations decreased. Nitrogen compounds remained at low values until July, but phosphate and silicate increased slowly with the approach of summer (without figure).

5.1.4.2 Phytoplankton

Phytoplankton in the backbarrier system of Spiekeroog is dominated by diatoms, which were found in variable numbers all over the year. Beside the diatoms, the haptophycea *Phaeocystis* spp. occurred regularly. Dinoflagellates, which often bloomed in the German Bight, were rarely found and occurred only in small numbers in the Wadden Sea.

In 1994, phytoplankton showed the following development (Fig. 5.1.3). As early as March, diatoms typical for a spring situation, occurred in the water, but only in low numbers. These were for example: *Asterionellopsis glacialis*, *Odontella sinensis*, *Guinardia delicatula*, *Thalassiosira rotula* and *Rhizosolenia shrubsolei*. The spring bloom was dominated by *Guinardia delicatula* and reached its maximum in April with the haptophyceae *Phaeocystis globosa* (*P. globosa*: 1.4×10^7 cells l^{-1}, *Guinardia delicatula:* 1.1×10^6 cells l^{-1}). In May, June and July cell numbers were low. In August a bloom of the diatoms *Leptocylindrus danicus* and *Leptocylindrus minimus* was observed.

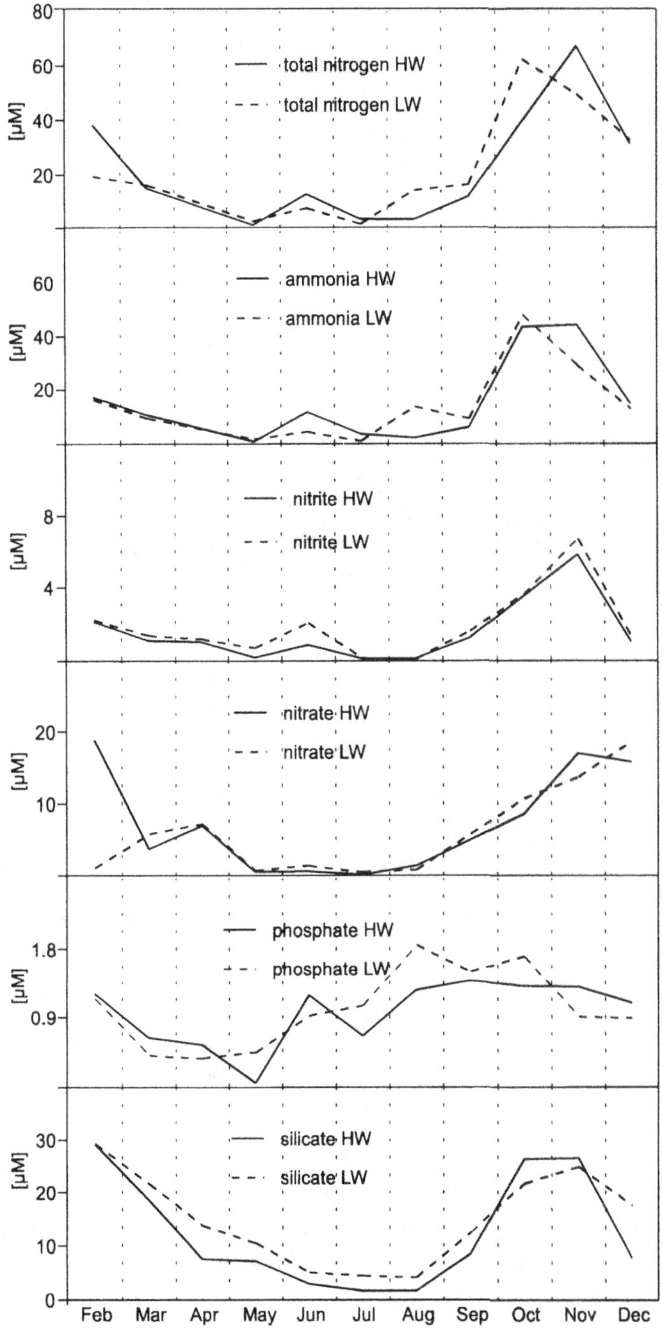

Fig. 5.1.2 Seasonal development of the dissolved nutrients in 1995. Values of high tide (line) and low tide (broken line) measurements are shown. (Liebezeit unpublished data)

In 1995 no bloom comparable in strength to that of the previous year was found (Fig. 5.1.3). However, the total amount of algal cells in the months from May to September was higher than in summer 1994 and corresponded to the cell numbers during the spring bloom of 1994. The summer of 1995 was characterized by different *Chaetoceros* species, which built the maximum in August. The missing spring bloom with the following high density of microalgae in summer 1995 was very unusual and the causes are not known.

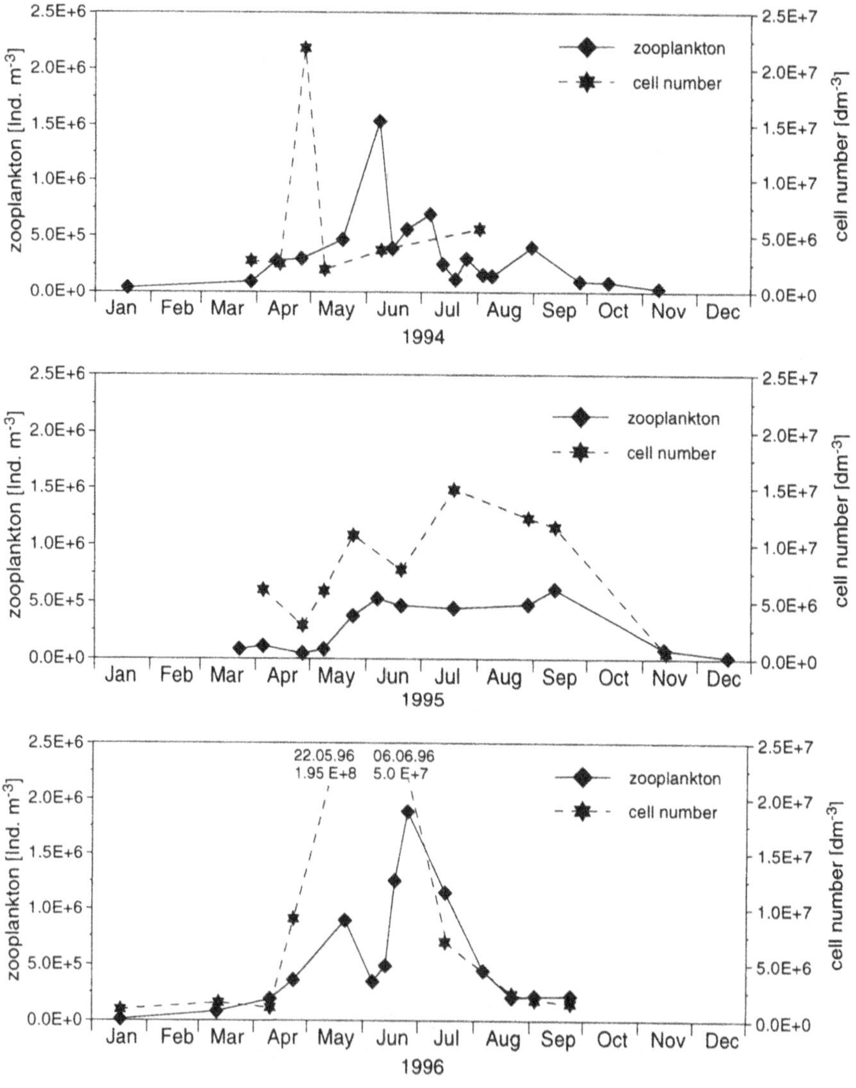

Fig. 5.1.3 Numbers of algal cells and zooplankton abundances in the years 1994, 1995 and 1996. High cell numbers in May and June 1996 are caused by an extraordinary strong bloom of *Phaeocystis pouchetii*. (Data from Niesel 1997 and Boysen-Ennen unpublished data)

The development of phytoplankton in 1996 was unusual (Fig. 5.1.3). Because of the cold winter and the extraordinary favourable light conditions in the Wadden Sea (Chaps. 3.2; 7) the large diatom *Coscinodiscus concinnus* bloomed. The bloom began in March and lasted until June. According to cell numbers, the bloom was inconspicuous (between 400 and 6400 cells l^{-1}), but its biomass can account for over 90 % of the total biomass. Diatom species, which normally occurred in the Wadden Sea at this time, were rarely found, except for small *Chaetoceros* species.

Simultaneously to the *C. concinnus* bloom, a bloom of the haptophyceae *Phaeocystis pouchetii* occurred. This bloom reached maximal cell numbers of nearly 200 millions cells per litre. Usually this species is found in northern latitudes and has a higher tolerance towards low temperature than *Phaeocystis globosa* (Jahnke 1987). The simultaneous appearance of the two cold-adapted algae *C. concinnus* and *P. pouchetii* are in line with the extraordinary low temperatures in spring 1996.

5.1.4.3 Zooplankton

The development of the zooplankton in the backbarrier system of Spiekeroog could be divided into four seasonal groups.

1. Winter plankton: It has few species and individuals and is dominated by copepods.
2. Spring plankton: Increasing amount of species and individuals with first meroplanktonic larvae and often hydromedusae in great numbers.
3. Early summer plankton: Maximal number of species with high individual numbers, leading to highest diversity in spite of high amounts of meroplanktonic larvae and young copepods.
4. Late summer/autumn plankton: The decrease of individual numbers and the increase of species numbers leads to a similar high diversity as in early summer.

In 1994, zooplankton reached maximum numbers in June, following the spring phytoplankton bloom (Fig. 5.1.3). The zooplankton maximum was mainly composed of copepods and larvae of molluscs and polychaetes. Whereas copepods occurred all over the year, larvae of molluscs were found only in May and June. Larvae of polychaetes were found irregularly until the end of August. After the bloom of the diatom *Leptocylindrus* ssp. in summer, an increase of the copepods was observed, who possibly profited from this bloom. In winter, zooplankton was dominated by copepods.

Compared with 1994, the total amount of zooplankton was much lower in 1995 and extended only 1/4 of the densities of the year before. The larvae of the polychaetes and molluscs reached low numbers. Only the numbers of copepods were similar to the year before. The total amount of zooplankton reached maximum values from June to September and then decreased to lower winter values.

In the year 1996, the zooplankton also showed differences with the years before (Fig. 5.1.3) The individual numbers were unusually high and showed two maxima. The first maximum appeared in the middle of May during the bloom of *C. concinnus*. It consisted of copepods and polychaete larvae. A second maximum was built in July by cirrepedia, polychaetes and copepods. In June large amounts of bivalve larvae were observed (3237 larvae 100 l^{-1}.)

The decrease of the zooplankton individuals in early June 1996 could have been caused by the lipid film, formed by the diatom *C. concinnus* (see above Sect. 5.1.3). Residues of the lipid film were observed as a white substance in the water and on the sediment surface in the Wadden Sea. Such fats can clog the respiration organs of organisms and cause their death.

5.1.5 Interannual Variability

In the Wadden Sea the variability between the years can be high. But some recurrent patterns can be observed nearly every year. The nutrient concentrations are high in winter, decrease in spring because of the phytoplankton bloom and usually a phosphate maximum is observed in summer (de Jonge & Postma 1974). Phytoplankton is characterized by a spring bloom, which is dominated by diatoms. The diatom bloom is followed by a bloom of flagellates, usually *Phaeocystis* spp. Often a second diatom bloom is considered in summer (Cadee 1986).

A comparison of the data taken over three years of observation showed a very high variability of plankton concentrations (e.g. no spring bloom in 1995). This high variability made it impossible to take one year as a reference year for the evaluation of stability properties. Rather, each year had to be treated as a special case.

Because of the different intensity and composition of the phytoplankton bloom in each year, the development of the zooplankton varied respectively. The year 1994 was characterized by a spring and summer bloom of diatoms. The spring bloom was followed by a bloom of *Phaeocystis globosa*. This temporal sequence is considered to be typical for the Wadden Sea (e.g. Hickel et al. 1992; Elbrächter et al. 1994). In the zooplankton, larvae of molluscs were scarcely represented, whereas copepods and polychaete larvae occurred with mean abundances.

The seasonal development of nutrient concentrations observed in 1995 followed a typical trend for this area (e.g. Gillbricht 1988). But instead of a spring bloom of phytoplankton a long summer bloom was observed. The zooplankton species reacted differently to the lack of the spring bloom and the high cell numbers in summer. Larvae of molluscs appeared later in the year and polychaete larvae were only found in small numbers.

The year 1996 was characterized by low temperatures in the beginning of the year, a bloom of *C. concinnus* and a mass occurrence of *Phaeocystis pouchetii*. Many copepod species reached the highest individual numbers of the study period in that year. Larvae of polychaetes and bivalves reacted differently depending on the species.

This large variability on the base of the food web implies that species on higher trophic levels have to face not only small scale variations of food abundances, but large scale variations between the years as well.

5.1.6 Relations Between Phytoplankton and Copepods

Copepods are the main group of the holozooplankton. In winter they dominate the zooplankton composition (see Sect. 5.1.4) and they reach maximum abundances in summer when nauplius and copepodite stages occur. The copepods *Acartia clausi*, *Temora longicornis*, *Centropages hamates* and *Pseudocalanus elongatus* were frequently recorded in the backbarrier system of Spiekeroog.

Copepods are omnivore (Valiela 1995). Their food consists of detritus, bacteria, algae, ciliates and nauplius stages of other copepods. Larvae of polychaetes and molluscs were also preyed upon. The feeding behaviour of the copepod is widely debated (Donaghay & Small 1979; Irigoien et al. 1996). Some species of copepods feed selectively on a specific particle size, others are non-selective with regard to particle size. The ability to switch between different prey sizes and prey species has also been recorded for some copepod species (Valiela 1995). Since not every potential food source of copepods could be studied in ELAWAT, only a relation between copepod numbers and algal densities is presented in the following. Because of the wide food spectrum of copepods, not the number of small algal cells, but the total amount of algal cells was chosen for the comparison.

The occurrence of copepods in the Wadden Sea appeared to be influenced by the amounts of algae. High individual densities were mostly recorded after the algal bloom. In 1995 the individual numbers of copepods were not lower than in the previous year, as was the case with polychaete and bivalve larvae and resulted from the lack of the spring bloom. Probably most of the copepods were able to grow well despite of the absence of an algal bloom, as they were able to profit from other sources of food. The copepods *Acartia clausi* could use the summer bloom, so that this species reached maximum densities and developed several generations later in the year (Fig. 5.1.4).

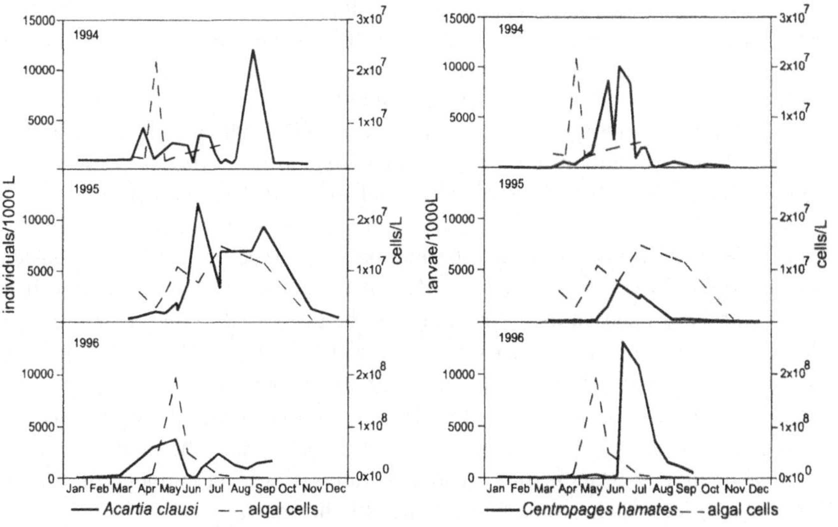

Fig. 5.1.4 Abundances of *Acartia clausi* and *Centropages hamates* and the numbers of algae cells during the time of observation (please note the different scales). (Data from Niesel 1997 and Boysen-Ennen unpublished data)

In 1996, most of the copepods (esp. *Temora longicornis*) could use *Coscinodiscus concinnus* as a food source and their numbers increased already during the bloom. The copepods already built fat droplets in spring, which was not observed at this time in the two preceding years, indicating a good utilization of *C. concinnus*. Only the numbers of *Centropages hamates* did not increase during the bloom of *C. concinnus*, as this species has a longer generation time and therefore occurs usually later in the Wadden Sea.

5.1.7
Relations Between Phytoplankton and Meroplanktonic Larvae

The amount of meroplanktonic larvae in the water column can be influenced by several factors, e.g. the intensity of spawning, abiotic conditions (e.g. water temperature), food supply and predation pressure. Spawning and predation were not assessed. For the remaining factors an evaluation of their influence of appearance of larvae will be given.

The food of bivalve larvae consists mainly of microalgae and bacteria. The size of food particles required by bivalve larvae is still under debate. Jørgensen (1981) states that particles should not exceed 2–5 μm in size, but Baldwin & Newell (1991) observed larval uptake of particles smaller than 30 μm. To assess the influence of food on the larvae, the algal fraction smaller 30 μm was considered here.

Larvae of polychaetes have a wide ranging food spectrum. They feed on small algae as the larvae of bivalves do. Some larvae of polychaetes are known to colonize aggregates in the water and feed on bacteria and detritus (Shanks & del Carmen 1997).

In order to estimate the food supply for meroplanktonic larvae the fraction of the small algae (< 30 μm) was calculated. To this fraction belong, among others, *Asterionellopsis glacialis*, *Skeletonema costatum*, small species of *Chaetoceros* and *Phaeocystis* spp. Including *Phaeocystis* spp. in this fraction (< 30 μm) can be disputed, because these algae also occur in colonies, which are substantially larger than 30 μm. However, in the backbarrier system of Spiekeroog these algae often occurred as individual cells and were therefore counted in this fraction.

Except for algal blooms, the fraction of small algae accounted for 75 to 90 % of total algal numbers. During the diatom bloom in spring 1994 the share was reduced to 15 %. At the beginning of the bloom of *Coscinodiscus concinnus* in 1996, the fraction of small algal cells was low, but reached nearly 100 % in April.

Cold winters with sea-ice formation in the Wadden Sea ("ice winters", Chaps. 3.2, 7) indirectly influence the numbers of larvae of the different species: Cold sensitive species as *Lanice conchilega* do not survive temperatures below 2 °C (Buhr 1991), so that their spawning can be decreased by the reduction of the adult population. According to Honkoop & van der Meer (1997) ice winters can also result in an increase of the amount of larvae of some species, if the egg development commences before the winter. Then, in a mild winter with poor nutrition, the eggs can be resorbed by the adult. But in a cold winter, the metabolism is reduced and all eggs can mature. The generally strong recruitment of *Macoma balthica* after ice winters could be explained by this mechanism (*loc. cit.*).

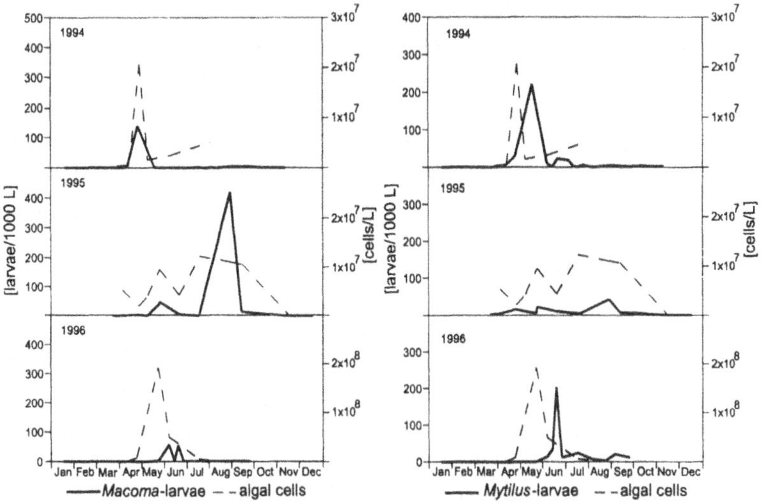

Fig. 5.1.5 Larval abundances of *Macoma balthica* and *Mytilus edulis* and the number of algae cells <30 µm during the study period (please note the different scales). (Data from Niesel 1997 and Boysen-Ennen unpublished data)

5.1.7.1 Bivalve Larvae

During the three years of observation larvae of *Mytilus edulis*, *Mya arenaria*, *Cerastoderma edule*, *Ensis directus* and *Macoma balthica* occurred regularly in the Wadden Sea. But abundances and timing differed.

Larvae of these species usually occur after the algal bloom. The late maximum of larval numbers in 1995 and the second spawning which was observed for some species in late summer, could be explained by the lack of the spring bloom and the long persistent high algal density in the summer of that year. A second spawning is known for *M. balthica* on its southern distribution limit (Gironde estuary). In the East Frisian Wadden Sea, however, it was not yet known to occur (Günther et al. 1998; Bachelet 1986; Chap. 5.3; Fig. 5.1.5)

A continuous spawning over summer was also observed for *M. edulis* in 1995, but the number of larvae was very small that year.

After the ice winter, the larval numbers of most species were comparable or higher in 1996 than in the previous year. Only *M. balthica* had low larval numbers after the ice winter, which was contradictory to the mechanism described above, after which this species should have a high recruitment rate after cold winters. An explanation cannot yet be given.

5.1.7.2 Polychaete Larvae

Polychaete larvae comprised the largest part of the zooplankton during the three years of observation. Especially high numbers were found for *Magelona papilli-*

cornis, *Lanice conchilega*, *Pygospio elegans*, *Polydora ligni*, *Spio martinensis* and further spionid polychaete larvae which could not be determined to species level. The timing and the abundances varied over the years of observation. (Fig. 5.1.6).

Larvae of some species reached rather high numbers before the algal bloom, e.g. *M. papillicornis* and larvae of spionids. These species could probably feed sufficiently on the few algae present, bacteria and detritus. The appearance of larvae of *L. conchilega* showed a close temporal connection to algal blooms in 1994 and 1995.

With the exception of *S. martinensis* the larval density of all other polychaete species was reduced in 1995. An explanation could be a nutrition limitation resulting from the lack of the spring bloom. The successful recruitment of *S. martinensis* could be explained by the late spawning of this species, which enabled it to profit from the summer bloom of algae in 1995.

After the ice winter, numbers of larvae of *L. conchilega* and *M. papillicornis* were severely reduced in 1996 (compare Chap. 5.3.6). Since the adults of both species are sensitive to cold temperatures, the low numbers of larvae are probably caused by the death of the adults. Other species, however, were not influenced by the cold temperatures, e.g. *P. elegans* and *P. ligni*. Their larvae reached record numbers in 1996.

The meroplankton data of the three years showed a high variability. While bivalve larvae appeared probably corresponding with the time of algal blooms and resulting from simultaneous spawning of several bivalve species, the occurrence of the polychaete larvae varied during the study period. This results from different reproduction strategies and the wide range of food spectrum of the larvae.

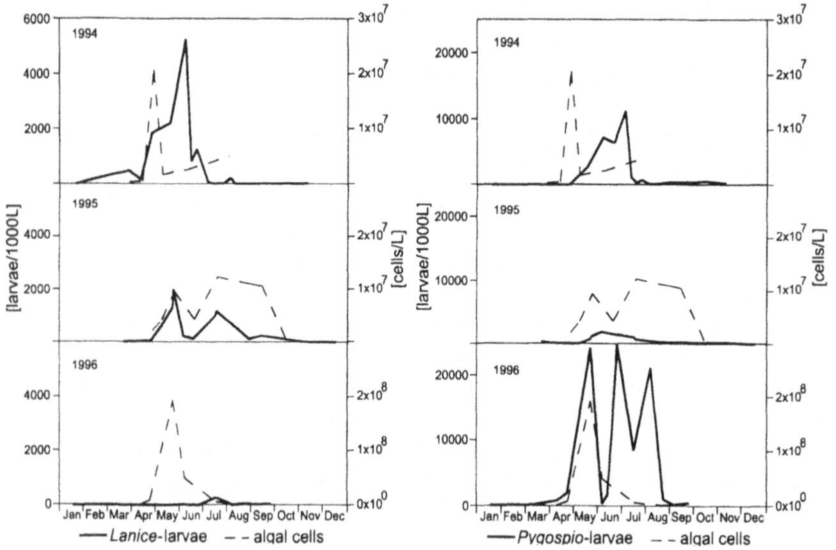

Fig. 5.1.6 Abundances of larvae of *Lanice conchilega*, *Pygospio elegans* and algal cells (< 30 µm) during the time of observation (please note the different scales). (Data from Niesel 1997 and Boysen-Ennen unpublished data)

5.1.8
Resuspension of Benthic Algae

Species of microalgae, living in and upon the sediment, belong to the microphytobenthos. The dominant group are diatoms (e.g. species genera *Amphora*, *Cymatosira*, *Melosira*, *Cylindrotheca*, *Navicula*, *Nitzschia*, *Odontella*), cyanobacteria (species of *Merismopedia* and *Oscillatoria*) were found only sporadically. According to their mode of living, the benthic diatoms were classified as a) sessile b) mobile c) solitary or d) colony-building types. The relative contribution of these groups to the microphytobenthic assemblage varied strongly.

Bloom-building diatoms, which occurred in the water body, were also found on the sediment (e.g. *Asterionellopsis glacialis*, *Coscinodiscus concinnus*, *Ditylum brightwellii*, *Eucampia zodiacus*). Diatoms, which usually occurred in the sediment, were also found in the water body and reached highest densities during flood- and ebb-flow (e.g. *Cylindrotheca closterium*, *Paralia sulcata*). During calm weather, sediment surface layers detached from the ground when the tide came in and floated as a film on the water surface. This film dissolved with the next high tide. Diatom species usually found in the surface films were *Cylindrotheca closterium*, *Cymatosira belgica* and species of the genera *Navicula*, *Nitzschia*, *Paralia*, *Odontella* and *Amphora*. A qualitative comparison of the diatoms in the sediment and in the detached surface film showed that both consisted mainly out of species which were dominant in the sediment. Since all categories of microphytobenthos (mobile, sessile, solitary and colony-building) were found in the floating material, it is assumed that the surface layer is mechanically detached and not made up by mobile diatoms actively entering the water column.

Altogether, 66 genera with 162 species (26 species thereof could not be identified) were found (Fig. 5.1.7). A comparison of the species lists of benthic and pelagic microalgae showed that more species and genera were dwelling in the sediment. 42 % of all genera found occurred in the water body as well as in the sediment. Genera which occurred only in the water column made up 14 % of the total microalgae, whereas genera, which were found only in the sediment accounted for 44 % (Table 5.1.4). The high number of genera occurring in both habitats indicates that a strict classification into benthic- or pelagic-living species or genera is only possible for a few microalgae in the Wadden Sea. The prevailing part of the genera seems to be able to survive in both habitats, at least for a short time.

This intensive exchange between the benthos to the pelagic and vice versa has various consequences for the microalgae. Algae, which enter the water body through resuspension, are exposed to an improved light situation in the water col-

Table 5.1.4 Number of diatom species and genera which were found in the pelagic and benthic habitats of the backbarrier system of Spiekeroog. (Data from Niesel 1997 and Keuker-Rüdiger unpublished data)

	Species	Genera
Pelagic	76	37
Benthic	124	56

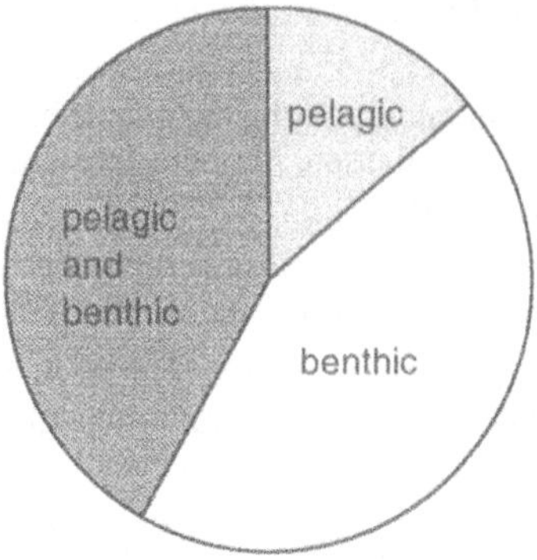

Fig. 5.1.7 Classification of algal genera found in the backbarrier system of Spiekeroog according to their benthic and/or pelagic habitat. The entire circle represents 100 %. (Data from Niesel 1997 and Keuker-Rüdiger unpublished data)

umn. On the other hand microalgae in the water column serve as food for suspension feeders (Marshall 1967; de Jonge & van Beusekom 1992). But the short-range transport in the water column can also be an advantage for the algae, because they can change their location on the sediment surface. Algae, which settle out of the water column onto the sediment, are exposed to an improved nutritional situation because of the higher nutrient concentrations in the pore water.

Furthermore, many algal species develop resting stages, that enable them to survive unfavourable seasons in or on the sediment (Boero 1994). These resting stages can also be resuspended into the water column by the current.

Acknowledgements

This chapter is based on the studies in ELAWAT by Bert Albers, Jan Backhaus, Alexander Bartholomä, Brigitte Behrends, Cord Bergfeld, Dietrich Blome, Michael Böttcher, Elisabeth Boysen-Ennen, Peter Brocks, H.-J. Brumsack, Monique Delafontaine, Burghard Flemming, Günther Hertweck, Andreas Hild, Udo Hübner, Ralf Kaiser, Margrit Kanje, Roswitha Keuker-Rüdiger, Kim Knauth-Köhler, Ingrid Kröncke, Wolfgang Krumbein, Thomas Leu, Gerd Liebezeit, Hermann Michaelis, H.D. Niemeyer, Verena Niesel, Georg Ramm, Jörn Reichert, Susanne Rolinski, Jürgen Rullkötter, Günther Sach, Ulrich Sommer, Monika Stamm, Jens Steuwer, T. Stoeck, Michael Türkay, Karl-Heinz van Bernem, Marlies Villbrandt, Frank Wolf and Ruth Zühlke. Their studies were published as cited in the text, and further unpublished data were kindly provided by Elisabeth Boysen-Ennen, Roswitha Keuker-Rüdiger and Gerd Liebezeit. The project was funded by the German Bundesministerium für Bildung, Wissenschaft, Forschung und Technologie (BMBF) under grant number 03F0112 A and B. The responsibility for the contents of the publication rests with the author. Thanks for the help in translating to Ute Fuhrhop and Wolfgang Wenzel. The authors wish to thank a reviewer for helpful comments on the manuscript.

References

Asmus R, Asmus H (1991) Mussel beds: limiting or promoting phytoplankton? J Exp Mar Biol Ecol 148: 215–232

Asmus R, Asmus H (1998) Bedeutung der Organismengemeinschaften für den benthopelagischen Stoffaustausch im Sylt-Rømø Wattenmeer. In: Gätje C, Reise K (Eds) Ökosystem Wattenmeer. Austausch-, Transport- und Stoffumwandlungsprozesse. Springer, Berlin Heidelberg New York. pp 257–302

Baars JWM (1979) Autecological investigations on marine diatoms. I. Experimental results in biogeographical studies. Hydrobiol Bull 13: 123–137

Bachelet G (1986) Recruitment and year-to-year variability in a population of *Macoma balthica* (L.). Hydrobiologia 142: 233–248

Baillie PW, Welsh BL (1980) The effect of tidal resuspenion on the distribution of intertidal epipelagic algae in an estuary. Estuarine and Coastal Marine Science 10: 165–180

Baldwin BS, Newell RIE (1991) Omivorous feeding by planktotrophic larvae of the eastern oyster *Crassostrea virginica*. Mar Ecol Prog Ser 78: 285–301

Bauerfeind E, Hickel W, Niermann U, Westernhagen HU (1990) Phytoplankton biomass and potential nutrient limitation of phytoplankton development in the southeastern North Sea in spring 1985 and 1986. Neth J Sea Res 25: 131–142

Boero F (1994) Fluctuations and variations in coastal marine environments. P.S.Z.N.I. Mar Ecol 15: 3–25

Buhr K-J (1981) Auswirkungen des kalten Winters 1978/79 auf das Makrobenthos der *Lanice*-Siedlung im Weser-Ästuar. Veröff Inst Meeresforsch Bremerh 19: 115–131

Cadée GC (1986) Recurrent and changing seasonal patterns in phytoplankton of the westernmost inlet of the Dutch Wadden Sea from 1969 to 1985. Mar Biol 93:281–289

Cadée GC, Hegemann J (1991) Historical phytoplankton data of the Marsdiep. Hydrobiol Bull 24 (2): 111–119

Dame R, Dankers N, Prins T, Jongsma H, Smaal A (1991) The influence of mussel beds on nutrients in the western Wadden Sea and eastern Scheldt Estuaries. Estuaries 14: 130–138

de Jonge VN, Postma H (1974) Phosphorus compounds in the Dutch Wadden Sea. Neth J Sea Res 8: 139–153

de Jonge F, Bakker JF, Dahl K, Dankers N, Harke H, Jäppelt W, Koßmagk-Stephan K, Madsen PB (Eds), (1993) Quality status report of the North Sea. The Wadden Sea. The Common Wadden Sea Secretariat Wilhelmshaven

de Jonge VN, v. Beusekom JEE (1992) Contribution of resuspended microphytobenthos to total phytoplankton in the Ems estuary and its possible role for the grazers. Neth J Sea Res 30: 91–105

de Jonge VN, v. Beusekom JEE (1995) Wind and tide induced resuspension of sediment and microphytobenthos from tidal flats in the Ems estuary. Limnol Oceanogr 40: 766–778

Donaghay PL, Small LF (1979) Food selection capabilities of the estuarine copepod *Acartia clausi*. Mar Biol 52: 137–146

Drebes G (1974) Marines Phytoplankton. Georg Thieme, Stuttgart

Drebes G, Elbrächter M (1976) A checklist of planktonic diatoms and dinoflagellates from Helgoland and List (Sylt) German Bight. Bot Mar 19: 75–83

Elbrächter M, Rahmel J, Hanslik M (1994) *Phaeocystis* im Wattenmeer. In: Lozán, JL, Rachor E, Reise K, von Westerhagen H, Lenz W (Eds). Warnsignale aus dem Wattenmeer. Blackwell, Berlin, pp 87–90

Gehm G (1996) Nährsalzverteilung in der bodennahen Grenzschicht des Wattenmeeres. Diplom-Arbeit, Universität Oldenburg

Gillbricht, M., (1988) Phytoplankton and nutrients in the Helgoland region. Helgoländer Meeresunters 42: 435–467

Grasshoff K, Ehrhard M, Kremling K (Eds), (1983) Methods of Seawater Analysis. Verlag Chemie, Weinheim

Grøntved J (1952) Investigations on the phytoplankton in the southern North Sea in May 1947. Medd Komm Danm Fisk Havund Ser Plankton 5,5: 1–49

Günther, CP, Boysen-Ennen E, Niesel V, Hasemann C, Heuers J, Bittkau A, Fetzer I, Nacken M, Schlüter M & Jaklin S, (1998) Observation of a mass occurrence of *Macoma balthica* larvae in midsummer. J Sea Res 40: 347–351

Hickel W, Berg J, Treutner K (1992) Variability in phytoplankton biomass in the German Bight near Helgoland, 1980-1990. ICES mar Sci Symp 195: 249–259

Himmelman JH (1975) Phytoplankton as a stimulans for spawning in three marine invertebrates. J Exp Mar Biol Ecol 20: 199–214

Honkoop PJC, van der Meer J (1997) Reproductive output of *Macoma balthica* populations in relation to winter-temperature and intertidal-height mediated changes of body mass. Mar Ecol Prog Ser 149: 155–162

Höpner T (1996) Schwarze Tage im Nationalpark Wattenmeer. Spektrum der Wissenschaft, 8/96: pp 16–22

Hüttel M, Ziebis W, Forster S (1996) Flow-induced uptake of particulate matter in permeable sediments. Limnol Oceanogr 41: 309–322

Irigoien X, Castel J, Gasparini S (1996) Gut clearance rate as predictor of food limitation situations. Application to two estuarine copepods: *Acartia bifilosa* and *Eurythemora affinis*. Mar Ecol Prog Ser 131: 159–163

Jahnke J (1987) The light and temperature dependence of growth rate and elemental composition of *Phaeocystis globosa* SCHERFFEL and *P. pouchetii* (HAR.) LAGERH. in batch cultures. Neth J Sea Res 23: 15–21

Jensen HS, Mortensen PB, Andersen, FO, Rasmussen E & Jansen A (1995) Phosphorus cycling in coastal marine sediment, Aarhus Bay, Denmark. Limnol Oceanogr 40 (5): 908–917

Jørgensen CB (1981) Mortality, growth, and grazing impact of a cohort of bivalve larvae, *Mytilus edulis* L. Ophelia 20: 186–192

Marshall N (1967) Some characteristics of the epibenthic environment of tidal shoales. Chesapeake Science 8: 155–169

Martens P, Elbrächter M (1998) Zeitliche und räumliche Variabilität der Mikronährstoffe und des Planktons im Sylt-Rømø Wattenmeer. In: Gätje C Reise K (Eds) Ökosystem Wattenmeer. Austausch-, Transport- und Stoffumwandlungsprozesse, Springer, Berlin Heidelberg New York. pp 65–80

Niesel V (1997) Populationsdynamische und ökophysiologische Konsequenzen des Wattaufenthaltes für Phytoplankter der Nordsee. Forschungszentrum Terramare - Berichte, 7: 1–85

Petri G, Vareschi E (1997) Utilization of *Phaeocystis globosa* colonies by young *Mytilus edulis*. Arch Fish Mar Res 45 (1): 77–91.

Postma H (1984) Introduction to the symposium on organic matter in the Wadden Sea. Neth Inst. Sea Res Publ Ser 10: 15–22

Roskam PT (1970) De verontreinigung van de zee. Chemisch Weekblad, pp 57–59

Schneider G, Hickel W, Martens P (1998) Lateraler Autausch von Nähr- und Schwebstoffen zwischen dem Nordsylter Wattgebiet und der Nordsee. In: Gätje C, Reise K (Eds) Ökosystem Wattenmeer. Austausch-, Transport- und Stoffumwandlungsprozesse. Springer, Berlin Heidelberg New York. pp 341–366

Shanks AL, del Carmen KA (1997) Larval polychaetes are strongly associated with marine snow. Mar Ecol Prog Ser 154: 211–221

Shum KT (1993) The effects of wave-induced pore water circulation on the transport of reactive solutes below a rippled sediment bed. J Geophys Res 98: 10284–10301

Small, AC & Twisk F (1997) Filtration and absorption of *Phaeocystis* cf. *globosa* by the mussel *Mytilus edulis* L. J Exp Mar Biol Ecol 209: 33–46

Starr M, Himmelman JH, Therriault J-C (1991) Coupling of nauplii release in barnacles with phytoplankton blooms: a parallel strategy to that of spawning in urchins and mussels. J Plankt Res 13: 561–571

Utermöhl H (1958) Zur Vervollkommnung der quantitativen Phytoplanktonmethodik. Mitt Int Verh Theor Angew Limnol 9: 1–38

Valiela I (1995) Marine Ecological Processes. Springer, Berlin Heidelberg New York

5.2 Biogeochemical Processes in Tidal Flat Sediments and Mutual Interactions with Macrobenthos

M. Villbrandt, A. Hild & S. Dittmann

Abstract: In this chapter results from investigations on biogeochemical processes and their interactions with macrofauna and bacteria in two tidal flats of the East Frisian Wadden Sea are presented. On the Swinnplate, the influence of a mussel bed (*Mytilus edulis*) on the surrounding sandflat, the inhabiting fauna as well as interactions between macrozoobenthos, microorganisms and geochemical parameters were examined. Samples were taken along a transect from the centre of a mussel bed towards the sandflat. Part of the biodeposits produced by the mussels was transported by the ebb current into the surrounding sediments. Thus, not only the grain size composition (mud content of the sediment) and the POC-content (particulate organic carbon) of the sediments changed along the transect, but also other physico-chemical sediment characteristics. These changes in the sediment in turn affected the composition of benthic assemblages.

The proportion of fine-grained material in the biodeposit-containing sediments was subject to seasonal fluctuations. During winter, the mud content decreased due to intensified erosion by increased wind and wave activity. In summer, the deposition of fine-grained biogenic sediments rich in POC increased due to higher biological activity and calm weather conditions. The chemical composition of the sediments was predominantly coupled to the particle size distribution and not directly to the activity of the mussels. During winter, the content of Mn, P and Cu in the suspended matter was coupled to the distribution of the clay minerals (Al-content). In contrast to that a strong fluctuation occurred in summer, which was attributed to early diagenetic processes due to microbial activity.

The input and decomposition of organic matter derived from plant material in the surface-sediments of the mussel bed could be documented and followed using specific organic biomarkers (fatty acids, sterols). Thus, it was possible to differentiate between signals of phytoplankton and macroalgal blooms or bacteria in the sediment of the mussel bed and to allocate these to specific events.

During the investigations on the influence of organic matter availability on biotic activities it turned out that not only the amount, but also its quality was important. Organic constituents of high quality such as photosynthetic pigments, hydrolyzable amino acids and fatty acids decreased with increasing distance from the mussel bed towards the sandflat. The composition of pigments and their degradation products (mainly phaeophorbides) in the biodeposits were related to seasonal inputs of diatoms and/or macroalgae. Some amino acids were enriched in the sediment underneath the mussel bed, but the relative contribution of amino acid-derived carbon to total POC was reduced, indicating a selective depletion of amino acids by the mussels. Investigations in a mesocosm confirmed that *M. edulis* selectively eliminated amino acids and pigments from the water column and converted them into biomass.

The spatial and temporal distribution of microbial biomass and activity was closely coupled to the availability of fresh, labile organic matter and followed the change of the macrofaunal distribution along the transect. Microbial activities in the sediment were strongly influenced by seasonal events such as the deposition of phytoplankton blooms in spring and autumn, mass development of macroalgae followed by fine-grained sediment accumulation in late summer, or by ice formation during the winter months.

The spatial distribution of the macrofauna was predominantly influenced by the content of mud and organic material in the sediment. In the sediments underneath the mussel bed, the macrofauna assemblage was dominated by small and fast-growing polychaetes (r-strategists), which were deposit-feeding. In the centre of the mussel bed, species number, abundance and biomass of the macrofauna were low. At a small distance to the mussel bed, the macrofauna assemblage became increasingly rich in species, biomass and individuals. With increasing distance from the mussel bed in the direction of the sandflat, which was devoid of mussel biodeposits, a reduction of macrofauna species, biomass and individual numbers was observed. Deep-dwelling, surface-deposit-feeding and predatory polychaetes were found here.

At the second study site, the Gröninger Plate, the response of a sandflat to an excessive input of organic matter was examined. Results showed that, for about a fortnight, oxygen and nitrate in the sediment were used up completely by the bacterial mineralization of organic substances. This was followed by an enrichment of reduction equivalents such as sulphides, which caused that the sediment turned anoxic up to the surface (black spots and areas) and that, under stagnant water conditions, anoxic waterbodies appeared on the sediment surface. The anoxic sediments emitted methane and hydrogen sulphide. Due to these unfavourable environmental conditions, most of the infauna disappeared.

5.2.1 Microbial and Geochemical Processes

5.2.1.1 Introduction

In the Wadden Sea, phytoplankton and microphytobenthos represent the primary producers and the first step of the nutrient and energy cycles (Asmus et al. 1998 a, b). A large portion of primary production enters the microbial loop. Another portion of unicellular algae serves as food for filtering or grazing molluscs, worms and crustacea (primary consumers). The plant biomass is converted into animal biomass and partly excreted. Especially suspension-feeders are important for benthic-pelagic coupling (Smaal & Prins 1993). Their effects on biogeochemical processes in the sediment and infaunal distribution patterns were studied in the framework of the ecosystem research project ELAWAT.

Blue mussels (*Mytilus edulis* L.) form dense beds on tidal flats (Chap. 3.6.1). Fundamental processes, which modify the sediments underneath a mussel bed and in the surrounding area, are the deposition of biodeposits (faeces and pseudofaeces), the passive capture of fine-grained material in the heterogeneous epibenthic structure of a mussel bed and the accumulation of POC and carbonate shells (Chap. 3.6.1; Postma 1988). Due to hydrodynamic effects, a part of the fine-

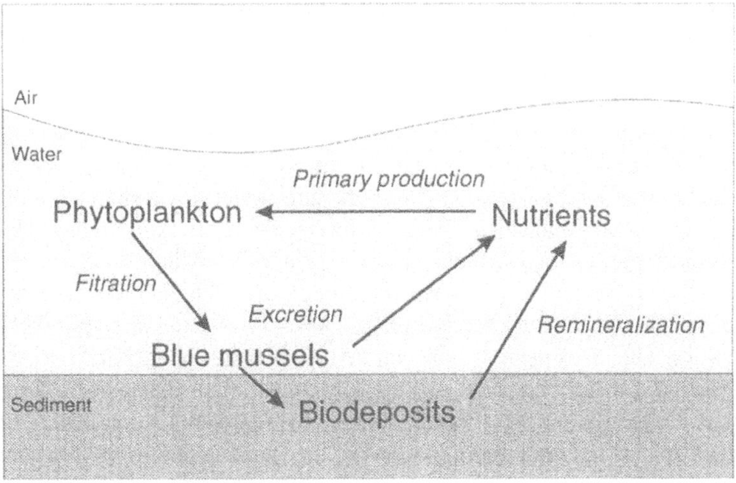

Fig. 5.2.1 The role of mussel beds in the nutrient cycle (Smaal, 1997 modified by Behrends).

grained material is exported from the mussel bed and can be mixed into the surrounding sandy sediments (Oost 1995). Apart from phytoplankton blooms (after active or passive sedimentation), dead benthic organisms and buried macroalgae, which are locally accumulated on a mussel bed, contribute to the input of organic matter to the sediment.

Organic matter, which is enriched in the sediments of a mussel bed, is degraded by bacteria. After microbial mineralization, the nutrients become available again for the food web (Fig. 5.2.1; Fenchel & Blackburn 1979). Freshly deposited phytodetritus leads to an increase in the mineralization activity within a short time (Daumas 1990; Graf 1992). Seasonal variations of primary production and the deposition of phytoplankton blooms (Middelburg et al. 1996), decaying macroorganisms as well as tidal and weather-induced modifications of sedimentation lead to a temporal heterogeneity of degradable biomass in the sediment and thus of the concentrations of the degradation products in the pore water lasting a few days or weeks. In addition, spatial variations occur, as detritus, which is deposited on the sediment surface out of the water column or derived from dead epiphytobenthos (seagrass, macroalgae), is reallocated by wave- and current action or distributed heterogeneously within the sediment by bioturbation (Postma 1988). Around the mussel bed in the backbarrier system of Spiekeroog Island, the decomposition of organic matter imported into the tidal flat sediment was traced with several techniques over space and time.

Several studies have shown that not only the quantity of organic matter, but also the quality is important for the spatial heterogeneity of nutrient concentrations and the availability as a food source for infaunal organisms (Middelburg et al. 1996; Dauwe 1999). The effects of the input of organic matter were studied in and around the mussel bed, looking at ways to describe the quality of organic matter and its modification by the mussels.

Further investigations dealt with the pore water chemistry in naturally occurring black spots (Höpner & Michaelis 1994) and sandflat sediments experimentally enriched with an overload of organic matter. The aim of these experiments was to determine a critical level for the input of organic matter into intertidal sediments and characteristic indicators for an excess of organic enrichments. The following possible effects were considered: the loss of the oxic sediment surface layer, a complete consumption of the sulphate in the pore water, toxic sulphide concentrations and increased methane emission. A further interest was to predict the regeneration ability of disturbed sediment surfaces to normal conditions.

The results of several studies carried out in close collaboration by a team of scientists in the ecosystem research project ELAWAT are summarized in the following sections. The investigations concentrated on two sites in the backbarrier system of Spiekeroog, the Swinnplate and the Gröninger Plate (Chap. 3.1). Of special importance were the input of organic matter, the remineralization of this material as well as spatial and temporal variations of the studied parameters. Microbial and geochemical processes and their interactions with macrozoobenthos are considered and discussed. The distribution of macrofauna under the influence of a mussel bed on the Swinnplate is described in Sect. 5.2.2.

5.2.1.2
Swinnplate

For the investigations on the Swinnplate (SP), sediment samples were taken at 6 sites along a transect from the centre of a mussel bed (*Mytilus edulis*) to a reference site in the sandflat (Chap. 3.1; Fig. 3.1.3). At the beginning of the investigations in spring 1994, the sites SP1 and SP2 were located in the area of the main mussel settlement. As a reference, site SP6 was situated on the sandflat. The direction of the transect was adjusted to the flow of the ebb current, with which the material discharge from the mussel bed took place. Sampling was carried out in intervals of one to two months. The range of methods applied covered the analysis of parameters of organic and inorganic geochemistry as well as microbiology.

The surface sediments of the transect sites were studied in order to estimate the influence of the mussel bed and temporal variations on the sediment composition. The conditions on the Swinnplate changed severely during the study period under the influence of weather and hydrodynamics. A larval spatfall in summer 1994 over the entire study area led to a new settlement of blue mussels south of SP5, which remained relatively stationary over the year 1995. Biodeposits produced there affected the reference site SP6. In August 1994 and 1995 macroalgae (*Enteromorpha* spp. and *Ulva* sp.) attached to the mussels on all sites. A summary of special events as well as spatial and temporal variations at the transect sites during the study period is given in Table 5.2.1.

Besides such single unpredictable events, the seasonal influences on the sediments of the transect were determined. In this context, both biological and physical variations had to be considered. The filtration activity of mussels varies with season (phytoplankton blooms) and the activity of microorganisms is in turn linked to the supply of organic matter and the ambient temperature. Physical modifications refer to wave energy, storms, temperature and further to the content of suspended matter (Krögel 1997) and the proportion of fine-grained material in the surface sediments (see Chap. 3.4).

Table 5.2.1 Events and observations at the transect on the Swinnplate during the study period

sampling plot location		**SP 1 old mussel bed**	**SP 2 mussel-patch**	**SP 3 *Lanice*-settlement**	**SP 4 muddy sandflat**	**SP 5 new settlement**	**SP 6 sandflat**
year	date\distance	0 m	260 m	360 m	960 m	1480 m	2330 m
1993	November 10	after a widespread distribution from SP1 to SP5 relocation of the mussels with a concentration in the East of the Swinnplate (sampling plots SP1 and SP2)					
1994	March 30					intensive spatfall of *M. edulis*	
	May 17					*M. edulis* established	partly influenced by SP5
	August 03	macroalgal cover (*Ulva* sp., *Enteromorpha* spp.) at sampling plots SP1 and SP2 (mussel bed)				new settlement of *M. edulis*, macroalgal cover (*Ulva* sp., *Enteromorpha* spp.)	
1995	April 27	predominantly old mussels with reduced faeces production	evolution of a dense mussel population, mainly young mussels with increased faeces-production, *Lanice*-settlement			new settlement of *M. edulis* established, 1995 stationary	
phytoplankton bloom from April to May							
	August 30	mud deposition at all transect sites					
	December 18	total destruction of the mussel bed	musselbed 3 to 4-years old			degeneration of the mussel bed	
ice winter							
1996	March 04	mussel bed totally eroded and destroyed			mud patches		
	July 17				new mud sedimentation		
	August 21	north of SP1 to SP4 development of a new mussel bed; with minor recruitment of *M. edulis*					

Sediment sorting

As discussed in chapter 3.4, the continuous reallocation of the surface sediments by wave action and tides produces a horizontal zonation of sediment types in the backbarrier tidal flats (sediment sorting). Deviating from this situation, a deposition of fine-grained material took place in mussel beds, where no fine-grained material would deposit by hydrodynamic factors only without the settlement and activity of the mussels (Chap. 3.6.1).

Differences in the sorting of the sediments were reflected in the elemental composition. Looking at the bulk of sediment composition, correlations of the element contents appeared which resulted from a high proportion of quartz and were largely caused by dilution effects (Hild 1997). The geochemical composition of individual components and of different grain size fractions (e.g. Al, heavy metals, etc.) showed that the element contents of different intertidal sediment types (e.g. mud flat, muddy sand) corresponded well with each other (Fig. 5.2.2).

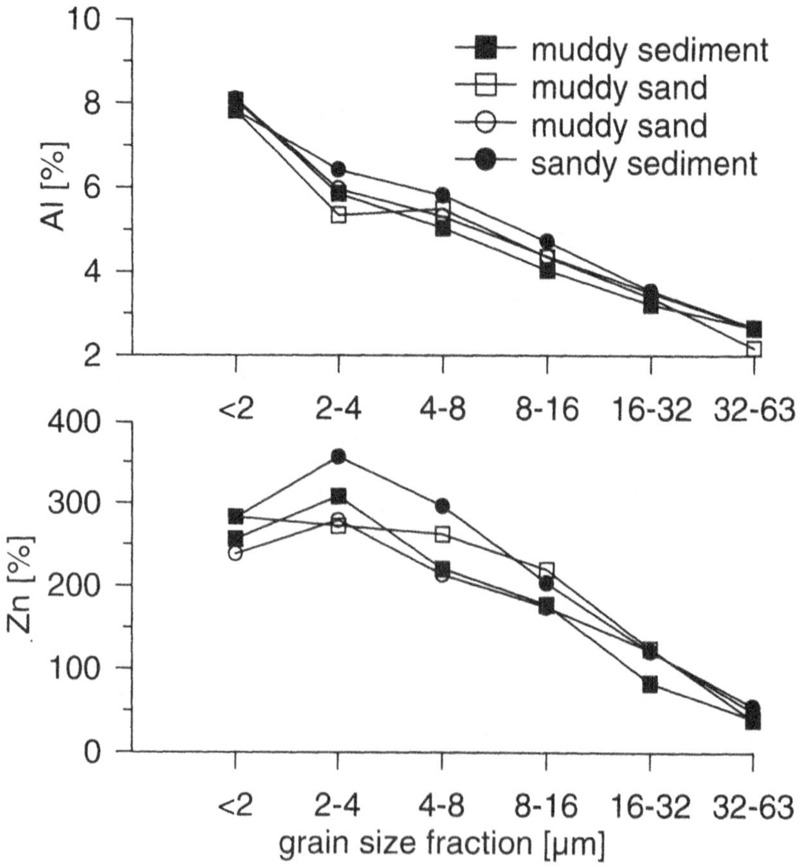

Fig. 5.2.2 Contents al Al and Zn in grainsize fractions of different sedimentary environments. (Data from Hild 1997)

Just like the element composition, the POC-content of the sediments was linked to the particle size distribution. The relationship between the POC-content (% of weight) and the mud content as well as the POC-concentration (% of volume) and the mud-content was elaborated by Delafontaine et al. (1996). These authors put special emphasis on the distinction between content (per unit of weight) and concentration (per unit of volume) for parameters determined in sediments and exemplified this for POC. Specifying a value as concentration takes the compaction and water content of the sediment into account. In the sediments of the backbarrier tidal flats, a linear correlation existed between the proportion of mud and the POC-content, but not between the proportion of mud and the concentration of POC. Contrary to the POC-contents, which had their maximum with a mud content of 100 %, the highest POC-concentrations occurred in sediments with a mud content of 60–70 %.

The finding of Delafontaine et al. (1996) is important for the "availability" of organic matter for organisms (but see section on quality of organic matter below) and for the calculation of mass balances on the base of sediment volumes. Their results showed that during the turnover of sand, substantial and previously underestimated amounts of POC were set in motion. The POC-contents were lower in sand than in mud, but due to the reallocated sediment masses in total, larger quantities of POC (approx. 40 %) were moved with the sand fraction.

Formation of gradients along the transect

The studies on the Swinnplate took place under the assumption that a gradient developed with the discharge of fine-grained material from the mussel bed with the ebb flow and a pronounced gradient indeed existed in 1994 at the beginning of the investigations. However, due to changes at single transect sites during the study period (macroalgal growth, mussel dislodgement or settlement), this gradient was not as pronounced later. In 1994, POC concentrations as well as mud and carbonate contents decreased continuously from SP1 (mussel bed) to SP6 (sandflat) (Flemming, Bartholomä, Delafontaine, Kröncke, unpubl. data). This situation lasted until the ice winter destruction of the mussel bed. After the ice winter 1995/96, the former gradient was destroyed by dislodgement of the mud packages and the death of the mussels (Table 5.2.2). Compared to the strongly fluctuating values in the sediment of the former mussel bed, the POC contents at the reference site in the sandflat (SP6) showed little variation over time (Kröncke, unpubl. data).

The contents of total extractable protein and amino acids also showed a continuous decrease along the transect in the direction of the sandflat in 1995 (Behrends 1997). During that year, the average protein contents increased. However, while the total contents of amino acids decreased along the transect from SP1 to SP6, the proportion of amino acid carbon (AAC) on the entire POC increased in the same direction (Behrends 1997). This result indicates that, relative to POC, the amino acids were not enriched in the sediments of the mussel bed, since the mussels take in particulate amino acids. An outline of the spatial and temporal variations of the parameters along the transect is given in Table 5.2.2.

Table 5.2.2 Summary of the results from the geochemical and microbial investigations at the transect on the Swinnplate

parameter	spatial variations (along the transect)	temporal variations (over the study period)
mud content	highest contents in biodeposits, lowest contents in the sandflat, decrease along the transect from sampling plot SP1 to SP6, good correlation with POC	fluctuating over the year, lowest contents in wintertime, following phytoplankton blooms, in August 1995 surficial mud cover of the whole transect
POC	highest values in biodeposits, lowest values in the sandflat	following the annual course of primary production, decrease in July, maxima in May/June and August
elemental distribution	coupled to the sorting of sediments (grain-size distribution)	phytoplankton blooms were characterized by high POC/metal-ratios in the sediments of mussel bed sites
proteins	decrease along the transect from SP1 to SP6	increase over the year till autumn/winter, comparing the single years decrease over the study period
amino acids	decrease along the transect from site SP1 to SP6, highest values in biodeposits, lowest values in the sandflat, percentage of amino acid bound C in POC increasing along the transect from SP1 to SP6	seasonal variations coupled to the development of phytoplankton blooms
lipids	highest values in biodeposits	highest values in 1994, 1996 after the ice winter generally lower than in previous years with a maximum in March (*C. concinnus*)
biomarkers	seasonal variations superimpose spatial differences	markers for macroalgae were high in August, maxima of diatom markers occurred in May and September
bacteria	more bacteria at mussel bed sites, relatively more copiotrophic und fermenting bacteria; in the sand flats less, with a higher share of oligotrophic bacteria, bacterial C-share of POC is highest in sandy sediments	increasing since May, maxima of total cell numbers in June and September
enzymes	decrease along the transect from SP1 to SP6	decrease in July, otherwise increase over the year (see bacteria)
Adenylate Energy Charge (AEC)	highest in the sandflat, lowest in the mussel bed	
RNA/DNA-ratio		increase till June, decrease in the summer and a second increase in October, signals follow phytoplankton blooms with a time delay
C/N-ratio	slight decrease along the transect from SP1 to SP6	slight variations, elevated in October 1995 at SP1

Decomposition of organic matter

Organic matter in marine sediments is mainly degraded via the activity of microorganisms (Fenchel & Blackburn 1979). Bacteria prefer fresh, easily degradable and high-quality material with high contents of fatty acids and proteins (Grenz et al. 1990). Thus they compete with other heterotrophic consumers for food. On the other hand, bacteria produce high-quality food in liquid (acids) and solid form, which can be used as food by meiofauna and protozoa (Bouvy & Soyer 1989). Both groups can further improve the quality of food for detritus-feeding organisms (Meyer-Reil & Faubel 1980). On a seasonal scale, an enhanced supply of fresh organic matter of high quality exists especially during phytoplankton blooms. When fresh organic matter is not available (e.g. during winter), some bacteria are able to use refractory organic matter (lignin, humic acids and mucopolysaccharides) and decompose high-molecular substances into smaller components, which in turn can be used by other organisms (Alongi & Hanson 1985). The low-molecular products and soluble inorganic compounds of C, N and P can also be released from the benthic system into the overlying water (Billen 1978; Klump & Martens 1981; Blackburn 1988; Hansen & Blackburn 1991). Apart from the mobilization of nutrients during the decomposition of organic matter, a release of Mn and Fe by reduction of Mn/Fe-oxi /hydroxides occurs (Sørensen & Jørgensen 1987; Canfield et al. 1993) as well as of trace elements. Released Fe is precipitated relatively fast in the form of Fe-sulphide complexes, or reoxidised. Main factors influencing mineralization processes in intertidal sediments are fauna, bioturbation, irrigation, tidal pumping, wave pumping, diffusion and temperature (Sørensen et al. 1979, Malcolm & Sivyer 1997).

Physiological groups of bacteria and substrate utilization

In ecophysiological investigations, bacteria are divided into physiological groups and taxonomic criteria are only of minor importance. Due to the different environmental conditions, distinct bacterial populations occurred in the sandflat and in the mussel bed sediments. Dividing bacteria into physiological groups, more copiotrophic organisms (adapted to high nutrient contents) and fermenting bacteria were found in the mussel bed sediments, whereas more oligotrophic bacteria occurred at SP6 (sandflat) (Leu unpubl. data). This showed that the bacterial populations were adapted to the environmental conditions at the tidal flat sites: to anoxic conditions and a sufficient supply of degradable substrates in the biodeposits and to oxic conditions and a lower (limited) supply of organic substrates in the sandflat.

Experiments on the utilization of different microbial substrates (BIOLOG™; Zak et al. 1994), in which several carbon sources were offered to the microflora, showed that the bacteria population in the mussel bed was able to transfer over 50 % of the offered substrates relatively quickly (within 12 h) (Albers unpubl. data). This fast reaction indicated an adaptation of the bacteria to a high nutrient supply. In contrast, the bacteria in the sandflat could not use the offered substrates until a time lag of 36 h had passed. However, after one day of incubation, the bacteria in the sandflat had developed the ability for the use of 90 % of all offered C-sources, while the bacterial flora in the mussel bed could use only approximately

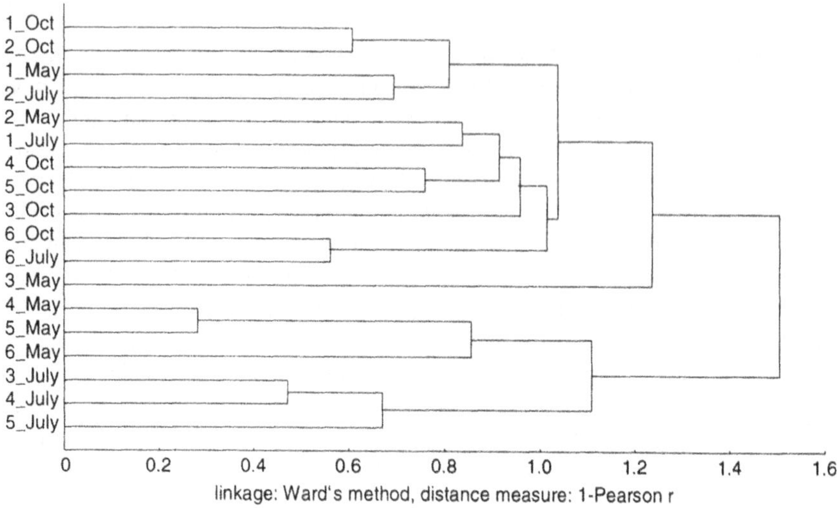

Fig. 5.2.3 Clustering of samples related to the usability or rather none-usability of 128 different C-sources (BIOLOG) by the microflora living in the sediments. SP1 to SP6: transect sites; SP1 and SP2 (mussel bed), SP6 (sandflat). Based on surface sediments sampled in May, July and October 1995. (Albers, unpublished data)

60 % (Albers unpubl. data). This means that, in contrast to the microflora in the mussel bed, the bacterial assemblage in the sandflat was not very specialized on particular substrates and could make use of a wider spectrum of potential C-sources. A cluster analysis of the BIOLOG™ data showed that the bacteria of the sites SP3, SP4, SP5 and SP6 were more similar to each other in the same months than to similar sites during different months (Fig. 5.2.3). This indicated a seasonal character of the studied parameters in response to events, such as phytoplankton blooms, sediment cover or macroalgal vegetation. In contrast, the bacteria in the mussel bed possess more site-specific characteristics of their utilization spectrum, i.e. a specialized microflora existed in the sediments underneath the mussel bed.

Adenylate Energy Charge (AEC)

The AEC-value can be calculated from the relation of (ATP + 0,5 ADP / (ATP + ADP + AMP) and is considered as a measure of the energetic state of the microbial population. Apart from a sufficient substrate supply, the availability of oxygen is especially important for a high biochemical energy content in the cells of the organisms. Thus, this parameter is influenced by the prevailing redox conditions. In comparison to the aerobic metabolism, the anaerobic metabolism is not so efficient for the cells, because the fermentation pathway delivers fewer energy (ATP) than the complete reduction of organic substance via aerobic respiration. This probably is a reason for the lower energy status of the biomass in the mussel bed than that in the sandflat (Albers unpubl. data). A comparison of the AEC-values of all transect sites showed increased values in the sandflat, i.e. the energy status of the biomass was generally more favourable there (Fig. 5.2.4). This is in agreement with the observation that the proportion of cultivatable bacteria (determined as colony forming units) on the total bacterial number (epifluorescene microscopic counting) was higher in the sandflat than in the mussel bed sediments (Fig. 5.2.5). Apart from these spatial variations, clear seasonal influences on the AEC-value were found also.

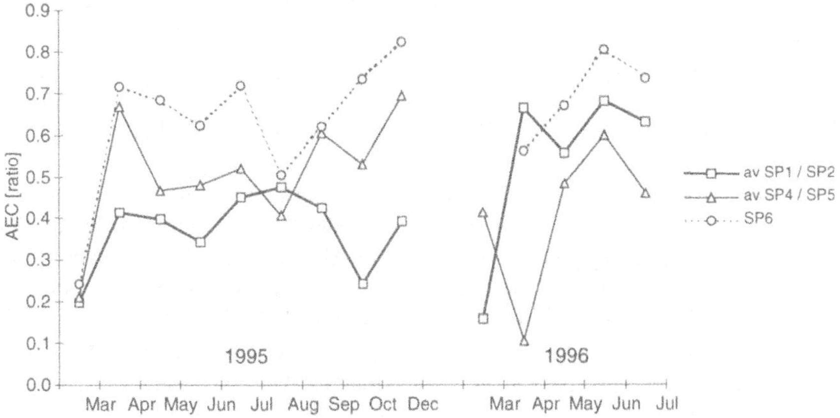

Fig. 5.2.4 Seasonal variations of the Adenylate Energy Charge (AEC) in sediments of the transect on the Swinnplate. Average values (av) are given for SP1 and SP2 (mussel bed) as well as for SP4 and SP5; SP6 (sandflat). (Data from Stoeck 1996)

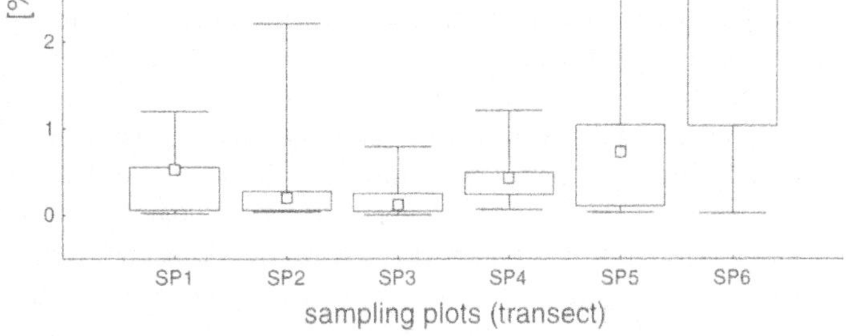

Fig. 5.2.5 Percentage of bacteria which can be cultivated (Most Probable Number, Zobell-medium) of the total bacterial number (microscopic count). SP1 and SP2 (mussel bed); SP6 (sandflat). Box and whisker plots based on values of April to October 1995. (Knauth-Köhler, unpublished data)

Bacterial exoenzym activities

The highest exoenzyme activities were measured in the sediments underneath the mussel bed, where the supply of degradable substrate for bacterial mineralization was highest and the largest surface area for bacterial colonization was present in form of fine-grained sediments. These sediments also contained the highest bacterial cell numbers (Leu unpubl. data). Within the sediment, extracellular enzymes were bound to particles and were therefore enriched in fine-grained (silty) sediments.

On a seasonal scale, the activity of exoenzymes (protease-/glucosidase-/phosphatase activity) showed a similar pattern to the seasonal development of the phytoplankton blooms (Leu unpubl. data) This was most pronounced at the reference site in the sandflat, and not obvious at the mussel bed.

On the basis the exoenzym activities it turned out that protein-containing high-quality substrate was taken up immediately after deposition and transferred into the food web. In vertical profiles, the highest bacterial exoenzym activities were found at the oxic sediment surface. The microbial turnover rates and exoenzym activities decreased with increasing sediment depth (Leu unpubl. data).

Like protease and glucosidase, which had similar seasonal patterns, the activity of phosphatase also showed a spatial behaviour in 1994 and decreased along the transect from the mussel bed to the reference site in the sandflat (Fig. 5.2.6). However, in contrast to the other two exoenzymes, the phosphatase activity rose slightly during the study period and differed in its seasonal development from the other two studied exoenzyme activities (Leu unpubl. data). This can be explained by the fact that increased activities of protease and glucosidase were related to an enhanced supply and mineralization of settled phytoplankton blooms, whereas high activities of phosphatase indicate a lack of dissolved phosphate.

Influence on the elemental composition

The knowledge of the element composition of different individual components and their dependence on the particle size distribution of the tidal flat sediments is important for the interpretation of biotic and abiotic geochemical processes. The influence of early diagenetic processes on the sediment geochemistry can be shown most clearly for the element Mn, since it reacts very sensitive to modifications of the redox environment. Under reducing conditions Mn is remobilized and released into the interstitial water. When this dissolved Mn reaches oxic conditions via upward diffusion, tidal pumping or other processes, it is precipitated as Mn-oxi-/hydroxide or adsorbed onto particles. In this case, the activity of bacteria plays a substantial role, since oxidation and reduction of Mn is mainly controlled by microbial processes. Thus, an enrichment of solid Mn oxide occurs frequently in the oxidized layers of the sediments in nearshore areas. But in the vertical sediment profiles taken on the Swinnplate this effect could not always be observed due to the occurrence of Mn-containing heavy minerals (ilmenite ($(Fe, Mg, Mn)TiO_3$)) (Fig. 5.2.7a). Only with consideration of the Mn-proportion bound to heavy minerals by a standardization on Ti (Fig. 5.2.7b), the typical enrichment of Mn at the sediment/seawater interface could be proven in all studied depth profiles. Seasonal

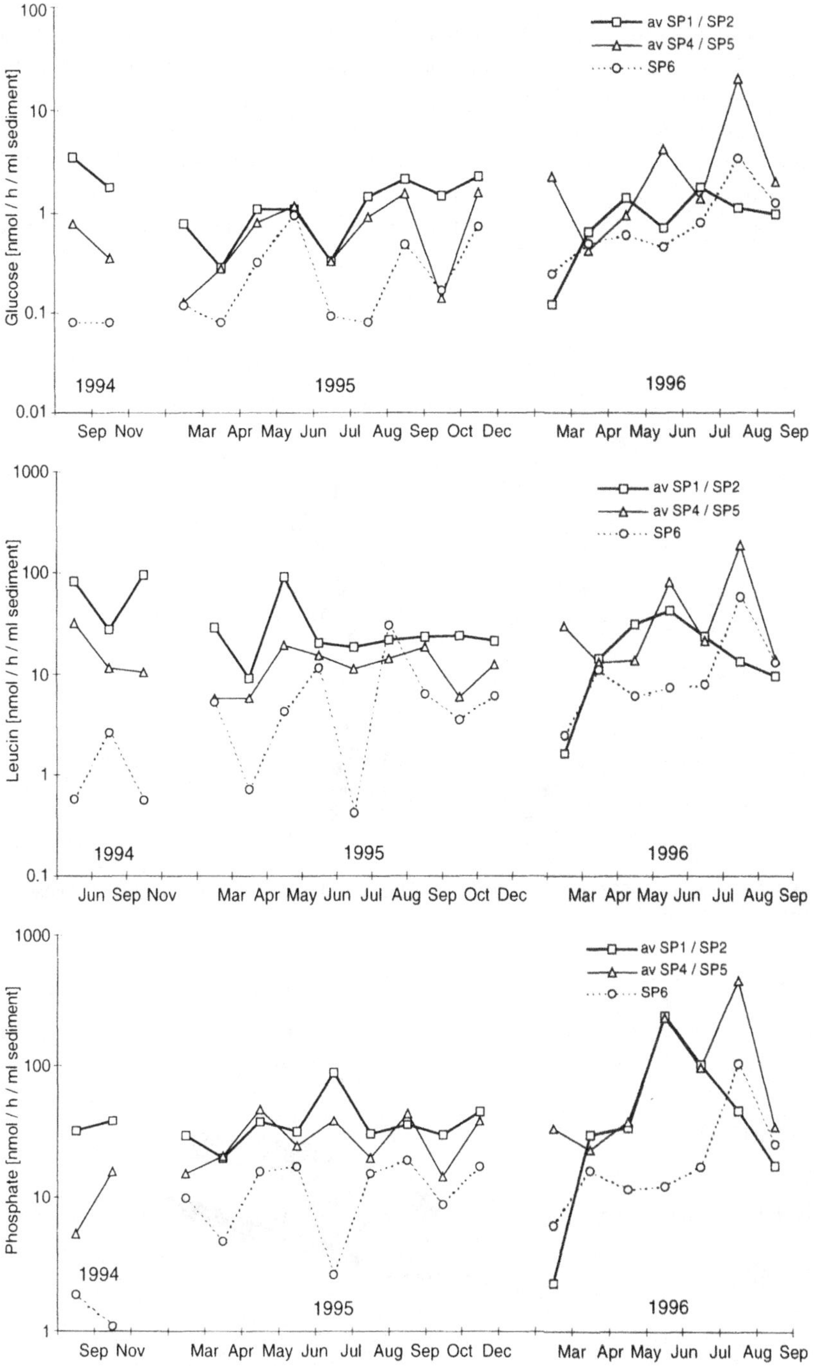

Fig. 5.2.6 Seasonal variations of the ß-D-Glucosidase- (a), Protease- (b) and Phosphatase- c) activities in the sediments. Average vlues (av) are given for SP1 and SP2 (mussel bed) as well as for SP4 and SP5; SP6 (sandflat). (Leu, unpublished data)

variations of the Mn-, Cu- and P-contents of suspended matter from the backbarrier area showed the influence of microbial degradation processes on the distribution of elements (Hild 1997). During the winter months, a good to very good correlation existed between these element contents and Al_2O_3-contents in suspended matter. Due to the dependence of Al_2O_3-contents on the fine-grained material proportion, this indicated a close coupling of these elements to detrital matter in winter. In summer, the Mn-values varied widely and were higher compared to winter values (Fig. 5.2.8).

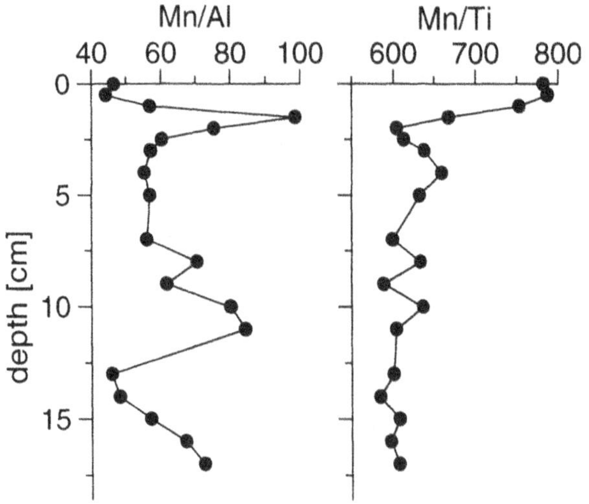

Fig. 5.2.7 Mn/Al- (a) and Mn/Ti- (b) depth profiles in sediments of the Swinnplate. (Data from Hild 1997)

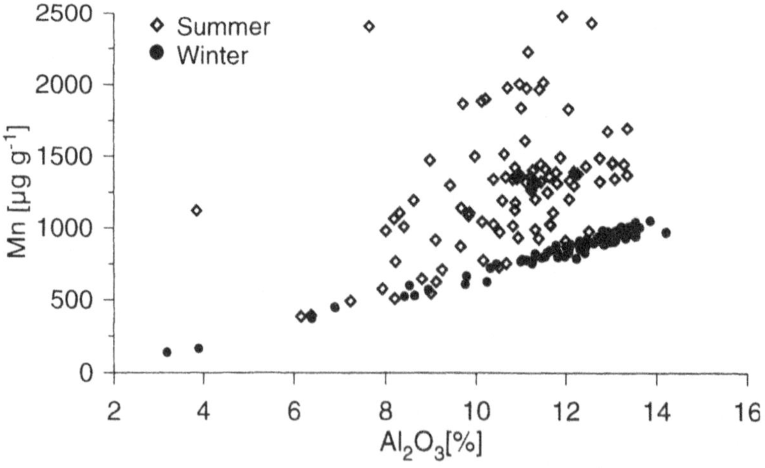

Fig. 5.2.8 Correlation of manganese versus Al_2O_3-contents in suspended matter from the Spiekeroog backbarrier system (summer and winter situation compared). (Data from Hild 1997)

This effect was attributed to an enhanced remineralization of organic matter during the summer months followed by Mn fluxes from the sediment. This interpretation was confirmed by the increase of dissolved phosphate concentrations in the interstitial water during summer (Sect. 5.2.1.3) as well as by seasonal variations of microbial activity, which were related to temperature and phytoplankton blooms.

Quality of organic matter

Coastal marine particulate organic matter (POC) consists for about 50 % of hardly degradable humic substances, while the remaining 50 % are composed of more easily degradable material: 10–15 % sugars, 5–10 % lipids and approx. 30 % amino acids (Liebezeit, pers. comm.). To study the effect of organic matter on sediment chemistry and macrobenthic distribution patterns, the determination of just the quantity of organic matter alone is not sufficient. The quality of organic matter can be defined from two perspectives: its source or its usefulness for other organisms.

Most authors define quality of organic matter as the availability or usefulness (potential to degrade, energetic use or nutritional value) as food source for organisms. Thereby, labile (easily degradable) and more refractory (hardly degradable or residual fraction) organic substances are distinguished often. This depends not on the molecular size (high-molecular or low-molecular substance), but on the type and stability of the bonds within the molecule. The evaluation of organic matter is usually carried out by the analysis of lipids, carbohydrates, proteins and pigments. In particular chlorophyll *a* and phaeopigments or fatty acids are used as substances to indicate labile organic material.

Quality and origin

From another perspective, the quality of the organic matter is defined by its different sources. This implies the question about the origin of the material, which can also be subject to seasonal fluctuations. In the context of the biomarker investigations, the fatty acid analyses showed only small proportions of terrestrial biomarkers and verified a predominant marine to estuarine origin of organic matter in the sediments of the Swinnplate (Brocks unpubl. data).

At all transect sites, higher contents of bacterial fatty acids were present from August to September than in spring (Brocks unpubl. data). While in spring and summer seasonal variations of fatty acids may have obscured possible spatial variations between the transect sites, clear differences existed in autumn between the mussel sites and the sandflat site.

The predominant marine origin of the organic matter could also be confirmed by carbon isotope analysis. The $\delta^{13}C$-values of the POC decreased continuously with increasing sediment depth from approximately -19 ‰ at the sediment surface to -25 ‰ at 25 cm depth (Böttcher unpubl. data). This may be explained by the different speed of early diagenetic degradation of terrestrial and marine organic matter. After Salomons & Mook (1981), particulate terrestrial organic carbon had $\delta^{13}C$-values of -27,4 ‰. However, the fresh organic matter of marine origin from the backbarrier area of Spiekeroog island had values between -16 and -18 ‰. Application of a two-component mixture model would lead to a marine proportion of approx. 80 % in the surface sediment, which decreased to a proportion of approx. 20–25 % at 25 cm depth. This would indicate a preference for degradation of marine organic carbon.

A further characterization of the origin of marine organic matter (phytoplankton, macroalgae and other marine organisms) with organo-geochemical methods is possible. For example, steroles are suitable to differentiate between the input of macroalgae and diatoms into the sediment (Volkman 1986). Sterols are components of the cell membrane in all organisms except bacteria. Their different molecular structures permit a specific allocation to the source organisms, a modification of the structure indicates diagenetic processes. The comparison of the concentrations of a steroid biomarker typical for diatoms (brassicasterol), with those characteristic for macroalgae (fucosterol), indicated how these sources contributed to the composition of organic matter in the sediment over the year. The results showed that an input of phytoplankton biomass into the sediment took place in spring, while in autumn a macroalgal occurrence could be distinguished from a following phytoplankton bloom (Brocks unpubl. data).

Effect of mussels on the quality of organic matter

The effect of blue mussels on organic matter could only be revealed when the quality of this material was considered. The quality was assessed in samples taken along the transect on the Swinnplate which were analysed for the proportion of amino acids on the POC. In an additional mesocosm experiment, the pigment composition was determined also (Behrends 1997). Amino acids are needed to cover the metabolic nitrogen demand of heterotrophic organisms and are thus of high qualitative importance.

The relative enrichment of THAA- (sum of the hydrolyzable amino acids) and POC-contents in the mud layers of the Swinnplate compared to the reference site in the sandflat showed the influence of mussels on the quality of organic matter (Fig. 5.2.9). A seasonal pattern appeared with a moderate THAA-enrichment in spring 1994 and 1995 and a fast increase in June (Behrends 1997). The enrichment of POC by the filtration of the mussels was relatively higher than the enrichment of THAA in the entire year of 1995. This was an indication for the selective consumption of amino acids by the mussels. The highest enrichment for POC was found during a macroalgal cover in August 1995. However, this was not the case for amino acids. It was concluded that a sedimentation of organic matter of lower quality had occurred by passive deposition. In contrast to that, the relative enrichment of THAA and POC were almost identical after the disappearance of the mussel beds due to the ice winter in 1995/96 (Behrends 1997). This suggests that, on the one hand, *M. edulis* reduced the quality of the organic matter and, on the other hand, that the influence of other heterotrophic organisms on the composition of organic matter is rather small.

The proportion of amino acid-carbon of total POC in the sediments of the mussel bed was clearly reduced in comparison to the sandflat sediments without mussels. This was also obvious from the lower enrichment of particulate amino acids relative to POC in the mussel beds than in the sandflat (Fig. 5.2.9).

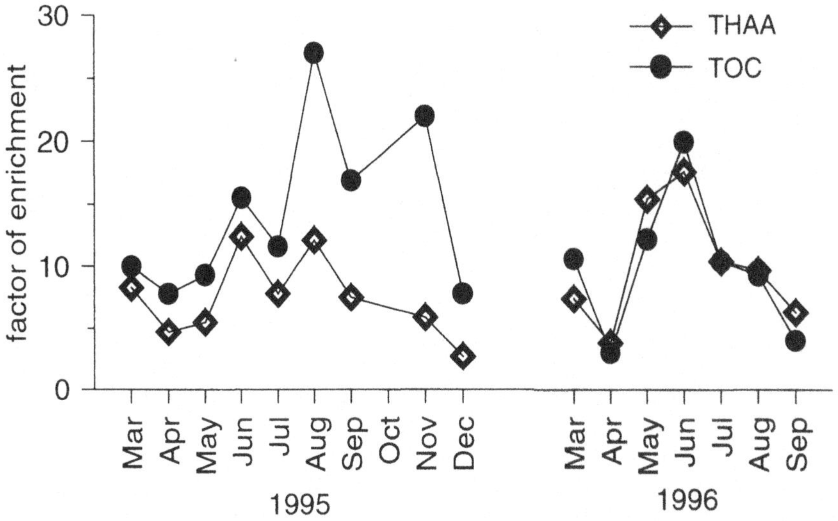

Fig. 5.2.9 Enrichment of THAA- (total hydrolyzable amino acids) and POC-contents in mud compared to sandy sediments; in 1995 influenced by the mussel beds and in 1996 after erosion of mussel settlement. (Data from Behrends, 1997)

A further effect became obvious by looking at the relative contents of amino acids. Obviously, a relation exists between the relative composition of amino acids and their total sum. In the samples from May 1995, the amino acids glycine and serine were positively correlated with the total sum of amino acids, while asparagine acid and glutamine showed a negative correlation with the THAA-content (Behrends 1997). It is not yet possible to infer a preferential use of these amino acids for the metabolism of *M. edulis*, although glycine was enriched in faeces, and the acidic amino acids asparagine and glutamine played a role in the synthesis of the mussel shells.

Investigations on photosynthetic pigments were carried out in a flow channel (mesocosmn) and showed a metabolic utilisation of chlorophyll *a* and a degradation of fucoxanthin by *M. edulis*. The mussels efficiently eliminated the two most frequent pigments chlorophyll *a* and fucoxanthin (58 % and 17 % of the suspended matter in the flow channel, Behrends 1997).

The average ingestion of chlorophyll *a* by the mussels (reduction of 37 % in the flow channel) was almost twice as high as the ingestion of POC, indicating a preferential ingestion of phytoplankton by the mussels (Behrends 1997). Particulate amino acids were also reduced by mussel filtration in the flow channel. The relative reduction of THAA with 39 % on average of the inflowing water was of about the same order of magnitude as that of chlorophyll *a*. The content of fucoxanthin in the suspended matter was reduced by the mussels by approx. 25 %. Fucoxanthin is converted to fucoxanthiol in the intestine of the mussels (Partiali et al. 1989). However, in comparison to the utilisation of chlorophyll *a*, this process is quantitatively of minor importance. Due to the use of the phytol group of the chlorophyll *a* by heterotrophic organisms, phaeophorbides are formed (Currie 1962; Volkman et al. 1980). The latter were excreted by *M. edulis* with their faeces and

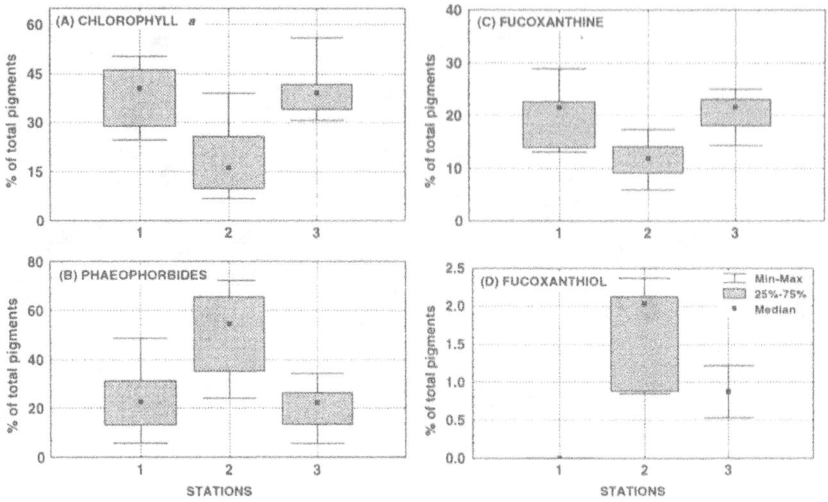

Fig. 5.2.10 Box and whisker plots of the proportional contents of Chlorophyll a (A), Phaeophorbides (B), Fucoxanthin (C) and Fucoxanthiol (D) in the sediments of the flume experiment (n = 8; station 1 in front of, station 2 in and station 3 behind the mussel bed). (Data from Behrends, 1997)

enriched in the sediment. However, also intact chlorophyll *a*-molecules were found in the biodeposits, which points to the fact that the material was not entirely used by the mussels. The degradation products – phaeophorbides and fucoxanthiol – were enriched in the faeces / pseudofaeces and in the sediment of the mussel bed compared to the sites outside the mussel bed (Fig. 5.2.10). Thus, besides the selective uptake of certain particles from the water column, the mussels actively changed the composition and quality of organic matter they ingested.

Quality and energy content

A further parameter to assess the quality of organic matter is its energy content (calorimetry). Since the energy content of phytoplankton declines with age (Graf 1987), this variable was used to indirectly estimate potential seasonal variations in the relative age of organic matter in mussel beds. This is relevant, as the age of POC is an important parameter for its use by benthic organisms (Rudnick 1989). Short-term enrichments of material with high POC content, which result e.g. from phytoplankton blooms, are degraded within less than one month time (Delafontaine 1995, 1996). Yet, overall these inputs should result in a rejuvenation of organic matter and be detectable by seasonal comparisons.

While the energy contents of POC in the sediments were relatively low (12–37 kJ g POC^{-1}) during spring 1995 and 1996, generally higher values were found in summer and autumn 1994 and 1995 (25–62 kJ g POC^{-1}). The energy content was not dependent on the POC- or mud contents. This result was attributed to seasonal effects of pulsed phytoplankton blooms (Chap. 5.1), which caused an increased energy content and a rejuvenation of the total organic matter in summer. The results are in good agreement with seasonal cycles of labile and refractable POC in surface sediments reported from Kiel Bight (Graf et al. 1983).

Indications of the quality of organic matter from microbial parameters

The microbial investigations supplied indirect evidence of the quality of organic matter. As part of these investigations, the RNA/DNA-ratio was analysed. This is a parameter for the average protein synthesis activity of a population (Paul & Voroney 1984) which, apart from the AEC-value, supplies further information about the fitness of the available biomass. A high RNA/DNA-ratio indicates an improved fitness of the biomass, since the protein biosynthesis and thus the division activity are high. This occurs when the cells have optimal growth conditions and are sufficiently supplied with nutrients.

In the seasonal development in the year 1995, an increase of the RNA/DNA-ratio was registered in June, followed by a collapse in summer from July to September and a second rise in October at the sites with mussel settlement (Albers unpubl. data). While the RNA/DNA-ratio in the old and young mussel bed sediments showed a similar seasonal pattern, the reference site in the sandflat without mussels hardly showed any seasonal fluctuation of this parameter at all.

The annual pattern of the RNA/DNA-ratio thus followed the typical seasonal development of the phytoplankton with a temporal delay of 1–2 months. However, the results can be biased by the sampling frequency, since fresh organic matter can be degraded within 4 weeks (see above).

Furthermore, the relation of the exoenzyme activities (glucosidase, protease, phosphatase) supplied indirect evidence of the composition of organic matter. While a high activity of protease indicated an increased protein content of the substrate and thus a good food quality, a high activity of glucosidase showed the presence of high-molecular sugars (low food quality). Only once after the ice winter and in contrast to the results from all other sampling dates, the activity of glucosidase was higher than the activity of protease.

Conclusion

The origin of the organic matter of the Swinnplate was characterized mainly by the input of material from marine and estuarine sources. The absolute POC-quantities were highest within and around the mussel bed, due to the filtration activity and the reduced current velocities by the epibenthic structure. The values decreased with increasing distance from the mussel bed. A gradient from the mussel bed into the sandflat resulted from the export of fine-grained material with the ebb current. The amount of deposited fine-grained material varied with the seasonal dynamics of phytoplankton, which could be confirmed by specific biomarkers in the sediment. The deposition of the organic matter through the filtration by the mussels was rapidly detectable in the sediment by increased contents of amino acids and lipid-biomarkers, whereas the reaction of microbial parameters was delayed. Seasonal variations were also found in the energy contents of the sediments with high values in summer and autumn. These fluctuations were also brought into connection with the input of fresh organic matter from phytoplankton blooms. Inspite of the increased POC-contents in the sediments of a mussel bed, the food quality of the organic matter for further heterotrophic organisms was reduced by the nutritional uptake of the mussels.

5.2.1.3 Gröninger Plate

The geochemical investigations on the Gröninger Plate were carried out in order to follow the reactions of the sediment- and pore water composition to an excessive input of organic carbon to the sediments. The emphasis of the investigation was on:

- monitoring the dynamics in the pore water chemistry of reference sites as background data for the experimentally enriched sediment sites
- field experiments on the enrichment of sediment with organic carbon
- improvement and use of a mathematical model on the development of black spots.

The fieldwork started on the Gröninger Plate in April 1994 and was continued until September 1996. Except for the winter months, samples were taken at least once a month mainly at the sandflat sites GP1 and GP3 (Chap. 3.1). During the entire investigation, the macrobenthos of the study area was dominated by *Arenicola marina* and *Lanice conchilega* (Chap. 3.6.2).

Reference dynamics

The field investigations concentrated mainly on the monitoring of the dynamics of dissolved pore water nutrients. Sediment profiles of phosphate, ammonia, nitrite, nitrate, silicate, sulphide, sulphate and DOC were measured to a depth of max. 50 cm. Samples were taken 2 to 8 times per month at station GP3 on the Gröninger Plate (Fig. 3.1.4), so that a relatively large number of sediment depth profiles was available, which were analysed with a variety of statistical procedures (Langner 1997).

The distribution of nitrate and nitrite was very variable. However, while the median values were close to the detection limit, enrichments occurred in spring and autumn, which could be explained by local peaks in single depth profiles. Compared to 1995, the values were generally lower in 1996 and showed little variation over the year. A decline of the values which was recorded in June/July 1995 could not be observed in 1996. Nitrite and nitrate occurred in sediment depths down to 25 cm although they are products of oxygen-dependent mineralization and consumed under anoxic conditions. Sporadic and small-scale maxima pointed to locally increased bioturbation activity, which facilitated an entry of oxygen down into deeper sediment layers.

With values below 300 µmol dm^{-3}, the sulphide concentrations of the pore water remained at low levels during 1994 and 1995. Unusually high sulphide concentrations (up to 1 mmol dm^{-3}), were detected in the year 1996, but most of the values were lower than 300 µmol dm^{-3} in that year as well.

The concentrations of ammonia, phosphate and sulphide as direct products of mineralization usually showed similar dynamics below the O_2 penetration depth, both in the overall pattern as well as in the single depth profiles. The annual variations indicated an intensified anoxic mineralization activity during the summer months, which could shift by one month or be prolonged until autumn. For example, the lowest downcore phosphate concentrations were recorded in spring 1994 and 1995, followed by an increase of the concentrations in summer and autumn, which was delayed in 1995 by around one month.

Immediately after the ice winter 1995/96 high pore water concentrations of ammonia, phosphate and sulphide were recorded up to mid February. Due to a stable weather condition with easterly winds, the inundation periods of the tidal flats on the Gröninger Plate were very short during that time, which led to a reduced pore water exchange and thus to an enrichment of mineralization products in the pore water. After a change in the weather conditions, the nutrient concentrations returned to a normal level. The development in 1996 was comparable to the preceding years. The maximum of the phosphate concentrations occurred in January and February, followed by a minimum in spring. In May and June 1996, simultaneously with the "black spot event" (Chap. 7) it came to an unusually high rise of the phosphate concentrations again, while the values measured in July and September 1996 were comparable to those of the previous years. A rise of the silicate concentrations in June 1996 could be an indication to the deposition and dissolution of a diatom bloom (*Coscinodiscus concinnus*). Also the phosphate-, sulphide and ammonia concentrations indicated an increased release of nutrients caused by mineralization processes during this time.

Enrichment experiments

Enrichment experiments were carried out in the years 1994 and 1995. In the course of these experiments, defined quantities of organic matter were buried in the sediment and the temporal development and change of the pore water chemistry was followed (Langner 1997, Oelschläger unpubl. data). In both years, additional control and reference sites were established and sampled at the same time as the experimentally polluted sites. In the control sites the sediments were treated in the same way as in the experimental sites, but no organic material was introduced into the sediment. Reference surfaces remained untreated.

In 1994, the experimental addition of organic matter took place by burying macroalgae (*Enteromorpha* sp.) in 10–15 cm sediment depth. The introduced algal material corresponded to a quantity of 0,83 kg C/m^2. During the initial phase of the experiment, daily samples were taken. Later on, sampling intervals were increased.

The input of organic matter into the sediment caused a quick reaction of the pore water chemistry. This was initially indicated by high sulphide concentrations, increased ammonia- and strongly decreased sulphate concentrations. In some profiles, the sulphide horizon reached the sediment surface and led to the formation of black spots. The vertical position of the maximum concentrations of S^{2-}, PO_4 and NH_4^+changed with time and moved towards the sediment surface. During the winter 1994/95, the sediment chemistry regenerated and the values of experiment and reference sites were comparable afterwards. This regeneration could have been due to an intensified exchange of pore water during storm events. However, higher sulphide-concentrations measured in summer 1995 still indicated the consequences of the enrichment.

Following the experiences of the previous year, starch (flour) was added to the sediment in different quantities (0.45 kg C/m^2 up to 2.7 kg C/m^2) in four experimental treatments in 1995 (Langner 1997, Oelschläger unpubl. data). Mesocosm studies, with different organic substrates had shown that starch caused the best agreement with a natural load e.g. by macroalgae. In these field experiments, the temporal development of total sulphur, dissolved sulphate and sulphide, iron (soluble and total), DOC, water-content, sediment density and POC were determined.

Directly after the experiment started, increased sulphide concentrations were detected at the three sites with the highest enrichment. After four weeks increased sulphide values were recorded at all experimental sites. Two months after the onset of the experiment, sulphide values decreased in the treatments with the lowest amounts of POC (S1, 0.45 kg C m^{-2}), whereas unchanged concentrations of sulphide were measured in the treatments S2 and S3 (medium load). At treatment S4 with the highest organic load (2.7 kg C m^{-2}) the rise of the sulphide concentration continued. After a decrease during the winter months one year after the beginning of the experiment the sulphide concentrations in treatment S4 increased again, to the level of the previous year. In treatment S3, the sulphide concentrations were slightly lower, while the values of the treatments S1 and S2 corresponded to those of the unpolluted reference sites. The results indicated that the regeneration times depended on the magnitude of the organic load. The released DOC quantities were also coupled to the amount of the load. The sulphate concentrations were reversely correlated to the sulphide-values - with increasing sulphide concentrations, the sulphate-values decreased. During the summer months, a complete exhaustion of the sulphate-pool was frequently observed at several sediment depths. Without an organic load, neither sulphide nor methane were emitted. Six weeks after the onset of the experiment a linear relation was found between the released CO_2-amounts and the degree of the load, while the emissions of sulphide and methane showed no linear correlation with the quantity of starch brought into the sediment. The production of sulphide was probably limited by the availability of sulphate. For methane, a clear difference was determined between day and night emissions, which did not occur for sulphide and CO_2.

In the context of the loading experiments effects on the isotope-geochemistry of solid and dissolved phases were examined as well (Böttcher unpubl. data). In contrast to the reference site, the DOC and ΣHS-concentrations in the sediments enriched with organic matter were increased (Fig. 5.2.11). Probably because sufficiently large quantities of reactive iron complexes were missing in the sandy sediments and/or the reaction rate of the iron-containing solid phases was too slow, the formed ΣHS could not be precipitated as iron-sulphide. In contrast to the open ocean, the dissolved residual sulphate was enriched with ^{34}S, as during the process of sulphate-reduction the light S-isotope changes preferentially into the reduced phase. Fractionation factors were between -5 to -25 ‰. The decrease of the fractionation factor was the result of an increase in the absolute rate of sulphate-reduction. The rate of sulphate-reduction is mainly a function of the availability of organic substances, pore water sulphate and the ambient temperature. Indeed, with increasing ambient temperature the isotope fractionation factor in pore water sulphate decreased.

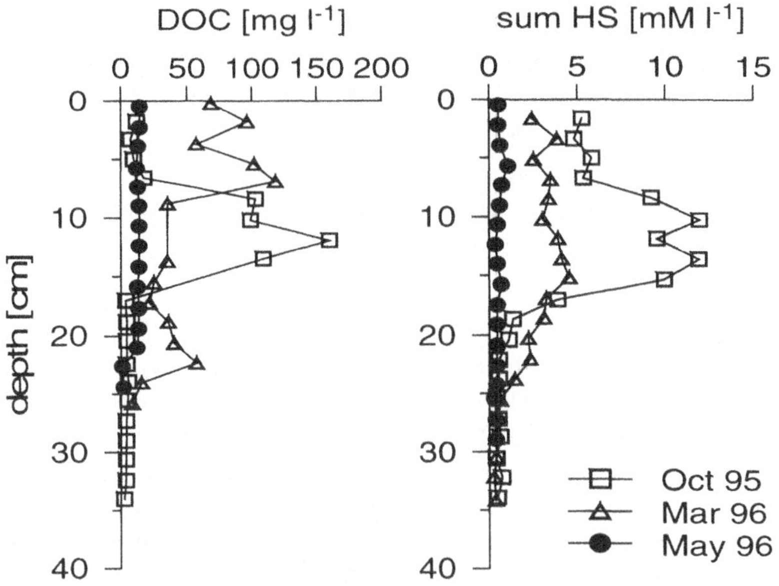

Fig. 5.2.11 Depth related pore water parameters of a C-loaded area on the Gröninger Plate as a function of time. (Böttcher, unpublished data)

Model for the development of black spots

The temporal development of the pore water chemistry after the input of organic matter was reproduced with a mathematical model (Ebenhöh 1996).

POC, DOC, CO_2, CH_4, O_2-consumption, nutrient efflux, H_2S and SO_4^{2-} were included as parameters in the model. It describes the change of the biogeochemical processes in the sediment, which were induced by an overload of POC. The data used for the calculations were derived from the results of the algal enrichment experiments in 1994 (see above). In general, a good agreement was found between model simulations and the field data of the algal enrichment experiments.

The calculated regeneration times after low enrichments were too long in the mathematical model compared to the regeneration in the field. This was attributed to the fact that bioturbation was not included as a parameter in the model-calculations. The processes simulated by the model predict an accumulation of sulphur in the form of iron-sulphide in the sediments after enrichment. Furthermore, the results show that not enough data were available to detect the more subtle effects of the loading with low POC quantities.

Conclusions

On the basis of the obtained results different scenarios were described, which could be regarded as diagnozing criteria for surpassing a critical level of organic input into intertidal sediments. Among these were:

- the occurrence of black spots at the sediment surface
- an increased release of trace gases (e.g. CH_4, H_2S) from the sediment
- the appearance of anoxic bottom water
- lack of bioturbation activity
- exhaustion of the buffer capacity of the sediments for sulphide (approx. 3 mmol dm^{-3})
- the accumulation of DOC in the pore water. While a value of approx. 10 mg dm^{-3} DOC is typical for unpolluted sediments, values of up to 4000 mg dm^{-3} DOC were measured in the experimental sites.

After the start of an enrichment experiment, the pore water concentrations increased in the sequence DOC, methane, sulphate(decrease), phosphate, sulphide, ammonia and decreased in the order DOC, sulphate(increase), ammonia, phosphate, sulphide again. The duration of the regeneration period depended on the degree of the load (concentration of the buried organic matter). The time requirement for sediment regeneration in the experimental sites was approximately one year, with the exception of the most heavily loaded site. However, from the mathematical model there were some indications that an increased sensitivity towards a renewed load (exhaustion of the sediment sulphide buffer capacity) can remain after the regeneration.

5.2.2 Distribution of Macrofauna Along the Transect

5.2.2.1 Introduction

The distribution pattern of benthic communities depends, among other factors, on physico-chemical sediment parameters and is subjected to modifications of POC- and mud contents. The classic model of Pearson & Rosenberg (1978) describes the succession of macrofauna due to POC-enrichment by organic pollution. An analogous distribution was found by Dittmann (1987) in sediments with experimental enrichments of biodeposits of the blue mussel *Mytilus edulis*, demonstrating that biodeposits can be a structuring factor for benthic communities. Her experiments showed that differences in the concentration of biodeposits influenced the small-scale distribution of benthic organisms.

In the context of ELAWAT, studies were carried out along a gradient with decreasing contents of mussel biodeposits. In order to examine the influence of the biodeposits of a natural mussel bed on the sediment chemistry as well as the subsequent interactions between physico-chemical parameters and the organisms, biogeochemists and benthologists collaborated at a common study site on the Swinnplate (Fig. 3.1.3). The filtration and deposition of organic matter by the mussels, its mineralization as well as its role in the structuring of microorganism

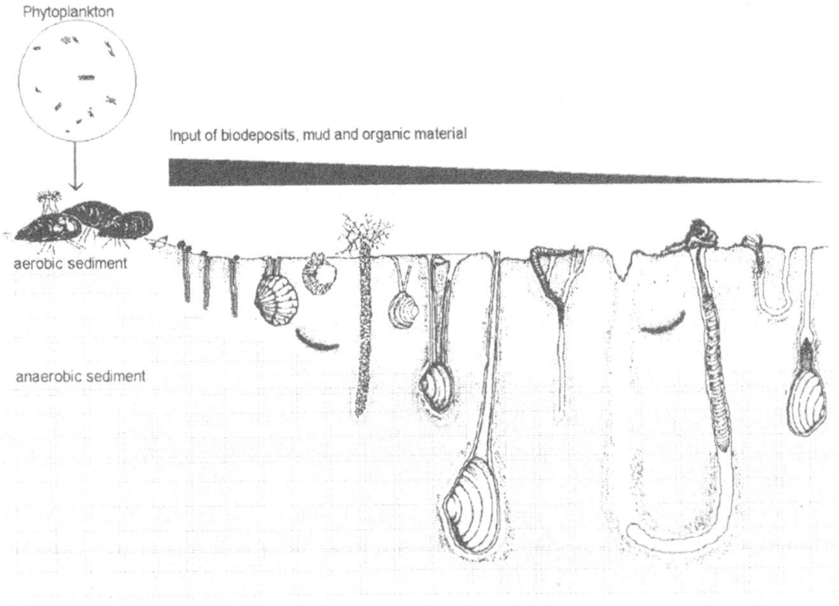

Fig. 5.2.12 Structural modifications in the benthic composition along gradients of organic enrichment (according to Pearson & Rosenberg 1978), physical disturbance (Rhoads et al. 1978) or particle size distribution (substrate gradient after Hertweck 1992).

and macrofauna communities due to modifications of the sediment characteristics were studied (Sect. 5.2.1.2).

The modifications in benthic assemblages described by Pearson & Rosenberg (1978) (increasing species diversity and higher proportion of deep-dwelling macrofauna with increasing distance from the source of contamination) are attributed to purely physical modifications according to other authors (Rhoads 1974; Rhoads & Boyer 1982; Rhoads et al. 1978). Hertweck (1992) assumed that a substrate gradient from the mudflat via muddy sandflats to sandflats is responsible for changes in the community structure and thus a possible factor controlling the macrofauna distribution. Therefore, the investigations in ELAWAT covered sedimentological parameters as well (Chap. 3.4; Sect. 5.2.1).

Epibenthic structures like mussel beds influence the exchange and transport of nutrients at the sediment surface (Fig. 5.2.12). A large part of the organic matter produced by primary production is taken up by the epibenthic organisms. Thereby, the labile organic matter is consumed and converted, e.g. partly into faeces having a different nutritional quality, affecting the availability of high quality POC for organisms living in deeper zones of the sediment. Bacterial biomass (cells, slime), attached to the surface of the sand grains or detritus, can serve as potential food for the fauna feeding in the sediment. The latter, however, delivers the food for epibenthic predators (Bouvy & Soyer 1989). Deposit-feeding macrofauna feeds on meiofauna and microorganisms, which are associated with faeces and pseudofaeces (Rhoads 1974).

In the sandflat, the macrofauna causes modifications of the sediment structure by its bioturbating activity and influences the concentration and mineralization of the organic matter (Anderson & Kristensen 1988). Bioturbation causes modifications of the physico-chemical sediment characteristics and may increase the metabolic activity of microorganisms (Yingst & Rhoads 1980).

The micro-topography of the redox-potential discontinuity (RPD-layer) in the sediments is changed by bioturbation- and ventilation-activities of the macrofauna. Without bioturbation, oxic mineralization of the organic matter is restricted to the sediment surface and the RPD-layer. Bioturbation causes a partial mixing of the sediment, which leads to relatively homogeneous depth profiles of microbial activities and bacterial cell numbers as well as of density gradients of particulate organic matter and nutrients. Thus, in sandflats the mineralization of the organic matter is not limited to the sediment surface, but also expanded to deeper zones of the sediment.

This chapter summarizes the studies carried out in ELAWAT by Kröncke, Bergfeld and colleagues. Detailed specifications of the geochemical reactions on the transect are given in Sect. 5.2.1.

5.2.2.2 Material and Methods

The macrozoobenthos was sampled in 2-monthly intervals at 6 different sites along a transect on the Swinnplate from May to November 1994, from March to December 1995 and March to September 1996 (Kröncke 1996 and unpublished; Chap. 3.1, 3.6). Eight replicate samples were taken per site with a corer (5 cm diameter, approx. 20 cm sediment depth) in an area of 0.25 m^2. The samples were sieved over 0.5 mm mesh size and fixed in 4 % buffered kohrsolin solution. The individuals were counted in the laboratory and determined to species level. For the determination of the macrofaunal biomass, the samples were dried for 24 h at 80 °C and combusted for 6 h at 480 °C. The biomass was expressed as mg ash-free dry weight (afdw) per 20 cm^2.

5.2.2.3 Results

The study area in the backbarrier system of Spiekeroog Island was characterized by extremely varying individual numbers (abundances) and a strongly fluctuating diversity of the macrofauna over the years (Bergfeld 1996). His survey of the whole Swinnplate yielded 70 different taxa of macrofauna, whereas only 43 different taxa were recorded at the transect sites. Polychaetes were the dominant group of organisms in terms of species. The typical sandflat community of macrofauna changed in composition under the influence of the biodeposits produced by the mussels (Kröncke 1996).

The sandflat was characterized by a poorly structured surface sediment and a high sediment mobility and bioturbation-activity. The macrofauna community was dominated by the polychaete *Arenicola marina*. Additionally, *Scoloplos armiger*,

Spio martinensis and *Nepthys hombergi* appeared frequently. The *Mytilus-Lanice*-community was more diverse in topography and habitat heterogeneity. The infauna here was dominated by oligochaetes, *Aphelochaeta marioni*, *Heteromastus filiformis* and *Capitella capitata*. Typical epifaunal organisms, as they were described from more stable mussel beds in the North Frisian Wadden Sea (Dittmann 1990), were missing. Table 5.2.3 summarizes the most important characteristics of the two major benthic communities distinguished on the Swinnplate (Bergfeld 1996).

Comparing the mussel bed (SP1) and sandflat reference site (SP6) showed distinct benthic assemblages differing in the species composition, abundance and biomass (Kröncke, unpubl. data). In multivariate analyses, the sandflat site (SP6) was always separated from the mussel bed site (SP1) for both the abundances as well as the biomass of the macrobenthos. In the mussel bed, small, fast-growing, opportunistic polychaetes (r-strategist) dominated the benthic community whereas deeper-dwelling polychaetes occurred in the sandflat.

Along the transect with increasing distance from the centre of the mussel bed, species numbers, abundances and biomass decreased in the direction of the sandflat. Compared to site SP3 (*Lanice*-site), quantitative parameters for macrofauna (species numbers, abundances and biomass) were lower at the old mussel bed (sites SP1+2). Abundances of macrofauna organisms were also lower at these sites in comparison to site SP5 with the new settlement of blue mussels. However, the initial trend along the transect was not as pronounced in all study years and was also subject to seasonal variations.

Table 5.2.3. Characteristics of the macrofauna surveyed on the Swinnplate in 1994/95 by Bergfeld (1996)

parameter	sandflat	mussel bed
concentration of organic matter	low	high
biogenic influence	bioturbation	biosedimentation
sediment surface and -structure	homogeneous, structure-poor	heterogeneous, structure-rich
sediment turnover	high	low-medium
macrofauna	poor in biomass, species and individuals	rich in biomass, species and individuals
community	*Arenicola*	*Mytilus-Lanice*
characteristic species	*Scoloplos armiger*, *Spio martinensis*, *Hydrobia ulvae*, *Eteone longa*, *Nepthys homgergii*, *Pygospio elegans*	Oligochaeta, *Aphelochaeta marioni*, *Capitella capitata*
feeding-type	surface deposit-feeder, carnivore polychaetes, grazers	subsurface deposit-feeder, omnivore polychaetes, filter feeder

During the study period, the benthic assemblages along the transect changed as a result of several events like dislodgement of the mussel beds, variations in mussel densities as well as a new settlement of *M. edulis* at site SP5 in 1994 (Table 5.2.1). During the same year, a strong spatfall of larvae of *Macoma balthica* occurred also. According to the results of the geochemical investigations, site SP5 was characterized by organic matter of high quality. During 1994 and 1995, biomass, average species number and abundances of the macrofauna were clearly higher at this site than at the other sites of the transect. However, the new settlement of *M. edulis*, which caused changes in the macrofauna, did not affect other parameters such as mud content, bacterial cell numbers or POC-concentrations in the sediment (Sect. 5.2.1). After the ice winter, the mussel beds had disappeared and no mussels were present until the summer 1996, when a new settlement of *Mya arenaria*, *Cerastoderma edulis*, *M. balthica* and *M. edulis* took place in the transect area.

Spatial and temporal variations of median species numbers, abundances and biomass of the macrofauna were high (Table 5.2.4). Median values for species numbers ranged from 2–11 species 20 cm^{-2} and varied at each site with season and between years. In 1994/1995, the median species numbers and abundances of the macrofauna were highest in spring (May). After the ice winter 1995/96, the highest species numbers and abundances were recorded later in the year (July) than in previous years. In 1994 and 1996, the highest values of macrofaunal biomass were determined in September, whereas in 1995, the maximum was detected in May.

Strong temporal fluctuations were also found for single species and taxonomic groups of the macrobenthos at each site. The mudsnail *Hydrobia ulvae* dominated at site SP6 in 1994. The dominating polychaetes and oligochaetes occurred with different frequencies at the sampling dates. However, no clear spatial or seasonal patterns were observed. After the ice winter, *Pygospio elegans* and *Aphelochaeta marioni* dominated the benthic assemblages in spring 1996 until summer.

A classification of the polychaetes into trophic groups showed a dominance of deposit-feeding species on almost all transect sites. Only in 1994 and 1995, the proportion of surface deposit-feeding polychaetes was smaller in the mussel bed sites SP 1+2 than at the other sites. During this time, subsurface deposit-feeders were more abundant in the mussel bed. The changing frequencies of the taxonomic and trophic groups were also reflected in a heterogeneous frequency distribution of the biomass at the sites in the course of the respective years (Table 5.2.4). Only in the spring of 1996, deposit-feeding polychaetes accounted for most of the benthic biomass at all sites.

The high temporal fluctuations of the abundances at the transect sites became also apparent in multivariate analyses of the assemblages. Multi-dimensional scaling (MDS) showed a clear gradient of benthic assemblages along the transect from the mussel bed to the sandflat in a few months only. Besides seasonal influences and the new settlement of mussels close of the reference site in the sandflat, the ice winter 1995/1996 affected the benthic assemblages also. In general, seasonal similarities of the transect sites were more pronounced than spatial similarities of the individual sites during the study period. MDS-analyses of the biomass data showed that the benthic assemblages of the transect sites were more similar to each other at most sampling dates. Seasonal similarities of all stations existed in May and September 1996.

Table 5.2.4 Ranges of spatial and temporal variations of species- and individual numbers as well as biomass of the macrofauna on the transect of the Swinnplate (Chap. 3.1). Values given are median 20 cm^{-2}, biomass is expressed as mg afdw 20 cm^{-2}. The numbers are the lowest and highest median for each case. The brackets indicate where (transect site SP 1–6, Fig 3.1.3) and when (months) the respective values were recorded. (Kröncke, unpublished data)

year	1994	1995	1996
spatial variation			
species number	4–11	4–10	2–8
	(SP2) (SP5)	(SP6) (SP3)	(SP6) (SP5)
abundances	27–167	8–109	7–138
	(SP2) (SP5)	(SP6) (SP2)	(SP6) (SP5)
biomass	4–278	5–109	3–88
(mg afdw)	(SP2) (SP5)	(SP6) (SP5)	(SP6) (SP5)
temporal variation			
species number	5–9	4–11	3–11
	(Sep.) (May)	(Oct.) (May)	(March) (July)
abundances	60–209	34–115	10–113
	(Nov.) (May)	(Dec.) (May)	(March) (July)
biomass	60–180	9–126	3–52
(mg afdw)	(Nov.) (Sep.)	(Oct.) (May)	(May) (Sep.)

5.2.3 Discussion

By the collaboration of several teams in ELAWAT, interactions between microbial and geochemical parameters with macrobenthos were investigated. At the studied transect sites on the Swinnplate, biogeochemical characteristics and certain benthic distribution patterns were determined, which partly influenced each other. Regarding the relevance of these interactions for the Wadden Sea, the results are discussed under the following aspects:

1. Effects of micro- and macrofauna on geochemical processes
2. Regulation of the benthic distribution patterns by biogeochemical parameters
3. Spatial and temporal variations of the interactions

5.2.3.1 Effects of Micro- and Macrofauna on Geochemical Processes

One aim of the investigations was a detailed study of the influence of blue mussels on sedimentation, transfer of organic matter from the water column into the sediment, transformation of the organic matter by filtration and digestion as well as the significance of microbial remineralization of the biodeposited organic matter.

The distribution of trace elements indicated a coupling of physical- and biological processes (early diagenesis, element mobilization and -fixation by microbial activity). However, no differences were found between the chemical composition in sediments of the same grain size fraction in mussel beds and in nearshore mudflats (Hild 1997). This indicated that the filtration and biodeposition of the mussels does not change the inorganic composition of the sediments.

The use of biomarkers could show the source of the material ingested and deposited by the mussels as well as seasonal variations in the input of organic matter to the sediment. However, estimating the origin of organic matter on the base of biomarkers supplies only an indirect impression of the degradability and availability for macroorganisms (Dauwe 1999, see Sect. 5.2.3.2).

Bacteria and mussels can derive a significant part of their nutritional requirements by poorly degradable material such as humic substances, chitin and cellulose (Kreeger et al. 1988; Kirchner 1995; Schmidt & Jonasdottir 1997). Kosfeld (1989) analysed the composition of organic matter in mussel faeces as approx. 60 % quickly degradable (within 6–8 weeks) and approx. 40 % of slowly degradable material (over 20 weeks). In mussel beds, the degradation of organic matter is accelerated, e.g. by the digestion of the mussel. The release rates of dissolved nitrogen and phosphorus components from the sediments of the mussel beds into the overlying water are enhanced (Nixon et al. 1976; Murphy & Kremer 1985; Doering et al. 1987; Dame & Dankers 1988; Dame et al. 1989; Asmus et al. 1990; Asmus & Asmus 1991; Smaal & Prins 1993). Particularly the release of nitrogen in the form of ammonia from the mussel bed exceeds that of other benthic communities in the Wadden Sea, by far (Asmus 1994). The increased release of nutrients from the mussel beds promotes the growth of macroalgae which attach to the mussel shells, while low N- and P-concentrations in the water column during summer limit the growth of phytoplankton (Asmus & Asmus 1991).

The effect of the mussels on the transformation of organic matter could be followed by comparing the patterns of amino acids and pigments before and after the ice winter (Sect. 5.2.1). Mussels used specific organic matter of high quality like amino acids for the production of their biomass. In the mussel bed, a higher content of fresh organic matter was available in comparison to the sandflat. However, this material was modified in its quality by the mussels through a depletion of amino acids and an enrichment of degradation products of pigments (Behrends 1997). The composition of amino acids in marine organic matter changes systematically in the course of the degradation of organic matter (Dauwe 1999). This author suggests that the special role of alkaline amino acids in the geochemistry of sediments and as part of the diet for heterotrophic organisms should receive more attention.

The input of POC-enriched biodeposits in the mussel bed sediments and the surrounding area increased the mineralization and composition of organic matter relative to the reference site. Due to the increased contents of fine-grained material and POC in the biodeposits and the associated intensified microbial mineralization, the anoxic depth horizon in the mussel bed sediment was closer to the surface than in sandflat sediments.

Biogenic structures are important sites for intensified bacterial decomposition (Forster 1991) and in turn the activity of the microorganisms influences the physico-chemical gradients in the sediment (redox processes). At the RPD-layer, a stimulation of benthic activities, both of microorganisms and of bacteria-feeding meiofauna, was observed by Fenchel & Riedl (1970). In sediments with a high supply of POC, such as mussel beds, a defined RPD-layer develops close to the sediment surface. In contrast, the RPD-layer of sandy sediments with a low supply of POC lies deeper within the sediment and is wider. Here, the microbial biomass and activity are primarily limited by the availability of the organic matter.

In the presence of oxygen, the organic matter could be oxidized completely via aerobic microbial respiration. Under anoxic conditions and with a lack of suitable electron acceptors, only a partial and incomplete degradation of the organic matter occurs. The products of fermentation are low-molecular organic acids, which could be used further by other groups of anaerobic bacteria as substrates for anaerobic respiration like sulphate-reduction or methanogenesis. The combination of both processes leads finally to the complete mineralization of the degradable organic matter and to the breakdown into its inorganic constituents. Anaerobic metabolic pathways are generally less efficient in their energy yield compared to aerobic pathways. For example, sulphate reducing bacteria have an about 80 % lower energy gain from the metabolized substrate than aerobic bacteria (Stumm & Morgan 1996).

Organic matter in the form of phytoplankton was clearly changed by the food uptake of the mussels. The mussels use predominantly those components of the filtered organic matter which are easy to degrade, like e.g. amino acids and pigments. Easily degradable, fresh organic matter of animal and plant origin was preferentially degraded by heterotrophic bacteria, whereas refractory material, could be stored and preserved in the sediment for longer time periods.

5.2.3.2 Regulation of Benthic Distribution Patterns by Biogeochemical Parameters

The dependence of benthic distribution patterns on geochemical parameters, food availability and quality of the organic matter was examined in detail in ELAWAT. It was shown that the production of biodeposits changed the benthos composition. In the sediments enriched with biodeposits the location of the boundary between oxic and anoxic sediment layers shifted due to the enrichment of POC and mud. This influenced the micro-, meio- and macrofauna living in the sediment. Most heterotrophic organisms (with the exception of anaerobic bacteria), depend on oxygen and cannot survive in an oxygen-free environment over a longer time period.

The distribution of microorganisms found in this study was mainly linked to the substrate supply. The highest total cell numbers and exoenzyme activities of bacteria were found in the mussel bed sediments. Here, apart from the sufficient supply of degradable substrates, the highest settlement surface for bacteria was available in form of fine-grained sediments. Laboratory experiments also indicated that the enzyme activity of the microorganisms was closely connected to the grain size of the sediment.

Ecological effects of low oxygen concentrations on macrobenthos were described by Diaz & Rosenberg (1995) and Levin et al. (1991). Not only the oxygen deficiency, but also the enrichment of metabolic endproducts (H_2S, NH_4^+), which are released into the pore water from the microbial mineralization, are toxic for macroorganisms at higher concentrations (Giblin et al. 1995) or, in the case of S^{2-}, even in trace levels.

Modifications of the organic and inorganic sediment composition entail certain distribution patterns of the benthos. The polychaete *Capitella capitata* is a typical representative of polluted coastal regions and an indicator for increased concentrations of organic matter (Weston 1990; Alongi & Tenore 1985). During the investigations on the Swinnplate, *Capitella* spp. occurred with higher abundances in the old mussel bed and the biodeposit-enriched sediments (Table 5.2.3). Probably, the natural habitat of this species-complex is POC-enriched sediment. Also the capitellid polychaete *Heteromastus filiformis* occurred more frequently at the edge of the mussel bed than in the sandflat. Thiermann et al. (1996) found a distribution of macrofauna which was related to the variation of the corresponding sulphide regime.

A classification of benthos into trophic groups revealed a modified composition of macrofauna assemblages along the transect. In the mussel bed on the Swinnplate, subsurface deposit-feeders were more frequent than in the sandflat. Predatory polychaetes were almost absent in the biodeposit-enriched sediments. A similar distribution of trophic groups was described by Dittmann (1990) as a functional interaction ("trophic group amelioration").

In the biodeposits of the mussel bed, bioturbation was strongly reduced, because no deep-dwelling macrofauna occurred here. Their abundance increased with increasing distance to the mussel bed. In the sandflat (reference site SP6) for example, *Arenicola marina* fulfils an important role as bioturbator (Reichert 1988). Almost half way between the centre of the mussel bed and the reference site in the sandflat, the conditions seemed to be most favourable for the macrofauna. These areas contained species- and individual-rich meio- and macrobenthic infauna with a dominance of polychaetes. Probably both the supply with organic matter as well as oxygen availability for aerobic respiration were sufficient here.

During the investigations on the Swinnplate, the content of organic matter turned out as an important factor for the distribution of macrofauna. However, together with the production of biodeposits, several other sediment characteristics changed (see Dahlbäck & Gunnarsson 1981). In the case of the biodeposits which were transported by the ebb current away from the mussel bed, changes in the content of organic matter were correlated with the proportion of fine-grained sediment material. Grain size distribution of the sediment itself can also influence the distribution of macrofauna, as it was shown e.g. for tubificoid oligochaetes (Giere & Pfannkuche 1982). Oligochaetes also dominated the infauna in mussel beds studied by Dittmann (1990) and Kröncke (1996).

However, not only the quantity, but also the quality of organic matter determines infaunal distributions. The availability of organic matter for macroorganisms can be roughly determined by leaching experiments (e.g. sum of the easily extractable macromolecules of the POC, Relexans et al. 1992). This labile organic substance represents the potentially available food for benthic consumers, however it usually makes up only a small part (less than 10 %) of the total carbon pool. Bulk parameters such as the content of organic carbon (POC), total nitrogen or the C/N-ratio are not enough to assess the quality of the organic matter. Furthermore, the traditional biogeochemical division of organic matter into big groups (protein, carbohydrates, fatty acids) is not sufficient as an indicator for the quality and availability of organic matter for macroorganisms (Dauwe 1999). A more precise characterization of the composition of this material is therefore necessary in the future.

5.2.3.3 Spatial and Temporal Variations of the Interactions

The input of POC-rich, fine-grained material into sandy sediments has substantial influences on the geochemical and biological processes (turnover rates of organic matter degradation, early diagenesis, element mobilization and fixation by microbial activity; Lochte 1993). These can be subject to seasonal variations and irregular special events (dislodgement of the mussels, new settlement or spatfall of larvae, ice winter, macroalgal development, mud deposition and phytoplankton blooms). For example, the proportion of biodeposits in the sediments was controlled by the filtration activity of the mussels, which was highest during phytoplankton-blooms from late spring to summer (Kautsky & Evans 1987). A further factor was the activity of microorganisms, which was linked to the supply of organic matter and the ambient temperature.

The parameters examined along the transect were characterized by high variations. The results of the MDS-analyses of the macrofauna investigations agreed with the data on biomarkers (Sect. 5.2.1). The reference site in the sandflat (SP6) was clearly distinguished from the other sites on the transect influenced by biodeposits. In the analysis of the biomarkers, the biodeposit-enriched sites resembled each other closely. Microbial tests on the functional diversity showed a high seasonal variation in the sediments outside of the mussel bed, whereas the bacterial populations in the mussel bed sediment had more site-specific properties.

The RPD-layer in coastal marine sediments shows seasonal dynamics (Jørgensen 1980), being millimetres to centimetres thick in winter, and in summer at times of increased POC deposition and elevated temperatures less than a few millimetres thin. Differences in the oxygen supply and sediment profiles between the transect sites were more clearly pronounced in the second half of the year, probably due to the increasing biogenic influences. The investigations on the Swinnplate showed that nutrient supply (e.g. in the form of phytoplankton blooms) leads to a direct stimulation of benthic activities and the signals of an input of organic matter were reflected in the geochemical parameters.

However, seasonal variations as well as small-scale fluctuations and deviations overshadowed the ideal pattern of an evenly decreasing gradient of POC- and mud-concentrations. For example, these deviations were caused by macroalgal growth particularly on the sites with mussels. In August 1995, large-scale deposition of mud occurred in the entire study area and was responsible for the similarity of oxygen conditions at all sites of the transect. In July, August and September 1995, mats of the macroalgae *Enteromorpha* spp. developed, which likewise exerted a more or less pronounced influence on the oxygen regime of the different transect sites, because they changed the benthic-pelagic exchange. Depending on the location where the algae occurred or where they were buried, they obstructed or blocked the oxygen diffusion across the sediment surface as well as the filtration activity of the mussels. Furthermore, the current velocity was reduced by the patchy macroalgal vegetation, which led to an additional physically-controlled sedimentation of POC-rich fine-grained material. When the macroalgae died, the microbial degradation and mineralization of the biomass caused an additional oxygen consumption.

5.2.4 Conclusions

The investigations at the transect on the Swinnplate aimed at bringing more light into interactions between biogeochemical parameters and the distribution patterns of macrobenthos, especially in relation to enrichments with organic matter (here in the form of biodeposits). Similar distributions were found as those already described by Pearson & Rosenberg (1978), Rhoads et al. (1978) and were also reported by Hertweck (1992), Dittmann (1987) and Kröncke (1996) for the Wadden Sea. The different succession models, giving more weight to either organic matter or sediment properties, do not really exclude or contradict, but can even support each other. As the results of the transect study show, mud- and POC-concentrations of the sediment were correlated and can both influence or regulate the distribution of macrofauna. Furthermore, parameters, which change simultaneously with variations of POC- and mud-contents (and partly being dependent on them), like oxygen- and sulphide-concentrations, redox-potentials and pH-values, cannot be ignored. In ecological systems, under natural conditions, the interaction of several environmental factors is usually decisive for the development of patterns and processes (Snelgrove & Butman 1994). The investigations on the Swinnplate had the difficulty that the location and density of the mussel bed changed from year to year. Thus, no continuous gradient of biodeposits was present on the studied transect. Future studies should check the interactions between POC, grain size and benthos under controlled laboratory conditions or in mesocosm experiments. Sulphide-and oxygen-profiles should be measured in these experiments as well. The specific investigation of the quality of organic matter should also be in the centre of attention, since the results showed, that quantitative determinations of bulk POC are not sufficient to explain its role as an environmental factor.

Acknowledgements

This chapter is based on the work and the data provided by several sub-projects of ELAWAT carried out by B. Albers, A. Bartholomä, B. Behrends, C. Bergfeld, M. E. Böttcher, E. Boysen-Ennen, P. Brocks, H.-J. Brumsack, M. T. Delafontaine, B. W. Flemming, G. Hertweck, A. Hild, T. Höpner, K. Knauth-Köhler, I. Kröncke, W. E. Krumbein, I. Langner, T. Leu, G. Liebezeit, B. Oelschläger, D. Rohjans, J. Rullkötter, T. Stoeck, M. Türkay, M. Villbrandt and F. Wolf. Their studies have been published as cited in the text and further unpublished data were kindly provided. We wish to thank two reviewers for their valuable and helpful comments on the manuscript. The project was funded by the German Bundesministerium für Bildung, Wissenschaft, Forschung und Technologie (BMBF) under grant number 03F0112 A and B. The responsibility for the contents of the publication rests with the authors.

References

Alongi DM, Hanson RB (1985) Effect of detritus supply on trophic relationships within experimental benthic food webs. II. Microbial responses, fate and composition of decomposing detritus. J Exp Mar Biol Ecol 88: 167–182

Alongi DM, Tenore KR (1985) Effect of detritus supply on trophic relationships within experimental benthic food webs. I. Meiofauna-Polychaete (*Capitella capitata* (Type 1) Fabricius) interactions. J Exp Biol Ecol 88: 153–166

Anderson FO, Kristensen E (1988) The influence of macrofauna on estuarine benthic community metabolism: a microcosm study. Marine Biology 99: 591–603

Asmus H (1994) Bedeutung der Muscheln und Austern für das Ökosystem Wattenmeer. In: Lozán LJ, Rachor E, Reise K, von Westernhagen H, Lenz W (Eds) Warnsignale aus dem Wattenmeer. Blackwell Wissenschafts-Verlag, Berlin, pp 127–132

Asmus H, Asmus RM, Reise K (1990) Exchange processes in an intertidal mussel bed: a Sylt-flume study in the Wadden Sea. Berichte der Biologischen Anstalt Helgoland 6: 1–79

Asmus H, Lackschewitz D, Asmus R, Scheiffarth G, Nehls G & Herrmann, J-P (1998b) Carbon flow in the food web of tidal flats in the Sylt-Römö Wadden Sea. In: Gätje C & Reise K (Eds). Ökosystem Wattenmeer. Austausch-, Transport- und Stoffumwandlungsprozesse. Springer, Berlin Heidelberg New York. pp 393–420

Asmus RM, Asmus H (1991) Mussel beds: limiting or promoting phytoplankton? J Exp Mar Biol Ecol 148: 215–232

Asmus R, Jensen MH, Murphy D & Doerffer, R (1998a). Primary production of microphytobenthos, phytoplankton and the annual yield of macrophytic biomass in the Sylt-Römö Wadden Sea. In: Gätje C & Reise K (Eds). Ökosystem Wattenmeer. Austausch-, Transport- und Stoffumwandlungsprozesse. Springer, Berlin Heidelberg New York. pp 367–391

Behrends B (1997) Aminosäuren in Sedimenten und Partikeln des Wattenmeeres. Dissertation, Universität Oldenburg

Bergfeld C (1996) Die Makrofauna-Besiedlung auf der Swinnplate im Rückseitenwatt der Insel Spiekeroog, Ostfriesisches Wattenmeer, Diplomarbeit, Universität Göttingen

Billen G (1978) A budget of nitrogen recycling in North Sea sediments off the Belgian coast. Estuarine Coastal and Marine Science 7: 127–146

Blackburn T H (1988) Benthic mineralization and bacterial production. In: Wiley J & Sons Nitrogen Cycling in Coastal Marine Sediments, Chichester, pp 175–190

Bouvy M, Soyer J (1989) Benthic seasonality in an intertidal mud flat at Kerguelen Islands (Austral. Ocean). Relationships between meiofaunal abundance and their potential food web. Polar Biol 10: 19–27

Canfield DE, Thamdrup B, Hansen JW (1993) The anaerobic degradation of organic matter in Danish coastal sediments. Geochimica et Cosmochimica Acta 57: 3867–3883

Currie RI (1962) Pigments in zooplankton faeces. Nature 193: 956–957

Dahlbäck B, Gunnarsson LAH (1981) Sedimentation and sulphate reduction under a mussel culture. Marine Biology 63: 269–275

Dame RF, Dankers N (1988) Uptake and release of materials by a Wadden Sea mussel bed. J Exp Mar Biol Ecol 118: 207–216

Dame RF, Spurrier JD, Wolaver TG (1989) Carbon, nitrogen and phosphorus processing by an intertidal oyster reef. Mar Ecol Prog Ser 54: 249–256

Daumas R (1990) Contribution of the water-sediment interface to the transformation of biogenic substances: application to nitrogen compounds. Hydrobiologia 207: 15–29

Dauwe B (1999) Organic matter quality in North Sea sediments. PhD thesis, Rijksuniversiteit Groningen

Delafontaine MT (1995) Organogene Bestandteile - Feinfraktionen. In: Krögel F (Ed) Ökologische Begleituntersuchungen zum Projekt "Europipe" – Teilprojekt Sedimentologie. Jahresbericht für 1994. Senckenberg am Meer Bericht 95: 59–60

Delafontaine MT (1996) Organogene Bestandteile – Feinfraktionen. In: Krögel F (Ed) Ökologische Begleituntersuchungen zum Projekt "Europipe" – Teilprojekt Sedimentologie. Jahresbericht für 1995. Senckenberg am Meer Bericht 96/1: 78–80

Delafontaine MT, Bartholomä A, Flemming BW, Kurmis R (1996) Volume-specific dry POC mass in surficial intertidal sediments: a comparison between biogenic muds and adjacent sand flats. Senckenbergiana marit 26: 167–178

Diaz RJ, Rosenberg R (1995) Marine benthic hypoxia: a review of its ecological effects and behavioural responses of marine macrofauna. Oceanogr. Mar Biol Ann Rev 33: 245–303

Dittmann S (1987) Die Bedeutung der Biodeposite für die Benthosgemeinschaft der Wattsedimente. Unter besonderer Berücksichtigung der Miesmuschel *Mytilus edulis*. Dissertation, Universität Göttingen

Dittmann S (1990) Mussel beds – amensalism or amelioration for intertidal fauna? Helgoländer Meeresunters 44: 335–352

Doering PH, Kelly JR, Oviatt CA, Sowers T (1987) Effect of the hard clam *Mercenaria mercenaria* on benthic fluxes of inorganic nutrients and gases. Mar Biol 94: 377–383

Ebenhöh W (1996) Schwarze Flächen im Watt – ein mathematisches Modell. UBA-Texte 64/96: 71–76

Fenchel T, Blackburn TH (1979) Bacteria and Mineral Cycling. Academic Press Inc., London

Fenchel T, Riedl RJ (1970) The sulphide system: a new biotic community underneath the oxidized layer of marine sand bottoms. Marine Biology 7: 255–268

Forster S (1991) Die Bedeutung biogener Strukturen für den Sauerstofffluß im Sediment. Ber Inst Meeresk, Kiel 206

Giblin AE, Foreman KH, Banta GT (1995) Biogeochemical processes and marine benthic community structure: which follows which? In: Jones CG, Lawton JH (Eds) Linking Species and Ecosystems. Chapman & Hall, Inc., pp 37–44

Giere O, Pfannkuche O (1982) Biology and ecology of marine Oligochaeta. A review. Oceanogr Mar Biol Ann Rev 20: 173–308

Graf G (1987) Benthic energy flow during a simulated autumn bloom sedimentation. Mar Ecol Prog Ser 39: 23–29

Graf G (1992) Benthic-pelagic coupling: a benthic view. Oceanogr Mar Biol Ann Rev 30: 149–190

Graf G, Schulz R, Peinert R, Meyer-Reil L-A (1983) Benthic response to sedimentation events during autumn to spring at a shallow-water station in the Western Kiel Bight. – I. Analysis of processes on a community level. Marine Biology 77: 235–246

Grenz C, Hermin M-N, Baudinet D, Daumas R (1990) In situ biochemical and bacterial variation of sediments enriched with mussel biodeposits. Hydrobiologia 207: 153–160

Hansen LS, Blackburn TH (1991) Aerobic and anaerobic mineralization of organic material in marine sediment microcosms. Mar Ecol Prog Ser 75: 283–91

Hertweck G (1992) Distribution patterns of benthos and lebensspuren in the Spiekeroog backbarrier tidal flat area (southern North Sea). In: Flemming BW (Ed) Abstr Vol Symp Tidal Clastics´92, Wilhelmshaven. Cour Forsch-Inst Senckenberg 151: 39–42

Hild A (1997) Geochemie der Sedimente und Schwebstoffe im Rückseitenwatt von Spiekeroog und ihre Beeinflussung durch biologische Aktivität. Forschungszentrum Terramare Berichte Nr. 5

Höpner T, Michaelis H (1994) Sogenannte "schwarze Flecken" – ein Eutrophierungssymptom des Wattenmeeres. In: Lozán JL, Rachor E, Reise K, v. Westernhagen H & Lenz W (Eds) Warnsignale aus dem Wattenmeer. Blackwell, Berlin, pp 153–159

Jørgensen BB (1980) Seasonal oxygen depletion in the bottom waters of a Danish fjord and its effect on the benthic community. Oikos 34: 68–76

Kautsky N, Evans S (1987) Role of biodeposition by *Mytilus edulis* in the circulation of matter and nutrients in a Baltic coastal ecosystem. Mar Ecol Prog Ser 38: 201–212

Kirchner M (1995) Microbial colonization of copepod body surfaces and chitin degradation in the sea. Helgoländer Meeresunters 49: 201–212

Klump JV, Martens CS (1981) Biogeochemical cycling in an organic rich coastal marine basin – 2. Nutrient sediment-water exchange processes. Geochimica et Cosmochimica Acta 45: 101–121

Kosfeld C (1989) Mikrobieller Abbau von Faeces der Miesmuschel (*Mytilus edulis* L.). Dissertation, Christian Albrechts Universität, Kiel

Kreeger DA, Langdon CJ, Newell RIE (1988) Utilization of refractory cellulosic carbon derived from *Spartina alterniflora* by the ribbed mussel *Geukensia demissa*. Mar Ecol Prog Ser 42: 171–179

Krögel F (1997) Einfluß von Viskosität und Dichte des Seewassers auf Transport und Ablagerung von Wattsedimenten (Langeooger Rückseitenwatt, südliche Nordsee). Dissertation, Universität Bremen

Kröncke I (1996) Impact of biodeposition on macrofaunal communities in intertidal sand flats. P.S.Z.N.I. Mar Ecol 17: 159–174

Langner I (1997) Statistische Klassifizierung der räumlichen und zeitlichen Heterogenität von Nährstoffkonzentrationen im Porenwasser von Wattsedimenten. Identifizierung von räumlichen Mustern und zeitlichen Entwicklungen. Forschungszentrum Terramare Berichte Nr. 6

Levin LA Huggett, CL, Wishner KF (1991) Control of deep-sea benthic community structure by oxygen and organic matter gradients in the eastern Pacific Ocean. J Mar Res 49: 763–800

Lochte K (1993) Mikrobiologie von Tiefseesedimenten. In: Meyer-Reil L-A, Köster M (Eds) Mikrobiologie des Meeresbodens. Gustav Fischer Verlag, Jena, pp 258–278

Malcolm SJ, Sivyer DB (1997) Nutrient recycling in intertidal sediments. In: Jickells TD, Rae JE (Eds) Biogeochemistry of Intertidal Sediments. Cambridge University Press, Cambridge, pp 84–98

Meyer-Reil L-A, Faubel A (1980) Uptake of organic matter by meiofauna organisms and interrelationships with bacteria. Mar Ecol Prog Ser 3: 251–256

Middelburg JJ, Klaver G, Nieuwenhuize J, Wielemaker A, de Haas W, Vlug T, van der Nat JFWA (1996) Organic matter mineralization in intertidal sediments along an estuarine gradient. Mar Ecol Prog Ser 132: 157–168

Murphy RC, Kremer JN (1985) Bivalve contribution to benthic metabolism in a California lagoon. Estuaries 8: 330–341

Nixon SW, Oviatt CA, Garber J, Lee V (1976) Diel metabolism and nutrient dynamics in a salt marsh embayment. Ecology 57: 740–750

Oost AP (1995) The influence of biodeposits of the blue mussel *Mytilus edulis* on fine-grained sedimentation in the temperate-climate Dutch Wadden Sea. Geologica Ultraiectina 126: 359–400

Partiali V, Tangen K, Liaaen-Jensen S (1989) Carotenoids in food chain studies – III. Resorption and metabolic transformation of carotenoids in *Mytilus edulis* (edible mussel). ComS Biochem Physiol 92B: 239–246

Paul EA, Voroney RS (1984) Field interpretation of microbial biomass activity measurements. In: Klug MJ, Redy CA (Eds) Current Perspectives in Microbial Ecology. ASM, Washington, pp 509–514

Pearson TH, Rosenberg R (1978) Macrobenthic succession in relation to organic enrichment and pollution of the marine environment. Oceanogr Mar Biol Ann Rev 16: 229–311

Postma H (1988) Tidal flat areas. In: Jansson B-O (Ed) Lecture Notes on Coastal and Estuarine Studies: Coastal - Offshore Ecosystem Interactions. Springer, Berlin Heidelberg New York, pp 102–121

Reichert W (1988) Impact of bioturbation by *Arenicola marina* on microbiological parameters in intertidal sediments. Marine Ecology 44: 149–158

Relexans J-C, Lin RG, Castel J, Etcheber H, Laborde S (1992) Response of biota to sedimentary organic matter quality of the West Gironde mud patch, Bay of Biscay (France). Oceanologica Acta 15: 639–649

Rhoads DC (1974) Organism-sediment relations on the muddy sea floor. Oceanogr Mar Biol Ann Rev 12: 256–300

Rhoads DC, Boyer LF (1982) The effects of marine benthos on physical properties of sediments. In: McCall SL, Tevesz MJS (Eds) Animal-Sediment Relations. Topics in Geobiology (FG Stehli, Ser. Eds) pp 1–51

Rhoads DC, McCall SL, Yingst JY (1978) Disturbance and production on the estuarine sea floor. Am J Sci 66: 577–586

Rudnick DT (1989) Time lags between the deposition and meiobenthic assimilation of phytodetritus. Mar Ecol Prog Ser 50: 231–240

Salomons W, Mook WG (1981) Field observations of the isotopic composition of particulate carbon in the southern North Sea and adjacent estuaries. Mar Geol 41: M11–M20

Schmidt K, Jónasdóttir SH (1997) Nutritional quality of two cyanobacteria: how rich is poor food? Mar Ecol Prog Ser 151: 1–10

Smaal AC, Prins TC (1993) The uptake of organic matter and the release of inorganic nutrients by bivalve suspension feeder beds. In: R.F. Dame. (Ed) Bivalve filter feeders in Estuarine and Coastal Ecosystem Processes. Nato ASI Series G, Springer, Berlin Heidelberg New York. Vol. 33: 271–298

Smaal AC (1997) Food supply and demand of bivalve suspension feeders in a tidal system. PhD thesis, Rijksuniversiteit Groningen

Snelgrove PVR, Butman CA (1994) Animal-sediment relationships revisited: cause versus effect. Oceanogr Mar Biol Ann Rev 32: 111–177

Sørensen J, Jørgensen BB (1987) Early diagenesis in sediments from Danish coastal waters: microbial activity and Mn-Fe-S geochemistry. Geochimica et Cosmochimica Acta 51: 1583–1590

Sørensen J, Jørgensen BB, Revsbech NP (1979) A comparison of oxygen, nitrate and sulphate respiration in a coastal marine sediment. Microbial Ecology 5: 105–115

Stoeck T (1996) Gesamtadenylate, DNA und Gesamtkeimzahl als Parameter zur Bestimmung der mikrobiellen Biomasse und Adenylate Energy Charge und RNA/DNA-Verhältnis als Parameter zur Bestimmung der mikrobiellen Aktivität in der biosedimentären Ablagerungsfahne einer Muschelbank (*Mytilus edulis*) im Niedersächsischen Wattenmeer. Diplomarbeit, Universität Kaiserslautern

Stumm W, Morgan JJ (1996) Aquatic chemistry. Chemical equilibria and rate in natural waters. Wiley Interscience Publications

Thiermann F, Niemeyer A-S, Giere O (1996) Variations in the sulphide regime and the distribution of macrofauna in an intertidal flat in the North Sea. Helgoländer Meeresunters 50: 87–104

Volkman JK (1986) A review of sterol markers for marine and terrigenous organic matter. Organic Geochemistry 9: 83–99

Volkman JK, Johns RB, Gillian FT, Perry GJ (1980) Microbial lipids of an intertidal sediment – I. Fatty acids and hydrocarbons. Geochimica et Cosmochimica Acta 44: 1133–1143

Weston DS (1990) Quantitative examination of macrobenthic community changes along an organic enrichment gradient. Mar Ecol Prog Ser 61: 233–244

Yingst JY, Rhoads DC (1980) The role of bioturbation in the enhancement of bacterial growth rates in marine sediments. In: Tenore KR, Coull BC (Eds) Marine Benthic Dynamics, pp 407–422

Zak J, Willig RM, Moorehead DL, Wildman HG (1994) Functional diversity of microbial communities: a quantitative approach. Soil Biol Biochem 26: 1101–1108

5.3 Settlement, Secondary Dispersal and Turnover Rate of Benthic Macrofauna

Carmen-Pia Günther

Abstract: In the three years of investigation, larvae of most benthic macrofauna species were observed in the plankton, but with variations in the order of magnitude of larval density and in the timing of peak abundance. For each of the investigated species initial settlement occurred regularly at specific main settlement sites. The abundance of settlers varied in dependence of larval supply, hydrodynamic conditions and (for *Lanice conchilega*) availability of hard-substrate. Especially after the ice winter 1995/96 the latter two factors appeared to be of relevance. In 1996 settlement of *Heteromastus filiformis* took place in wide areas of the Gröninger Plate, thus increasing the area of initial settlement compared to the preceding years. Recruitment of *L. conchilega* was limited in 1996 by the low numbers of larvae available and the lack of suitable substrate for settling.

In the polychaetes under investigation, zonation patterns of adults developed by initial settlement, while those of adult bivalves developed by postlarval dispersal. After extreme conditions, such as cold winters with ice formation, secondary dispersal may have some importance in determining zonation patterns of polychaetes as well.

The turnover rate (i.e. the turnover of benthic organisms on a specific site due to emigration, resuspension, initial and secondary settlement) was extremely high even for sedentary polychaetes (e.g. *Pygospio elegans*). The importance of this result for recolonization experiments (Chap. 6) in which this species dominated in all years of investigation, has still to be evaluated in future studies. The *in situ* turnover rate of soft sediment macrofauna can also be valuable in estimating the energy flow through the macrobenthic fauna more precisely.

The order of magnitude of the reproductive output, settlement patterns and settlement behaviour as well as mobility of macrofauna demonstrate that the studied species react resilient to disturbances in the Wadden Sea by balancing their population level.

5.3.1 Introduction

To persist in a given area, a macrozoobenthic population has to develop mechanisms, which lead either to a high probability of survival for an individual over a certain period of time, or to a regular replacement of the individuals at this place. The latter is achieved by colonization processes such as initial settlement (settlement of larvae) or secondary settlement (settlement of postlarvae or adult stages). Both processes contribute in undisturbed as well as disturbed benthic habitats to develop and maintain stocks of macrozoobenthic species.

Due to the relatively short study period and the strong interannual variability in seasonality, no reference dynamics of settlement and secondary dispersal of macrofauna could be determined in the context of ELAWAT. Thus, in the following chapter the attempt will be made, to identify successive temporal and spatial patterns at species level concerning settlement and secondary dispersal, to relate these processes to environmental conditions and to estimate the turnover rates of macrozoobenthic organisms due to secondary dispersal.

5.3.2 What Determines Year-Class Strength in Intertidal Soft Bottom Benthic Macrofauna?

Recruitment of intertidal benthic macrofauna is depending on certain abiotic and biotic factors which occur every year in a more or less sequential order but with variable intensity.

The amount of gametes produced each year depends on the age rsp. size structure of the adult population, its abundance and the food availability during gametogenesis. In bivalves, eggs produced by boreal species, such as *Macoma balthica*, may be resorbed during mild winters with low food availability, to meet the energy requirements of the adult. During cold winters, the metabolism of the adults is assumed to be so much reduced that eggs are not resorbed and a high number of eggs can be released at the next spawning event (Honkoop & van der Meer 1997). The authors conclude that rich spatfalls of bivalves after cold winters are the result of the reduced metabolism of adults in the preceding winter. But this does not explain, why recruitment success after cold winters varies among bivalve species. For example, after the ice winter 1995/96 the soft-shell clam *Mya arenaria* had a strong recruitment in the Spiekeroog backbarrier tidal flats (Jaklin unpublished data). In the Sylt-Rømø area *C. edule* (Strasser pers. comm.) recruited best of all infaunal bivalve species. After a preceding ice winter of 1985/86 *M. balthica* had a strong recruitment in the tidal flats of the island of Borkum (Günther 1990). The latter observations were in agreement with the recruitment pattern observed in long-term series from Balgzand (Beukema 1979), which showed very good recruitment of this species after cold winters. In the Spiekeroog backbarrier system the opposite pattern was observed after an ice winter in 1996: only low densities of *M. balthica* larvae were recorded in the water column (Chap. 5.1) and subsequently the abundance of initial settlers was low. The reason for this variability in recruitment success after cold winters is not yet known.

In general, the order of magnitude of bivalve larvae in the water column seemed to be related to the order of magnitude of initial settlers on the Gröninger Plate (Boysen-Ennen, Jaklin unpublished data). A more precise prediction of the year class strength from larval densities will not be possible, as initial settlement is determined by small-scale factors, which vary interannually and depend on the local conditions (Armonies 1998). For most polychaete species this information is not available. Results obtained during ELAWAT prove that for the tube-worm *Lanice conchilega* no relationship exists between larval density and abundance of initial settlers (see Sect. 5.3.6).

5.3.3 Development of Spatial and Temporal Patterns by Initial Settlement and Secondary Dispersal

Initial settlement of benthic macrofauna is mainly determined by two factors: (1) active substrate selection and (2) hydrodynamics (Butman 1987). According to this author both factors may co-occur in the same species but function on different spatial scales. Thus, for soft sediments such as those investigated in ELAWAT (Gröninger Plate and Swinnplate, see Figs. 3.1.3 and 3.1.4) the place of initial settlement results from a combined effect of species specific behaviour and local hydrodynamics.

The benthic assemblages in the sandflats of the Gröninger Plate were dominated by polychaetes. For most polychaete species, general data on settlement, recruitment and secondary dispersal are sparse or completely lacking.

Table 5.3.1 Objectives of the investigations carried out from 1994–96 for the studied species. The location of the sampling stations is given in Fig. 3.1.2

	1994	1995	1996
Period of investigation	beginning of April until middle of September	end of April until the middle of August, one additional sampling in December	end of April until middle of September, one additional sampling each in January and March
Lanice conchilega	small and large-scale distribution of initial settlers	settlement experiments comparison of two adult habitats	effects of the ice winter
Heteromastus filiformis	-	large scale distribution of initial settlers	large scale distribution of initial settlers
Scoloplos armiger	-	initial settlement, small scale variability	-
Macoma balthica	year class strength of initial settlement, secondary dispersal, 2-5 stations	initial settlement, secondary dispersal, 2-4 stations	initial settlement, secondary dispersal, 2-4 stations
Ensis americanus	secondary dispersal, 2-5 stations	initial settlement, secondary dispersal, 2-4 stations	initial settlement, secondary dispersal, 2-4 stations
Cerastoderma edule		initial settlement, secondary dispersal, 2-4 stations	initial settlement, secondary dispersal, 2-4 stations
Mya arenaria		initial settlement, secondary dispersal, 2-4 stations	initial settlement, secondary dispersal, 2-4 stations

Thus, investigations in ELAWAT focused on the polychaete species *Lanice conchilega*, *Scoloplos armiger* and *Heteromastus filiformis*, with *S. armiger* as a representative for polychaetes with an exclusively benthic mode of reproduction. Furthermore, each of these species is of major importance or a representative for one of the different macrofauna associations of the Gröninger Plate: *L. conchilega* for the so-called *Lanice* association, *S. armiger* is common in *Arenicola marina* sandflats and *H. filiformis* dominates in muddy sands. For the Wadden Sea, no information was available on the long-term population trend of *L. conchilega*, but increasing abundances were recorded for *S. armiger* and *H. filiformis*, with the latter species reaching untypically high abundances in sandy sediments (Obert 1982; Beukema 1989).

In the course of ELAWAT, investigations were carried out on the Gröninger Plate in the backbarrier system of Spiekeroog island. In the years 1994, 1995 and 1996 sampling was carried out from spring to autumn with different objectives (Table 5.3.1).

Sampling was carried out with a corer of 33 cm² area (penetration depth 20 cm, 5–7 replicates) in time intervals of 1–2 weeks during the main sampling phase. Samples were fixed intact with 5 % buffered formalin. In the laboratory the samples were sieved using a sieve column of 500, 250 and 125 µm mesh size. Each of the fractions was analysed separately. If necessary, specimens of the 500 µm fraction were separated into age groups.

The method of drift fauna sampling was developed by Armonies (1992). First results from the Gröninger Plate were published by Jaklin & Günther (1996).

5.3.3.1 Spatial Variability of Initial Settlement of Selected Species

The tube building polychaete *L. conchilega* was found to be a hard substrate settler, selecting preferentially tubes of conspecies adults (Heuers, 1998). Single juveniles were found to settle on eroded shells of cockles (*Cerastoderma edule*) and soft-shell clams (*Mya arenaria*). Beside the blue mussel (*Mytilus edulis*) *L. conchilega* is one of the rare hard substrate settlers of importance in the soft bottom environment of the Wadden Sea. In contrast to Bayne (1964) who described that *M. edulis* settles exclusively attached to hard substrate, early postlarval stages of this species were observed in the soft sediments on the southern edge of the study area (GP6). But early *L. conchilega* were never observed in the sediment. A metamorphosis of the aulophora larva without available hard substrate was only described to occur under experimental conditions and was apparently due to low current velocities (Bhaud & Cazaux 1990). More detailed results on the population dynamics of *L. conchilega* will be presented in Chap. 5.3.6.

M. balthica, *Ensis americanus*, *C. edule*, *M. arenaria* and *H. filiformis* settle on soft substrates (Jaklin, Heuers unpublished data; Heuers 1998). In 1995 these species settled mainly in the area of the tidal channel at the southern border of the Gröninger Plate (GP6, Fig. 5.3.1). As bivalve larvae are known to settle passively (Table 5.3.2; Günther 1990; Armonies 1998), this pattern indicates that the hydrodynamics on this site facilitated settlement. Interannual variations in spatial

Table 5.3.2 Characteristics of the main settlement sites of the macrofauna species under investigation

	Substrate	Currents	Chemical cues	Physical cues
Lanice conchilega	hard substrate, preferentially a vertical structure	strong currents, probably turbulence due to tube structures	no	contact with hard substrate
Heteromastus filiformis	mud and muddy sand	no information available	probably microbially-derived	
Cerastoderma edule *Ensis americanus* *Macoma balthica* *Mya arenaria*	sand and muddy sand	sedimentation area, turbulence?	according to current knowledge: no	according to current knowledge: no

settlement pattern as observed between 1995 and 1996 were either resulting from variations of hydrodynamics in the period of settlement (different current velocities, varying wind speed and direction, Armonies 1998) or modified sediment characteristics. While the first is of importance for bivalve settlement, the latter probably affected settlement of *H. filiformis* in 1996.

Considering only postlarval stages of the size close to metamorphosis (i.e. animals of the fraction 125 < animal < 250 µm) in 1995 and 1996, the maximum abundance of settlers was found a week earlier at site GP2 than at GP6. Both sites are flooded by water coming from the nearest channel (Landbalje). There was a temporal shift of highest larval densities from the outer towards the inner part of the Landbalje as the example of *L. conchilega* aulophora larvae shows (Fig. 5.3.2). This phenomenon can be explained by the tidal transport pattern of sediment particles and organisms in the backbarrier system (Chap. 3.3). Nevertheless, at station GP3 the same temporal pattern was observed as in the Otzumer Balje and there was no indication of temporal variation between different sites at the Gröninger Plate with respect to *L. conchilega* larvae (Fig. 5.3.3). Due to behavioural differences, transport of competent bivalve larvae may follow different rules than aulophora larvae of *L. conchilega*.

In total the share of initial settlers within the 0-group of bivalves was low when compared to results from Günther (1991, 1992). This probably indicates that not all postlarvae found at the main settlement site have undergone metamorphosis there.

After the ice winter 1995/96, initial settlement of *H. filiformis* took place all over the study area Gröninger Plate, with sandflats areas more densely colonized than muddy sand areas. Sedimentbound chemical cues may have been the reason for the changed settlement pattern. For the closely related species complex *Capitella*, H_2S was assumed to be a settlement cue (Cuomo 1985; Butman et al.) and cannot be simply transferred to field conditions. Neither *Capitella* sp. nor *H. filiformis* reacted with increased abundances in the defaunation experiments (Chap. 6) in which the H_2S content of the sediment increased due to the experi mental conditions. Most likely, these polychaete species are not attracted by H_2S,

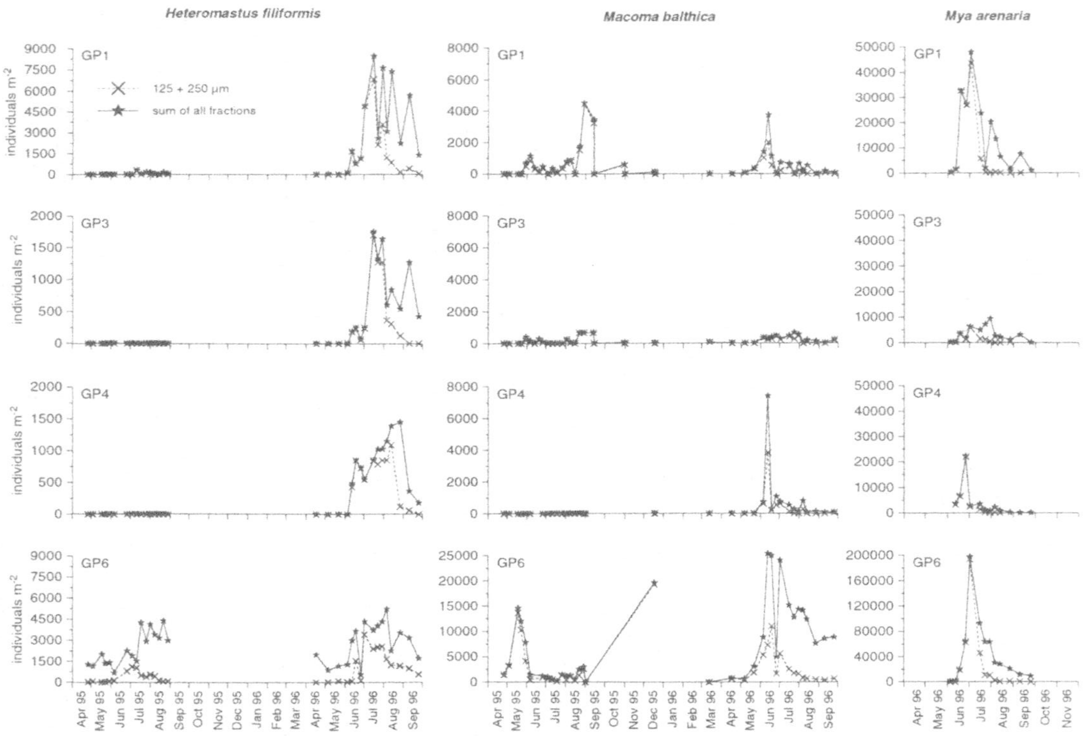

Fig. 5.3.1 Mean abundance of *H. filiformis*, *M. balthica* and *M. arenaria* at four sampling sites on the Gröninger Plate, East Frisian Wadden Sea (Heuers, Jaklin unpublished data). Presented are the individual numbers per m^2 based on all size fractions (solid line) and the smaller sized specimens retained on the 125 µm and 250 µm mesh (dotted line). Please notice the different scales

but by metabolic products of microorganisms. Until now examples for such relationships are only known from macrofauna species settling on hard substrate (compare Rodriguez et al. 1993 and references therein). Future investigations should test, whether chemical cues play a role in initial settlement of *H. filiformis*.

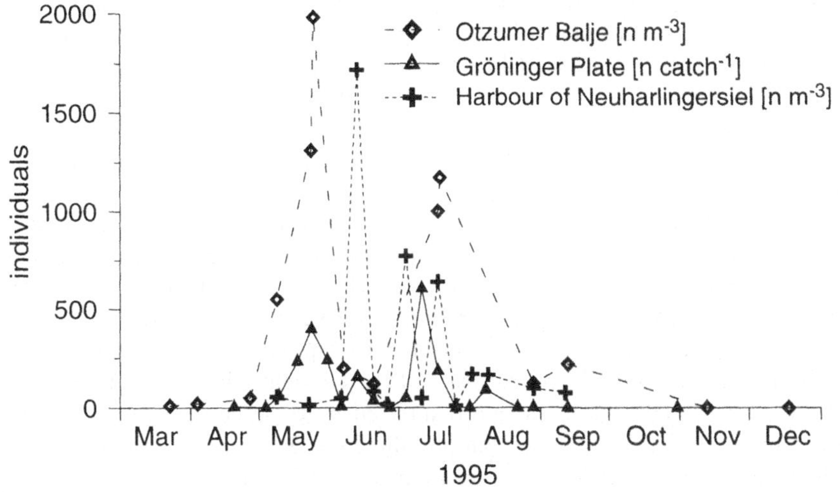

Fig. 5.3.2 Seasonal development of larval numbers of *Lanice conchilega* at several sampling sites in the backbarrier system of Spiekeroog island (tidal inlet Otzumer Balje, sandflat Gröninger Plate (GP3) and harbour of Neuharlingersiel, see Fig. 3.1.2), high tide samples 1995 (unpublished data from Boysen-Ennen, Jaklin and Heuers)

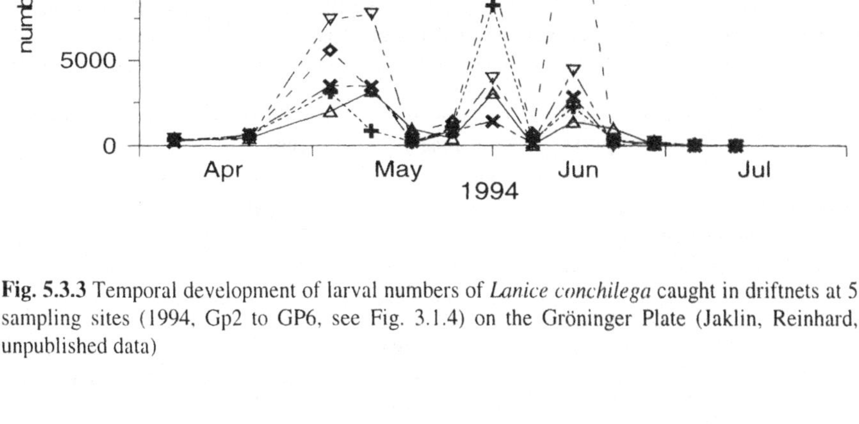

Fig. 5.3.3 Temporal development of larval numbers of *Lanice conchilega* caught in driftnets at 5 sampling sites (1994, Gp2 to GP6, see Fig. 3.1.4) on the Gröninger Plate (Jaklin, Reinhard, unpublished data)

5.3.3.2
Temporal and Spatial Variability in the Benthos vs. Water Column (Secondary Dispersal)

Water column dispersal of postlarval *L. conchilega*, *H. filiformis* and *S. armiger* was observed only occasionally (Table 5.3.3). Armonies (1998) reported that the occurrence of *S. armiger* in drift nets and sediment traps correlated with weather conditions and could be explained by passive resuspension. This can also be assumed to be the case for *H. filiformis,* although (as e.g. for *Pygospio elegans* Sect. 5.3.4) bedload transport cannot be excluded for this species (Smith & Brumsickle 1989). Heiber (1988) described a phase of secondary water column transport of juveniles in spring.

A general spatial and temporal pattern could be observed for the actively migrating bivalves (*M. balthica*, *E. americanus*) with a temporal shift in maximum abundances (driftfauna as well as benthos samples) between stations close to the southern tidal channel and those situated more in the centre of the Gröninger Plate. Starting from the area of initial settlement (GP2 rsp. GP6) small postlarvae migrate with the tides towards central parts of the Gröninger Plate. Due to the fact that the western and central part are dominated by flood currents (Brandt et al. 1995; Chap. 3.3) the resulting net transport direction should be eastwards, as long as the main current direction is not superimposed e.g. by easterly winds. Thus, the bivalves could either accumulate in the eastern part of the Gröninger Plate, which was not investigated, or be transported into nearby subtidal areas.

Because driftfauna nets of the type developed by Armonies (1992) catch only a true water column transport of benthic stages of the macrofauna, an epibenthic net was additionally used in 1995 (exposition height: directly above the sediment surface, height of the net 16 cm; mesh size 500 µm; unpublished data from Steuwer, Jaklin & Heuers). Comparing the results from both methods are interesting indications on the emergence of some species. The polychaete species *Eteone longa*, which was observed to be a good swimmer, was mainly found in the near bottom

Table 5.3.3 Maximum numbers of drifting juveniles of *L. conchilega*; *H. filiformis* and *S. armiger* in the years 1993-1996. In the different years the period of investigations as well as number of stations varied (1993 and 1994: GP2 - GP6; 1995 and 1996: GP3 and GP6). The information has the following sequence: (1) maximum number of individuals caught at one date, (2) frequency of occurrence in driftnets during the study period and (3) station. -: no animals caught, ?: no data available.

	1993	1994	1995	1996
Lanice conchilega	-	-	2,2,GP3	1,1,GP3
Heteromastus filiformis	1,1,GP6	-	?	
Scoloplos armiger		67, 2, GP	?	
		1, 2, GP3		
		5, 2, GP4		
		1, 1, GP5		
		1, 1, GP6		1, 1, GP6

water. Crustaceans belonging to the genus Gammarus (*G. locusta*, *G. salinus*) were also found more closely to the bottom, while *Urothoe poseidonis* moved into higher parts of the water column. There is still a need for more information concerning the vertical distribution of benthic macrofauna in the water column. Special attention of future research should be given to the question which life stages may reach which height above ground and what this means in relation to the potential transport distance.

5.3.4
In situ Turnover Rates of Juvenile Macrofauna in Soft Sediments

A consequence of secondary dispersal of juvenile macrofauna, either by active migration or passive resuspension, implies – at a given site at a given area – a turnover of organisms. Regular sampling of benthos documents the change of abundance over time, but does not give any information whether either input or output of organisms or the relation between these factors had caused the change.

Beside benthos samples further methods have to be applied, such as emergence or sediment traps, to analyse the turnover of organisms. Armonies (1994) was the first to estimate the turnover rate by a combination of benthos and sediment trap-sampling. A combination of driftfauna nets and benthos data was not suitable for the quantitative evaluation of turnover rates, as the driftfauna catches only record the occurrence of juveniles and adults in the higher water column but not close to the bottom (Armonies, pers. com.; Jaklin unpublished data). Detailed questions on population level, e.g. concerning the dependence of the turnover rate from (1) the abundance, (2) size of the organisms or (3) abiotic conditions, could not be answered by applying the methods mentioned above.

For this reason an *in situ* staining technique was developed using neutral red as marker (Jaklin unpublished data). Staining was carried out by pouring seawater with dissolved neutral red onto fenced areas in the field. The fence was made of plastic penetrating several cm into the sediment as well as protruding several cm above the sediment surface. The stained seawater penetrated during the emergence period into the sediment and stained the biota there. Surplus of neutral red was washed away with the next flood period, after which the fence was removed. Sampling in the plots was done on a daily base, the technique is described in Sect. 5.3.3. Juveniles of the bivalves *M. balthica*, *E. americanus*, *C. edule* and *Tellina* sp. but also of the polychaetes *Nereis* spp., *E. longa* and *P. elegans* were clearly visible stained. Thus, juveniles of several abundant species could be successfully stained, but this technique has its specific limitations similar to sediment or emergence traps. Polychaetes such as *S. armiger* or *Aphelochaeta marioni* were not stained sufficiently or the red colour was not visible due to the reddish colour of the worms themselves. The same accounted for adult *Hydrobia ulvae*. According to laboratory experiments the staining did neither affect behaviour nor increased mortality of the macrofauna. In polychaetes the red colour persisted for a period of 2–4 weeks, depending on the species stained and individual variability (Jaklin, unpublished data).

First results obtained by staining experiments in 1995 implied that at the site GP3 the local stock of the tube building polychaete *P. elegans* was completely exchanged within 6 days. The turnover period of *M. balthica* at a site close to the southern tidal channel exceeded 10 days. Though the Gröninger Plate is more exposed compared to the Oddewatt (a sheltered tidal flat at the island of Sylt) the turnover rate of this species was lower as estimated by Armonies (1994) for the latter area. The reason for this observed difference can be manifold: Armonies obtained his value probably from smaller individuals, and the methods applied were different. Repeated experiments at the Gröninger Plate in late summer 1995, when the mean individual size of the bivalves was higher, suggest the turnover rate to be size dependent. Thus, turnover of an entire O-group of one species is not continuous over time; e.g. 50 % of the *M. balthica* postlarvae had disappeared from the experimental plot within 2 days, but even after 10 days stained specimen could still be observed.

5.3.5 Small-Scale Temporal Variability in Benthic Distribution Patterns: Patchiness vs. Mobility

Considering the high mobility of juvenile benthic macrofauna the question arose whether adults disperse also due to active behaviour or passive resuspension. For this purpose, local hydrodynamics, short term fluctuations in macrofaunal abundances and water column dispersal of adults were investigated simultaneously to test for external factors driving spatial variability and variations in abundance of macrofauna (Niemeyer, Michaelis, Steuwer, Kaiser, Fleßner unpublished data). In spring and autumn of 1994 and 1995 two campaigns of 14 days duration each were carried out in which the above mentioned parameters were measured on a daily basis. In addition import experiments were carried out to determine the local colonization by adult macrofauna.

The results of the benthos sampling yielded high fluctuations in abundance between days, but in most cases without significant changes (Steuwer unpublished data). Specifically high variability was observed for *Urothoe poseidonis* (median between 50 and 1500 ind m^{-2}) and *Capitella capitata* (median between 50 and 1200 ind m^{-2}), and a low daily variability for *Arenicola marina*. Applying the time series analysing method developed by Ibanez & Dauvin (1988), the analysis of the 1994 spring data gave no indication for cyclic patterns in the changes of abundance of selected macrofauna species related to the spring tide rhythm.

For some species (e.g. *Phyllodoce maculata, P. elegans, S. armiger*) spring and autumn abundances were significantly different depending on the yearly recruitment success.

The general large deviation from the mean abundance indicates that in spite of the high number of samples (9–16 parallel samples, each of 177 cm^2 area) some of the species were not sufficiently sampled and temporal signals due to discrete events were superimposed by the "noise" of spatial inhomogeneity. Subsequently, for example, predators feeding on these organisms face a high variability of their prey items in time (scale of tides) and space (scale of km^2). This could affect the food web, making estimates of energy flux which do not address the proper scales erroneous.

During the sampling periods no extreme event such as a storm occurred, so that the influence of hydrodynamics on variations in macrofauna abundance could not be analysed. Only twice a deviation from normal hydrodynamic conditions was observed, with no or only a short slack water around high tide. In these periods the abundances of several macrofauna species (*Crangon crangon*, *L. conchilega*, *U. poseidonis*, *C. capitata*) showed a short-term reduction. This reaction was observed with a delay of about 1–2 days and was possibly due to deposit-feeders responding to a short-term shortage of food related to the changed hydrodynamic conditions. Specific hydrodynamic parameters such as current velocities as well as tidal levels at slack tide and inundation periods could not be related to the occurrence of driftfauna in the water column (Steuwer, Kaiser unpublished data). But a link between the number of abundant species and the residual current was assumed to exist. In one case in May 1994 the vector of the residual current at the Gröninger Plate (GP3) pointed westwards, affecting the net transport direction of the animals in the water column.

For the import experiments carried out in 1995, azoic sediments were exposed either for one day or one week. Comparable to the results from driftfauna studies these experiments indicate that a part of the variability in macrofauna abundance is caused by mobility. Unfortunately, this part could not be quantified due to the generally low abundance of macrofauna in 1995. Plots exposed for one week were not colonized in higher abundances as plots exposed for one day. As the *in situ* staining experiments this implies a high turnover rate of benthic macrofauna at the study site.

5.3.6 *Lanice conchilega*

L. conchilega is a polychaete species with separate sexes. In its life history this species passes through two larval phases, of which the last one, the so-called aulophora larva, lives about 4–6 weeks in the water column (Kessler 1963). The hypothesis that the distribution of larvae at high tide above the Gröninger Plate determines the distribution of juveniles in the sediment was falsified by the field data. The larvae showed a random distribution pattern, whereas the distribution of juveniles was determined by the distribution of tubes of adult *L. conchilega*.

Experimental results as well as data from field studies revealed that in the Wadden Sea the existence of hard substrate, preferentially tubes of adults is a requirement for initial settlement of *L. conchilega* larvae (Heuers 1998). Thus, *L. conchilega* is one of the few dominant benthic organisms in the Wadden Sea settling on hard substrate. The preferred settlement sites are adult tubes, tube-like structures as they were used in the experiments and, to a small extent, shells (Heuers et al. 1998). As after the ice winter 1995/96 initial settlement took place in the absence of adults in the intertidal or even shallow subtidal, intraspecific settlement cues are not in operation in this species in contrast to the tube-building polychaete *Phragmatopoma californica*. Adults of this species release chemical cues which are of importance in initiating settlement behaviour and metamorphosis of the larvae (Pawlik 1986).

The distribution of juvenile *L. conchilega* was determined by a combination of settling behaviour of the larvae and the small and large scale distribution of adults. Modifications were due to varying abundances of adults and variations in local hydrodynamics (Heuers, 1998). Harvey & Bourget (1995) observed that in a dense tube lawn the current velocity has to reach a critical value upon which turbulence occurs facilitating settlement. In case the current velocity is lower than this value no turbulence will develop and the probability of larval settlement is reduced.

At which current speed turbulence develops depends on the density of tubes, their diameter and the surface roughness in combination with currents. In the natural habitat as well as under experimental conditions higher abundances of tubes had positive effects on settlement of larvae. Most likely the probability for small-scale areas with low currents rises with increasing abundances of tubes (Carey 1983, 1987). The intensity with which dense tube lawns may affect the small scale hydrodynamics was demonstrated in experiments with densely packed artificial tubes leading to a sediment rise of several cm above the surrounding sediment surface (Zühlke et al. 1998). A similar accumulation of sediment was observed in dense patches of *Polydora* sp. (Daro & Polk 1973).

From the investigations carried out at the Gröninger Plate it appeared that recruitment including initial settlement into habitats with an adult population and subsequent detachment of juvenile worms from the adult tube to build their own solitaire tube maintains the spatial distribution of this species, provided the juveniles do not disperse far from their place of initial settlement. The low numbers of juvenile *L. conchilega* in the driftfauna nets as well as in the epibenthos net suggest water column transport to be limited, though bedload transport cannot be excluded.

In the summer 1996 after the ice winter patches of 20 to 70 *L. conchilega* developed on the Gröninger Plate without hard substrate locally available. The spatial distribution of the aggregates was similar to the zonation patterns of this species in the previous years. But how they developed could not be completely explained. The temporal gap between larval supply and the appearance of juveniles in the benthos presumably indicates that recolonization of the Gröninger Plate was achieved by secondary dispersal of juveniles. Their initial settlement may have taken place on shells of bivalves, filamentous algae or in dense *Pygospio* lawns, where they have been observed to settle in 1996.

In 1996 recruitment of *L. conchilega* must have originated from regions relatively far from the Spiekeroog backbarrier system. In the ice winter 1995/96 not only the intertidal part of the population disappeared completely but also subtidal ones down to 10 m water depth in the tidal inlets and seaward off the East Frisian islands (Reichert pers. comm.).

Strong differences in growth of juvenile *L. conchilega* at the different sites on the Gröninger Plate in 1996 point to the importance of local differences in food supply, which may affect the large-scale spatial distribution of this species on the long run. High growth rates of juveniles in the central part of the Gröninger Plate imply a good food availability at this site of a former adult habitat. Food supply will be the main advantage for the aulophora larvae to settle on adults tubes, accordingly competition between adults and newly metamorphosed juveniles is supposed to be low. This is a contradictory life history pattern to that of *Arenicola marina*. Habitat changes of juvenile lugworms are interpreted as avoidance of intraspecific competition between adults and juveniles (Beukema & DeVlas 1979; Reise 1985).

Acknowledgements

This chapter is based on the studies in ELAWAT by A. Bartholomä, C. Bergfeld, E. Boysen-Ennen, M.T. Delafontaine, B.W. Flemming, G. Hertweck, I. Kröncke, M. Stamm, M. Türkay, F. Wolf, W.E. Arntz, C.-P. Günther, C. Hasemann, J. Heuers, S. Jaklin, K. Reinhard, R. Kaiser, M. Michaelis, H.D. Niemeyer, J. Steuwer. Their studies were published as cited in the text, and further unpublished data were kindly provided by Elisabeth Boysen-Ennen, Jens Heuers, Sandra Jaklin and Jürgen Steuwer. The project was funded by the German Bundesministerium für Bildung, Wissenschaft, Forschung und Technologie (BMBF) under grant number 03F0112 A and B. The responsibility for the contents of the publication rests with the author. The author would like to thank a reviewer for valuable comments to the chapter.

References

Armonies W (1992) Migratory rhythms of drifting juvenile molluscs in tidal waters of the Wadden Sea. Mar. Ecol Prog Ser 83: 197–206

Armonies W (1994) Turnover of postlarval bivalves in sediments of tidal flats in Königshafen (German Wadden Sea). Helgoländer Meeresunters 48: 291–297

Armonies W (1998) Driftendes Benthos im Wattenmeer: Spielball der Gezeitenströmungen? In: Gätje C, Reise K (Eds) Ökosystem Wattenmeer. Austausch-, Transport- und Stoffumwandlungsprozesse. Springer, Berlin Heidelberg New York, pp 473–498

Bayne BL (1964) Primary and secondary settlement in *Mytilus edulis* L. (Mollusca). J Anim Ecol 33: 513–523

Beukema JJ (1979) Biomass and species richness of the macrobenthic animals living on a tidal flat area in the Dutch Wadden Sea: effects of a severe winter. Neth J Sea Res 13: 203–223

Beukema JJ (1989) Long-term changes in macrozoobenthic abundance on the tidal flats of the western part of the Dutch Wadden Sea. Helgoländer Meeresunters 43: 405–415

Beukema JJ, de Vlas J (1979) Population parameters of the lugworm *Arenicola marina*, living on tidal flats in the Dutch Wadden Sea. Neth J Sea Res 13: 331–353

Bhaud MR, Cazaux CP (1990) Buoyancy characteristics of *Lanice conchilega* (Pallas) larvae (Terebellidae). Implications for settlement. J Exp Mar Biol Ecol 141: 31–45

Brandt G, Fleßner J, Glaser D, Kaiser R, Karow H, Münkewarf G, Niemeyer HD (1995) Dokumentation zur hydrographischen Frühjahrs-Meßkampagne 1994 der Ökosystemforschung Niedersächsisches Wattenmeer im Einzugsgebiet der Otzumer Balje. Berichte zur Ökosystemforschung – Hydrographie Nr. 8

Butman CA (1987) Larval settlement of soft-sediment invertebrates: The spatial scales of pattern explained by active habitat selection and the emerging role of hydrodynamical processes. Oceanogr Mar Biol Ann Rev 25: 113–165

Butman CA, Grassle JP, Webb CM (1988) Substrat choices made by marine larvae settling in still water and in a flume flow. Nature 333: 771–773

Carey DA (1983) Particle resuspension in the benthic boundary layer induced by flow around polychaete tubes. Can J Fish Aquat Sci 40 (Suppl. 1): 301–308

Carey DA (1987) Sedimentological effects and palaeoecological implications of the tube-building polychaete *Lanice conchilega* Pallas. Sedimentology 34: 49–66

Cuomo MC (1985) Sulphide as a larval settlement cue for *Capitella* sp. I. Biogeochemistry 1: 169–181

Daro MH, Polk P (1973) The autecology of *Polydora ciliata* along the Belgian coast. Neth J Sea Res 6: 130–140

Dubilier N (1988) H_2S - A settlement cue or a toxic substance for *Capitella* sp.I larvae? Biol. Bull 174: 30–38

Günther C-P (1990) Zur Ökologie von Muschelbrut im Wattenmeer. Dissertation, Universität Bremen

Günther C-P (1991) Settlement of *Macoma balthica* on an intertidal sandflat in the Wadden Sea. Mar Ecol Prog Ser 76: 73–79

Günther C-P (1992) Settlement and recruitment of *Mya arenaria* in the Wadden Sea. J Exp Mar Biol Ecol 159: 203–215

Harvey M, Bourget E (1995) Experimental evidence of passive accumulation of marine bivalve larvae on filamentous epibenthic structures. Limnol Oceanogr 40 (1): 94–104

Heiber W, (1988) Die Faunengemeinschaft einer großen Stromrinne des Wurster Wattengebietes (Deutsche Bucht). Dissertation, Universität Bonn

Heuers J (1998) Ansiedlung, Dispersion, Rekrutierung und Störungen als strukturierende Faktoren benthischer Gemeinschaften im Eulitoral. Dissertation, Universität Bremen

Heuers J, Jaklin S, Zühlke R, Dittmann S, Günther C-P, Hildenbrandt H, Grimm V (1998) A model on the distribution and abundance of the tube-building polychaete *Lanice conchilega* (Pallas, 1766) in the intertidal of the Wadden Sea. Verhandlungen Ges Ökologie 28: 207–215

Honkoop PJC, van der Meer J (1997) Reproductive output of *Macoma balthica* populations in relation to winter-temperature and intertidal-height mediated changes of body mass. Mar Ecol Prog Ser 149: 155–162

Ibanez F, Dauvin J-C (1988) Long-term changes (1977-1987) in a muddy fine sand *Abra alba - Melinna palmata* community from the Western English Channel: multivariate time-series analysis. Mar Ecol Prog Ser 49: 65–81

Jaklin S, Günther C-P (1996) Macrobenthic driftfauna at the Gröninger Plate. Senckenbergiana marit 26: 127–134

Kessler M (1963) Die Entwicklung von *Lanice conchilega* (Pallas) mit besonderer Berücksichtigung der Lebensweise. Helgoländer Wiss Meeresunters 8: 425–476

Obert B (1982) Bodenfauna der Watten und Strände um Borkum - Emsmündung. Jber. 1981, Forsch.-Stelle f. Insel- und Küstenschutz, Norderney 33: 139–162

Pawlik JR (1986) Chemical induction of larval settlement and metamorphosis in the reef-building tube worm *Phragmatopoma californica* (Sabellariidae: Polychaeta). Mar Biol 91: 59–68

Reise K (1985) Tidal flat ecology. An experimental approach to species interactions. Springer, Berlin Heidelberg New York

Rodriguez SR, Ojeda FP, Inestrosa NC (1993) Settlement of benthic marine invertebrates. Mar Ecol Prog Ser 97: 193–207

Smith CR, Brumsickle SJ (1989) The effects of patch size and substrate isolation on colonization modes and rates in an intertidal sediment. Limnol Oceanogr 34: 1263–1277

Zühlke R, Blome D, van Bernem KH, Dittmann S (1998) Effects of the tube-building polychaete *Lanice conchilega* (Pallas) on benthic macrofauna and nematodes in an intertidal sandflat. Senckenbergiana marit 29: 131–138

5.4
Modelling the Spatial and Temporal Distribution of *Lanice conchilega*

Volker Grimm

Abstract: A preliminary grid-based simulation model is presented designed to explore the mechanisms responsible for the large differences in density of *Lanice conchilega* in the backbarrier tidal flats of Spiekeroog. Patches with low or high density occur in areas with low or high overall velocity respectively of the near-bottom flow. In the model, the relationship between near-bottom flow, local density of tubes, and local recruitment is described phenomenologically. The results of the model lead to the testable hypothesis: observed differences in density are not due to different larval supply.

Two reasons make the spatial and temporal distribution of *Lanice conchilega* particularly suitable for spatially explicit modelling (cf. Chap. 8): firstly, in contrast with almost all other macrozoobenthic species of the Wadden Sea, *Lanice* adults are sessile, and secondly, as has been revealed by studies under ELAWAT, there is a direct relationship between local abundance and recruitment (Chap. 5.3).

Areas with low and high densities of *L. conchilega* (Chap. 3.5) are described by Hertweck (1995) from the backbarrier tidal flats of Spiekeroog Island. After ice winters during which almost all worms are killed, these differences in density are restored within two to four years (Chaps. 3.6 and 7). Here we present a preliminary grid-based model (see Chap. 8 for an introduction to grid-based modelling) which was designed at the end of the ELAWAT project to explore the mechanisms responsible for the large differences in density of *L. conchilega* observed in the field (for a more detailed description of the model, see Heuers et al. 1998).

The model describes a transect of one to three metres in length with a depth of 10 cm. This transect is divided into virtual grid "cells" (cf. Chap. 8) each 1 cm wide. The model is essentially one-dimensional, because the depth of the transect is not considered explicitly but only to obtain an idea of the carrying capacity of a cell. Each cell may contain between zero and ten tubes of adult *L. conchilega*. Time is not described continuously in the model, but proceeds in discrete one–year steps. During a time step, the number of tubes within each cell is updated twice. The first update is due to recruitment, the second one summarizes the effects of mortality throughout the whole year.

Concerning recruitment, for each grid cell a number of *maxL* (for example, 20) larvae have the chance to settle in accordance with the individual recruitment (or settlement) probability *Ps*. In the programme which implements the model, a random number is drawn from a uniform distribution for each individual larva (i.e., from the interval [0, 1]). Only if this number is smaller than *Ps* the larva settles in the grid cell. *Ps* is calculated as the sum of the spontaneous, very small probability *Psp*, and probability *Pd* which depends on the local density of the adult's tubes. *Psp* describes recruitment which occurs irrespective of the adult's tubes.

To calculate *Pd*, first the local tube density, *N*, is determined as the mean of the number of tubes in the cell and its four neighbouring cells. In this way the model takes into account the fact that settlement occurs in two stages. First, aulophora larvae of *L. conchilega* are attached to the tubes of adults and build their first tube. Sooner or later the juvenile *L. conchilega* then leaves the adult's tube and plants its own tube into the sediment. How far the juveniles move from their initial site of settlement is unknown; in the model, we assume a maximum distance of 5 cm. Thus, to calculate recruitment in a certain cell, a neighbourhood of 5 cm on either side has to be taken into account.

Now, the calculation of *Pd* depends on assumptions of how local tube density determines local recruitment. The most simple approach would be to assume a linear relationship between *Pd* and *N*, which has been tested in the model (Heuers et al. 1998). With this assumption, it was not possible to reproduce two empirical findings simultaneously: the emergence of patches with low or high density of *L. conchilega* and the short time span which is needed to establish these densities. The model thus suggests that simply the amount of hard substrate provided by the adult's tubes is not sufficient to explain the spatial and temporal distribution of *L. conchilega* observed in the field.

Therefore, a relationship between *Pd* and *N* is assumed which in phenomenological terms takes into account the effect of local tube density on near-bottom flow. It must be emphasized that flow is not modelled explicitly, i.e. a hydrodynamical model is not used. The following assumption is used:

$$Ps = Psp + Pd = Psp + (\exp(vscale\ N) - 1) / scale$$

In addition, *Pd* is bound to the interval [0, 0.5]. The parameter *vscale* describes how the overall velocity of near-bottom flow influences the exponential relation-

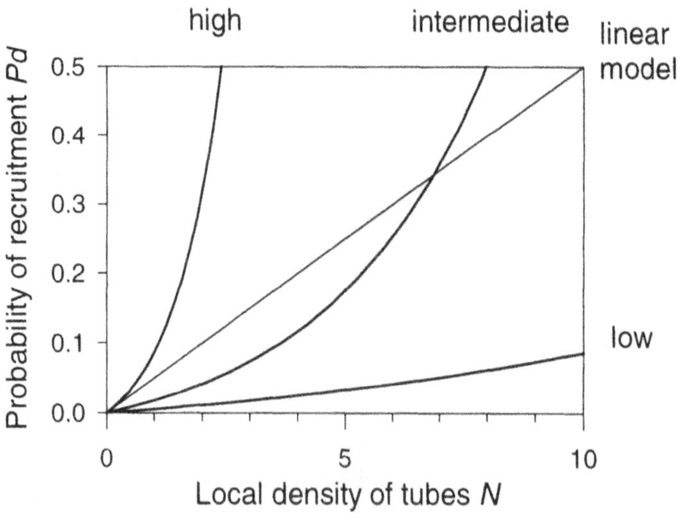

Fig. 5.4.1. Relationship between local tube density, *N*, and probability of recruitment, *Pd*, in a grid cell of the model for three different flow velocities (*vscale* = 0.1, 0.3, and 1.0). The straight lines describes the linear model of Heuers et al. (1998).

ship between density-dependent recruitment probability, *Pd*, and local density, *N* (Fig. 5.4.1). *Scale* is a scaling parameter modifying the slope of the curves in Fig. 5.4.1.

Ps is a probability, i.e. a number in the interval [0, 1]. The formula for calculating *Pd* means that for a high overall velocity of the near-bottom flow, even a relatively low local tube density leads to high recruitment, whereas for a low velocity of the near-bottom flow even high local tube densities will only allow for low recruitment (cf. Carey 1983; Eckman 1983; Butman 1986; Harvey et al. 1995; Friedrichs 1996). The idea behind this assumption is that for a given flow velocity a critical tube density is necessary to generate a flow skimming over the tube patch leaving a layer within the patch with reduced net flow velocities. Within this layer, larval settlement may be facilitated or larvae may actively select for such flow conditions. It is assumed that the critical tube density decreases with increasing overall flow velocity.

Local density within a grid cell is limited by a local carrying capacity (for the results presented here, ten tubes). If after calculating recruitment more than ten tubes stand in a grid cell, the number of tubes is reduced to ten. After calculating

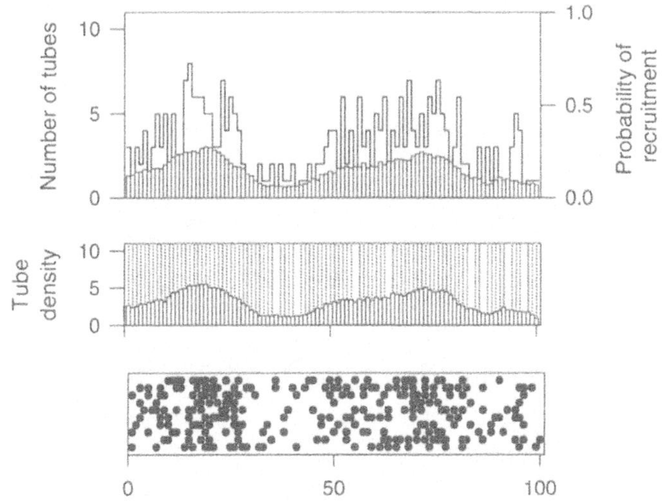

Fig. 5.4.2. Three different ways of visualizing the spatial distribution of tubes as produced by the model. In all three cases, the x-axis denotes location on the transect (in cm) or the number of the grid cell respectively. Upper panel: Actual number of tubes (line plot) and probability of recruitment, *Pd*, calculated from the distribution of tubes (bars). Middle panel: Local density of tubes (bars; here, the mean is only taken over two neighbouring cells) and vertical lines indicating whether at least one tube is present in a grid cell. Lower panel: The model does not explicitly consider the location of tubes within a grid cell. Here, to produce a top view of the distribution, each tube has been randomly assigned to a certain location within a cell.

recruitment and, if necessary, density dependence, mortality is taken into account. Each individual (including those newly established) has the probability *Pmort* (= 0.5) of dying.

In the following, the results of simulating a one-metre-long transect (i.e., a hundred grid cells) are presented. Simulations start with five randomly distributed tubes. Fig. 5.4.2 shows three ways of visualizing the distribution of the tubes. In contrast with the linear model mentioned above, this nonlinear, flow-dependent model enables the production of populations with low or high tube density within a few years. These differences in density persist for ten years or more (Fig. 5.4.3). The parameter generating these long-lasting density differences is *vscale*, which describes the effect of the near-bottom flow on the relationship between local density and recruitment.

To further understand the relationship between near-bottom flow, local density, larval supply and recruitment, more detailed field records are required from differ-

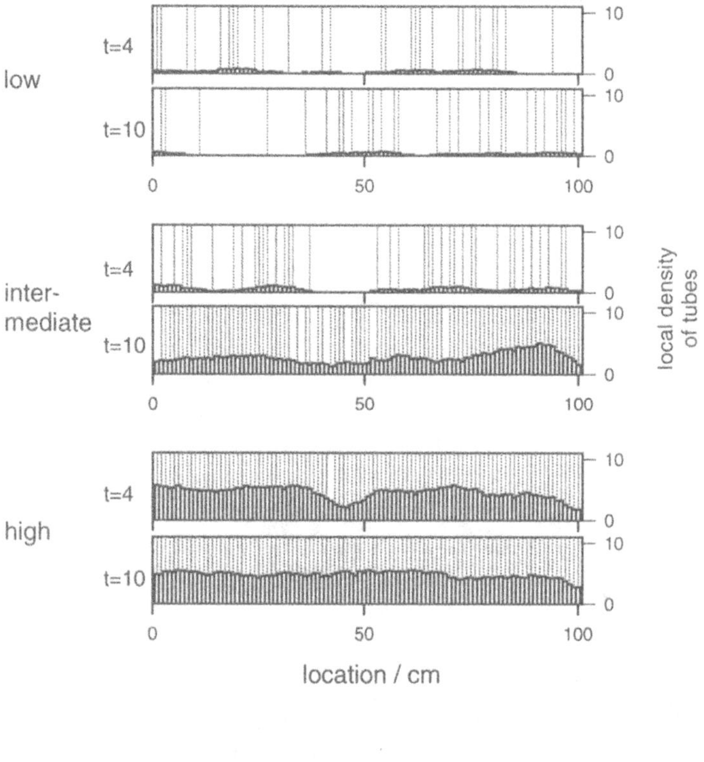

Fig. 5.4.3. Spatial distribution as produced by the model (see Fig. 5.4.2) for three different flow velocities (low, intermediate, high) in the fourth and tenth year after start of the simulation (Parameters: *vscale* = 0.1, 0.3, and 1.0; *maxL* = 20, *Psp* = 0.005, *scale* = 20, *Pmort* = 0.5). Note that three different transects in the three panels are meant to represent different parts of a sandflat with differing mean flow velocities.

ent areas with different densities of *L. conchilega*. These can help to validate the model assumption that the *local density of* L. conchilega *is strongly influenced by the overall velocity of the near-bottom flow: low density occurs in areas where low velocities prevail, whereas high densities occur where high velocities prevail.* Measurements of flow velocity, which were performed in June 1994 on the Gröninger Plate, indicated a mean velocity of about 10 cm/s in an area with a low density stand of *L. conchilega*, and 20 cm/s near high density patches (Brandt et al. 1995). These results tentatively confirm the model assumption. For a real test, however, much more detailed records of the flow regime are required. It must be emphasized that in our phenomenological approach "low velocity" is a metaphor for one or more aspects of the near-bottom flow which are not yet known (cf. Friedrichs 1996).

We do not claim that the model provides a detailed reflection of reality, as it merely takes into account those factors which are thought to be essential for the spatial and temporal distribution of *L. conchilega*. The model assumptions are extremely simple because more detailed information is unavailable (but see Friedrichs 1996) and because unnecessary detail would make it more difficult to explore and understand what the model does. However, this all means that the model cannot be tested directly. Instead, it allows testable hypotheses to be formulated. The model gave us an idea of what might be essential in reality, and suggests the following hypothesis:

The regional distribution pattern of L. conchilega *is not caused by local differences in larval supply*. This hypothesis goes back to the linear model (Heuers et al. 1998), which could not reproduce the spatial and temporal dynamics recorded by Hertweck (1995) by solely varying larval supply (*maxL*). The records of larval abundance and phenology taken in the ELAWAT project give no indications that larval supply on different parts of the Gröninger Plate were different. To further test this hypothesis, driftnets should be located in areas with low or high tube density respectively.

The model can also be used to formulate a verbal model which explains the striking spatial pattern of mounds and depressions emerging in some parts of the sandflats. The mounds are densely populated with tubes of *L. conchilega*, whereas the depressions, which are filled with water even at low tide, are free of tubes. The typical span of the mounds and depressions is one to three metres, i.e. larger than the mounds-and-depressions pattern which also emerges in some parts of the sandflats, but without being populated by *L. conchilega* (Heuers et al. 1998). The verbal model states that dense aggregations of tubes deflect the near-bottom flow. This may lead to such high lateral flow velocities that a further lateral expansion of the patches as predicted by the model is prevented. Similar phenomena have been reported for macroalgae in rivers (Sand-Jensen & Mebus 1996).

Although models cannot supplant field studies and experiments, they help to integrate the knowledge available and to identify important gaps in this knowledge. The results presented here merely represent some initial, preliminary runs of the model. Nevertheless, we believe that models of the kind presented here are an ideal tool to assist the exploration of spatial and temporal dynamics of *L. conchilega* and other benthic species.

Acknowledgements

The model presented in this section was developed together with C.-P. Günther, J. Heuers, S. Jaklin, R. Zühlke, S. Dittmann and H. Hildenbrandt. I thank the reviewer for helpful comments on the manuscript. The project was funded by the German Bundesministerium für Bildung, Wissenschaft, Forschung und Technologie (BMBF) under grant number 03F0112A and B. The responsibility for the contents of the publication rests with the author.

References

Brandt G, Fleßner J, Glaser D, Kaiser R, Karow H, Münkewarf G, Niemeyer HD (1995) Dokumentation zur hydrographischen Frühjahrs-Meßkampagne 1994 der Ökosystemforschung Niedersächsisches Wattenmeer im Einzugsgebiet der Otzumer Balje. Berichte zur Ökosystemforschung - Hydrographie Nr. 8, Nieders. Landesamt für Ökologie - Forschungsstelle Küste, Norderney

Butman CA (1986) Larval settlement of soft-sediment invertebrates: Some predictions based on an analysis of near-bottom velocity profiles. In: Nihoul JCJ (Ed) Marine Interfaces Ecohydrodynamics. Elsevier, pp 487–513

Carey DA (1983) Particle resuspension in the benthic boundary layer induced by flow around polychaete tubes. Can J Fish Aquat Sci, 40 (Suppl. 1): 301–308

Eckman JE (1983) Hydrodynamic processes affecting benthic recruitment. Limnol Oceanogr 28: 241–257

Friedrichs M (1996) Auswirkungen von Polychaetenröhren auf die Wasser-Sediment-Grenzschicht. Diplomarbeit, Universität Kiel

Harvey M, Bourget E, Ingram RG (1995) Experimental evidence of passive accumulation of marine bivalve larvae on filamentous epibenthic structures. Limnol Oceanogr 40: 94–104

Hertweck G (1995) Verteilung charakteristischer Sedimentkörper und Benthossiedlungen im Rückseitenwatt der Insel Spiekeroog, südliche Nordsee. I. Ergebnisse der Wattkartierung 1988-92. Senckenbergiana marit 26: 81–94

Heuers J, Jaklin S, Zühlke R, Dittmann S, Günther C-P, Hildenbrandt H, Grimm V (1998) A model on the distribution and abundance of the tube-building polychaete *Lanice conchilega* (Pallas, 1766) in the intertidal of the Wadden Sea. Verhandlungen der Gesellschaft für Ökologie 28: 207–215

Sand-Jensen K, Mebus JR (1996) Fine-scale patterns of water velocity within macrophyte patches in streams. Oikos 76: 169–180

5.5 Biotic Interactions in a *Lanice conchilega*-Dominated Tidal Flat

Sabine Dittmann

Abstract: The tube-dwelling polychaete *Lanice conchilega* provides a biogenic structure in the sandflats of the Wadden Sea, which has direct or indirect effects on other benthic organisms. In order to separate the effect of the tube-worm from mere effects of the tubes, experiments were carried out using artificial tubes. Both patches with living *L. conchilega* as well as patches with artifial tubes showed increased abundances and species numbers of benthic organisms compared to ambient sediment without this biogenic structure. In both cases (natural and artifical tubes) an increased settlement of juvenile mussels occurred on the tubes. Predatory polychaetes were also more numerous in both types of tube fields. Spionid polychaetes were more numerous in patches with a high density of *L. conchilega*. However, the only true commensalistic association found was between *L. conchilega* and the polychaete *Malmgreniella lunulata,* living in the worm tubes. The density and distribution of *L. conchilega* varied throughout the study period. The tube-worms formed localized patches of 1–2 m in size with a high density of *L. conchilega*. Over time, the individual numbers declined following the severe ice winter of 1995/96.

5.5.1 Introduction

Biogenic structures of polychaete tubes can affect the distribution pattern of benthic organisms by modifying the sediment (Eckman et al. 1981; Carey 1987) and hydrographic parameters (Eckman 1983; Friedrichs 1996), or by interactions between species (Woodin 1978). The interaction between the tube-dwelling polychaete *Lanice conchilega* (in the following text referred to as *Lanice*) and the benthos were studied by Zühlke et al. (1998) in a sandflat of the East Frisian Wadden Sea, the Gröninger Plate (Chaps. 3.1 and 3.6.2). To investigate the interactions with other fauna, the benthic associations of sandflats with different *Lanice* densities were analysed. Therefore, additional samples were taken on the Dornumer Nacken, a backbarrier tidal flat behind the island of Langeoog, where *Lanice* attained densities twice as high as on the Gröninger Plate. Furthermore, experiments were set up with artificial tubes to study the effect of the tube structure as such (Zühlke et al. 1998). The *Lanice*-community was studied over the entire duration of ELAWAT, allowing an assessment of natural variations in the density of this polychaete and its effects on associated fauna. The investigations, carried out by a number of teams in ELAWAT, are summarized in this chapter.

5.5.2 Methods

The benthic fauna in the sandflats and especially in *Lanice*-patches was studied with several approaches in ELAWAT. Each study had a specific question which required respective methods. These are briefly outlined below with respect to their relevance for this chapter. The main study areas were the Gröninger Plate and Dornumer Nacken in the backbarrier tidal flats of the islands of Spiekeroog and Langeoog (see Chap. 3.1). Especially two sampling sites at the Gröninger Plate with different *Lanice* densities (Table 5.5.2) are often referred to in the text (GP1 and GP2, see Fig. 3.1.4).

1. Spatial distribution: Macrobenthos was sampled with a corer of 100 cm^2 surface area and penetration depth of 20 cm. To assess small-scale distributions, a 5x5 multicorer of 25 corers with 100 cm^2 surface area each was used in addition. The samples were sieved through 1 mm mesh. To assess the smaller sized infauna, a corer of 32 cm^2 surface area was used to 10 cm sediment depth. These samples were sieved through 0.5 mm mesh. The fauna was sorted from unpreserved samples, determined to the lowest possible taxonomic unit and counted. See Zühlke et al. (1998) for details. Spatial patterns of harpacticoid copepods were assessed with 50 corers of 2x2 cm^2 each arranged to a transect of 1 m length (Sach & van Bernem 1996). Nematodes in the vicinity of *Lanice*-tubes were sampled with corers of 0.2 cm^2 surface area to a depth of 5 cm. The corers were inserted in a row up to 2 cm distance from the tubes and additional samples were taken in 5 and 10 cm distance (Zühlke et al. 1998).
2. Short-term variability: Macrobenthos samples were taken daily during sampling campaigns on the Gröninger Plate each spring and autumn from 1994 to 1996 (Steuwer, unpubl.). The benthos was sampled with a corer of 177 cm^2 surface area to 30 cm sediment depth and sieved through 0.5 mm mesh. The samples were treated alive. In addition, driftnets (after Armonies 1992) were installed in the vicinity of the grid to assess dispersal of benthic fauna. These nets (opening diameter 20 cm, 0.5 mm mesh) were set up 20 cm above the sediment surface and could rotate with the tide.
3. Experiments: To test for the effect of the tube structures as such, worm tubes were experimentally simulated by inserting metal sticks of about the same size as real *Lanice*-tubes into the sediment (Zühlke et al. 1998). In six experimental treatments, 1000 of these metal sticks (4 mm wide, 25 cm long) were implanted in each of the 1 m^2 sized plots, protruding 2–3 cm above the sediment surface. In three additional treatments, the same number of metal sticks was implanted, but level with the sediment surface. These plots served as controls for a possible effect of the metal on benthic organisms. Seven reference plots of 1 m^2 size were left totally untreated. The sites for the three treatments were arranged at random in the field.

5.5.3
Species Composition and Macrobenthos Abundances

58 macrobenthic species were recorded during these studies in the tidal flats of the Gröninger Plate and the Dornumer Nacken (Table 5.5.1). Polychaetes contributed the majority of the species (52 %), followed by crustacea (22 %) and molluscs (19 %). The average species density at site GP1 was 4 species per 100 cm^2 sediment. Diversity (gamma-index [0,3]; Pfeifer et al. 1995) varied between 2.64 and 6.7, whereby the lowest values were recorded after the ice winter (Steuwer, un-

Table 5.5.1 Species list of macrobenthic species recorded on the studies sites of the Gröninger Plate and Dornumer Nacken, East Frisian Wadden Sea, in the years 1992–1996 (Steuwer, van Bernem et al. unpublished data)

Cnidaria	
Anthozoa	
Actinothoe anguicoma	(Price)
Mollusca	
Gastropoda	
Hydrobia ulvae	(Pennant)
Bivalvia	
Mytilus edulis	L.
Mysella bidentata	(Montagu)
Cerastoderma edule	(L.)
Petricola pholadiformis	(Lamarck)
Macoma balthica	(L.)
Angulus tenuis	(Da Costa)
Scrobicularia plana	(Da Costa)
Abra alba	(Wood)
Ensis americanus	(Binney)
Mya arenaria	L.
Annelida	
Polychaeta	
Malmgreniella lunulata	(Delle Chiaje)
Harmothoe sarsi	(Kinberg)
Pholoe minuta	(Fabricius)
Eteone longa	(Fabricius)
Phyllodoce maculata	(L.)
Phyllodoce mucosa	(Oersted)
Eumida sanguinea	(Oersted)
Eumida bahusiensis	Bergström
Pirakia punktifera	(Grube)
Microphthalmus sczelkowii	Mecznikow
Microphthalmus aberrans	(Webster & Benedict)
Autolytus prolifer	(O.F.Müller)
Nereis diversicolor	(O.F.Müller)
Nereis virens	Sars
Nereis longissima	Johnston
Nephtys hombergii	Savigny
Nephtys incisa	Malmgren
Scoloplos armiger	(O.F.Müller)
Spio filicornis	(O.F.Müller)
Polydora cornuta	Bosc
Pygospio elegans	Claparede
Spiophanes bombyx	(Claparede)
Malacoceros tetraceros	(Schmarda)
Marenzelleria viridis	(Verril)
Magelona mirabilis	(Johnston)
Aphelochaeta marioni	(De Saint-Joseph)
Capitella capitata	(Fabricius)
Heteromastus filiformis	(Claparede)
Arenicola marina	(L.)
Lanice conchilega	(Pallas)
Oligochaeta	
Tubificoides benedeni	(Udekem)
Tubificoides pseudogaster	(Dahl)
Crustacea	
Cumacea	
Bodotria arenosa	(Goodsir)
Iphinoe trispinosa	(Goodsir)
Cumopsis goodsiri	(Van Beneden)
Isopoda	
Idotea sp.	
Amphipoda	
Gammarus locusta	(L.)
Gammarus salinus	Spooner
Urothoe poseidonis	Reibisch
Bathyporeia pilosa	Lindström
Microprotopus maculatus	Norman
Corophium volutator	(Pallas)
Corophium arenarium	Crawford
Decapoda	
Crangon crangon	(L.)
Carcinus maenas	(L.)
Echinodermata	
Asteroida	
Asterias rubens	L.

publ.). Individual densities of macrobenthos varied in the three study years between 300–53 500 individuals per m^2 and the lowest values were also recorded immediately after the ice winter. The highest values occurred in summer 1996, due to massive recruitments of several polychaete and bivalve species (Chaps. 5.3 and 7). Macrobenthos abundances varied not only between years and seasons, but also daily. Especially abundant species with patchy distribution patterns (e.g. the amphipod *Urothoe poseidonis* and the polychaete *Capitella capitata*) varied in numbers between days. Yet, in most cases the variance was higher between subsamples taken on the same day than between samples from consecutive days. The high variability of benthos abundances was not related to the occurrence of organisms of the respective species in the driftfauna.

A classification of trophic groups showed that their shares varied over the years in relation to variations in the abundances of single species and their respective recruitment (Chap. 5.3).

5.5.4 *Lanice conchilega*: Density and Distribution

Lanice occurred at different population densities at the study sites of the Gröninger Plate and Dornumer Nacken and also showed small-scale density variations at each of the sites (Table 5.5.2). The density of *Lanice* also varied between years, with a decline both inter- and subtidally (Reichert pers. comm.) after the ice winter of 1995/96.

Lanice had a small-scale distribution pattern with patches of 1–2 m in size, alternating with ambient tidal flat areas, where it was absent or occurred with lower densities. Multicorer analyses revealed that an increase in density was combined with a higher small-scale aggregation. In the studied backbarrier system, *Lanice* occurred mainly in tidal flats of intermediate or higher elevation. *Lanice*-dominated tidal flats often had a small-scale topography with low mounds around the *Lanice*-patches and slight depressions in between (van Bernem et al. unpubl.).

Size classes of *Lanice* were assessed by measuring the diameter of the thoracic segments (Reichert pers. comm.). Adult *Lanice* had an average thorax diameter of 0.3 cm. At the end of the year 1995, over 50 % of the *Lanice*-population at the Dornumer Nacken were adults with a thorax diameter of 0.3–0.45 cm. After the ice winter 1995/96, no living *Lanice* were recorded in the study area until June 1996. In July, juvenile *Lanice* with a thorax diameter of 0.05–0.2 cm appeared again on the tidal flats, but only a few individuals had grown to the size of 0.3 cm by the end of the year.

The tubes of *Lanice* are a biogenic structure reaching about 20 cm into the sediment and protruding 2–3 cm above the sediment. Dense aggregations of polychaete tubes can affect the currents at the benthic boundary layer (Friedrichs 1996) and thus alter small-scale habitat conditions. Thus, the topography of mounds and depressions described above develops. After the ice winter 1995/96 and the recolonization by *Lanice* in the following summer, this topography was not pronounced, as the juvenile worms were obviously too small to exert the same

Table 5.5.2 Individual densities (ind. m^{-2}, minimal, maximal and mean values given) of *Lanice conchilega* in the studied tidal flats of the backbarrier system of Spiekeroog and adjacent tidal flats. The small-scale densities refer to values within *Lanice*-patches (van Bernem et al., unpublished data)

site	min	max	mean
Large scale			
Gröninger Plate: GP 3			
1992			162
1994	50	156	103
Gröninger Plate: GP 1			
1994	0	1036	295
1995	312	930	671
Small scale			
Gröninger Plate: GP 1	630	1790	1303
Dornumer Nacken	2480	4500	3356

effect on the sediment as the adults. Also, the patches with a higher density of the tube-worms were of much smaller diameter (20–30 cm in size).

Furthermore, *Lanice* affects the density of the oxic sediment layer. The thickness of the oxidized sediment layer varied between 0.3–2.5 cm over summer 1996 in *Lanice*-dominated tidal flats, with lowest values in depressions without *Lanice* and highest values in sites with artificial tubes (see below). The shear-stress was measured once and showed no difference between patches with or without *Lanice* nor artificial tube fields (1.1–1.5 g cm^{-2}).

5.5.5 Benthos Associated with *Lanice conchilega*

Next to the lugworm *Arenicola marina, Lanice* was the dominant macrobenthic species known to affect the benthos in the tidal flats through its biogenic structure (Reise 1985). Thus, the interactions between this prominent macrobenthic species and microphytobenthos, meiofauna and macrofauna were studied in detail in ELAWAT (Zühlke et al. 1998). Sampling sites with a high density of *Lanice* also had the highest densities and species numbers of macrobenthos compared to sites with a lower density or no *Lanice*. In May 1995, infaunal densities (653 ind. m^{-2}) recorded in a *Lanice*-patch at site GP1 were twice as high as at the other sampling sites with less or no *Lanice*. Multivariate analyses showed a distinct association in this *Lanice*-patch compared with the other sites.

In the following, the reaction of the benthos to *Lanice*, its biogenic structure and the small-scale topography are distinguished.

Reaction of benthos to Lanice

The infauna associated with *Lanice*-patches occurred with higher individual abundances and higher species numbers than infauna in ambient sites without *Lanice*

(Zühlke et al. 1998). At both study areas, the Gröninger Plate and the Dornumer Nacken, benthos abundances were significantly enhanced within the *Lanice*-patches at all sampling dates (Fig. 5.5.1). Juvenile bivalves (*Mya arenaria, Mytilus edulis, Macoma balthica*) were more frequent in patches with *Lanice* than without and settled especially on the tentacle crown of the worm tubes. Polychaetes, which made up the majority of the benthos, occurred with more species and individuals in *Lanice*-patches. In particular, abundances of predatory polychaetes (*Eteone longa, Nephthys hombergii, Nereis diversicolor*) were higher. Independent of the *Lanice* density, the commensal polychaete *Malmgreniella lunulata* was more abundant with than without *Lanice*. Higher numbers of spionids (especially *Pygospio elegans*) were only recorded in the *Lanice* site at the Dornumer Nacken. Seasonal and annual variations of polychaete abundances were, however, common (see above). Multidimensional scaling (MDS) of the benthos associations revealed distinct faunal groups at the sampling sites in each year. Samples taken in February 1996, straight after the ice winter, had a very different benthos composition from all other dates in the MDS. In the course of the recolonization of *Lanice* in the summer of 1996, a benthos association comparable to the previous years developed. Only the commensal polychaete *M. lunulata* and predatory polychaetes did not yet achieve densities as before. Contrary to the previous years, abundances of *Polydora cornuta* and *Phyllodoce mucosa* were enhanced in the *Lanice*-patches.

The meiofauna distribution was also affected by *Lanice*. Studies on the small-scale distribution of copepods showed that two species (*Harpacticus obscurus* and *Halectinosoma gothiceps*) were restricted in their occurrence to the *Lanice*-patches (Sach & van Bernem 1996). Modelling their small-scale distribution with a modified Thomas-process (a two-phase stochastic point process relating spatial distributions of so-called parent and daughter points, Pfeifer et al. 1996) confirmed that these species aggregate around the tubes of *Lanice* and only few individuals occur in the space between the worm tubes. A cluster analysis indicated a greater similarity of the copepod fauna among sites with *Lanice* than among sites without. Copepods could use the biogenic structure provided by *Lanice* as a shelter from high current velocities. Their aggregation around the tubes could, however, also be related to their reproductive cycle, as many of the copepods found near the worm tubes were in pre-copula (Sach & van Bernem 1996).

Nematodes were three times as abundant at the worm tubes than at 1.5–5 cm distance (Zühlke et al. 1998). The majority of the nematodes close to the *Lanice* tubes were DOM (dissolved organic matter)-feeders, following the classification of trophic groups given in Giere (1993). At 2 cm distance from the worm tubes, diatom-feeders dominated amongst the nematodes. A relation between small-scale distributions of food sources and nematodes was also confirmed with multicorer data, where nematodes of the same trophic group had similar small-scale distributions (Blome et al. 1999). In total, 72 nematode species of 25 families were recorded in *Lanice*-patches out of 235 nematode species known from the East Frisian Wadden Sea and adjacent estuaries (Blome 1996).

The colony-forming benthic diatom *Amphipleura rutilans* was more abundant within the *Lanice* patches. This species was absent in the spring of 1996 after the ice winter (Ramm pers. comm.).

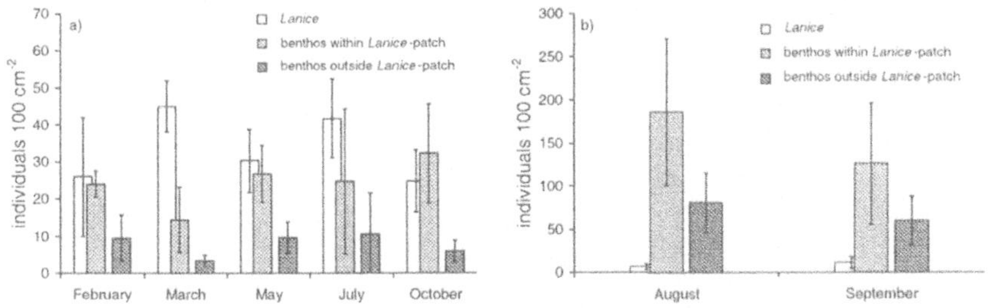

Fig. 5.5.1 a, b Comparison of benthos abundances (ind. $100cm^{-2}$, mean values with standard deviation, n=2-10) in 1995 at sites with and without *Lanice conchilega*. a) Dornumer Nacken, b) Gröninger Plate. The samples taken at the Dornumer Nacken were sieved through 1 mm mesh size, those from the Gröninger Plate through 0.5 mm (van Bernem et al., unpublished data and data from Zühlke et al. 1998)

Reaction of benthos to the tube structure

In the course of the experiment, about 1–3 cm of sediment accumulated around the artificial tubes in the experimental treatment, just like in natural *Lanice*-patches. Samples taken 1–2 months after the set-up of the experiment yielded increased species- and individual numbers in the experimental treatment compared to control and reference sites (Fig. 5.5.2). There was no significant difference in the benthos of control and reference sites. Juvenile bivalves (*M. arenaria, M. edulis*) settled on the artificial tubes. Abundances of spionid polychaetes (*P. cornuta, P. elegans*), *N. diversicolor* and *Capitella capitata* were significantly higher in the experimental treatments. Multivariate data analysis showed distinct assemblages within and outside of the artificial tube fields. In fact, the benthos association around the artificial tubes could not be distinguished by multivariate techniques from those in natural *Lanice*-patches sampled at the same time (Zühlke et al. 1998). These results show the relevance of the habitat heterogeneity provided by the worm tubes for the benthos distribution. Thus the benthos association in *Lanice*-patches is mainly an effect of the biogenic structure and not of direct species interactions between *Lanice* and associated fauna. The only exception is the commensal *M. lunulata*, which was missing in the artifical tube fields.

Reaction of benthos to the small-scale topography

Tidal flats occupied by *Lanice* often have a pronounced surface structure of shallow mounds and depressions. The effect of this small-scale topography on the benthos, with and without *Lanice*, could be analysed after the ice winter 1995/96, when the local population of *Lanice* had declined (Chap. 7) and a ripple structure resembling the mounds and depressions had developed. In June 1996, when *Lanice* was still absent from the tidal flats, no differences were recorded in the benthos abundances between mounds and depressions. Two months later, *Lanice* started to recolonize the area and was found in both the hydrodynamically induced mounds and depressions. Twice as many benthic organisms were found in mounds with *Lanice* compared to those without or to depressions with and without *Lanice*.

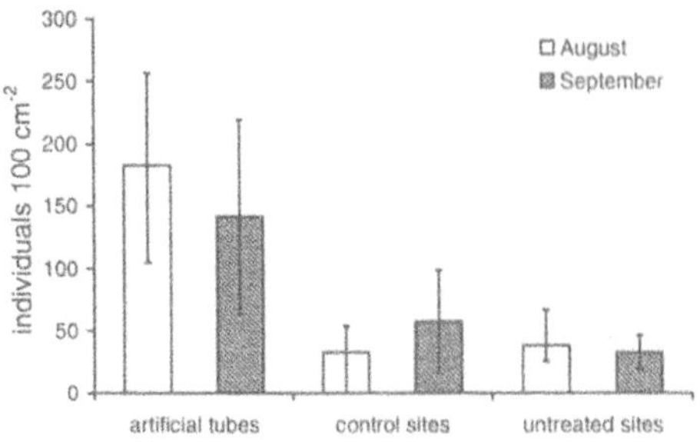

Fig. 5.5.2 Abundances (ind. 100 cm⁻², mean values with standard deviation) of macrobenthos in the experimental treatment (artificial tubes protruding above the sediment surface, n=6), control sites (artificial tubes implanted level with the sediment surface, n=3), and reference sites that were left untreated (n=3). The experiment was set up in July 1995 and samples were taken on 22. August and 12. September 1995 (data from Zühlke et al. 1998)

Mainly juvenile polychaetes (*P. elegans*) and bivalves (*M. arenaria, M. edulis*) had settled in the mounds with *Lanice*. Classifying the infauna into trophic groups showed that the densities of deposit-feeders and predators were higher. Four of the polychaetes (*Eteone longa, N. diversicolor, C. capitata* and *Heteromastus filiformis*) occurred in higher abundance in mounds regardless of the presence of *Lanice*. The small-scale distribution of benthic organisms is thus a complicated result of reactions to biogenic structures of worm tubes and the habitat heterogeneity provided by small-scale topography.

5.5.6 Discussion

The benthos in tidal flats characterized by dense aggregations of *Lanice conchilega* has a temporary and optional association with the biogenic structure of this tube-building polychaete. The habitat provided by *Lanice* is temporary in itself, as *Lanice* is a cold-sensitive species, whose intertidal part of the population dies in ice winters (Chap. 7). The sediment accumulation around the worm tubes, which caused the small-scale topography of mounds and depressions, developed also around empty *Lanice* tubes and artifical tubes.

The increased species- and individual numbers recorded in *Lanice*-patches and patches with artificial tubes are also known to occur around the tubes of other species of polychaetes (Woodin 1978). In contrast to previous studies, relative abundances of some species differed between natural and artificial tubes in the study by Zühlke et al. (1998). Juvenile bivalves had attached to both types of tubes with their byssus threads, but the abundances of the species differed in each case. Also, opportunistic polychaetes (spionids, *C. capitata*) were more abundant in artificial tube fields, showing the typical response of early colonizing species (cf. Chap. 6). And the commensal polychaete *Malmgreniella lunulata* (Hartmann-Schröder 1996) was absent without *Lanice* tubes. Beukema (1979) also recorded a decline of *M. lunulata* after the die-off of *Lanice* following an extremely cold winter in the Dutch Wadden Sea.

The empty *Lanice* tubes still present after the ice winter 1995/96 had no effect on the benthos. But six weeks after the recolonization of *Lanice* in the tidal flats in summer 1996, increased benthos abundances were recorded, showing a high elasticity of this distribution pattern. The colony-forming diatom *Amphipleura rutilans* and green algae were frequent that summer on sites where *Lanice* had settled and on the artificial tube fields. Both probably facilitated the settlement of juvenile bivalves and polychaetes (Dauer et al. 1982).

The benthic colonization of *Lanice*-patches can be active (Armonies 1994) or passive (Eckman 1979). A passive accumulation was confirmed by the experiments using artificial tubes. The effect of small-scale topography as such on benthos distributions was shown by the higher benthos abundances in mounds with or without *Lanice* compared to depressions. The biogenic structure of the worm tubes can reduce current velocities near the sediment surface and thus create favourable hydrodynamic conditions for the settlement of larvae as well as dispersing macro- and meiobenthic organisms (see also Chap. 5.3 and 5.4). Positive correlations between the structural complexity of worm tubes and meiofauna distributions were found by Bell (1985) and Sach & van Bernem (1996). Mangrove-derived tannins (Alongi 1987) and other natural bromophenol sources (Steward et al. 1992) can have inhibitory effects on associated benthic nematodes. No such effect was demonstrated for bromophenols potentially leaching from *Lanice* tubes (Goerke & Weber 1990).

The higher densities of infaunal predators in *Lanice*-patches could be a response to the higher meio- and macrobenthic densities at these sites (Zühlke et al. 1998). While previous studies discussed the increased benthos abundances in natural and artificial tube fields as a result of a reduced rate of epibenthic predation or an increase of shelter (Woodin 1978; Dauer et al. 1982), the studies in ELAWAT indicate more complex mechanisms. The association of benthic organisms to polychaete tubes also varies over time (Bell & Woodin 1984). The interactions and trophic relations between *Lanice* and its associated fauna should thus be studied further.

Acknowledgements

This chapter is based on the studies in ELAWAT by Hans-Peter Bäumer, Carlo van Bernem, Dietrich Blome, Michael Grotjahn, Hermann Michaelis, Roswitha Keuker-Rüdiger, Heidrun Ortleb, Dietmar Pfeifer, Georg Ramm, Jörn Reichert, Günther Sach, Ulrike Schleier, Jürgen Steuwer and Ruth Zühlke. Their studies were published as cited in the text, and further unpublished data were kindly provided by Dietrich Blome, Jörn Reichert, Jürgen Steuwer and Ruth Zühlke. I thank the reviewer for helpful comments on the manuscript. The project was funded by the German Bundesministerium für Bildung, Wissenschaft, Forschung und Technologie (BMBF) under grant number 03F0112 A and B. The responsibility for the contents of the publication rests with the author.

References

Alongi DM (1987) The influence of mangrove-derived tannins on intertidal meiobenthos in tropical estuaries. Oecologia 71: 537–540

Armonies W (1992) Migratory rhythms of drifting juvenile molluscs in tidal waters of the Wadden Sea. Mar Ecol Prog Ser 83: 197–206

Armonies W (1994) Drifting meio- and macrobenthic invertebrates on tidal flats in Königshafen: a review. Helgoländer Meeresunters 48: 299–320

Bell SS (1985) Habitat complexity of polychaete tube-caps: Influence of architecture on dynamics of a meioepibenthic assemblage. J Mar Res 43: 647–671

Bell SS, Woodin S (1984) Community unity: experimental evidence for meiofauna and macrofauna. J Mar Res 42: 605–632

Beukema JJ (1979) Biomass and species richness of the macrobenthos animals living on a tidal flat area in the Dutch Wadden Sea: effects of a severe winter. Neth J Sea Res 13: 203–223

Blome D (1996) An inventory of the free-living marine nematodes of the East Frisian Wadden Sea including the estuaries of the rivers Ems, Jade and Weser. Senckenbergiana marit 26: 107–115

Blome D, Schleier U, van Bernem K-H (1999) Analysis of the small-scale spatial pattern of free-living marine nematodes from tidal flats in the East Frisian Wadden Sea. Marine Biology 133: 717–726

Carey DA (1987) Sedimentological effects and palaeoecological implications of the tube-building polychaete *Lanice conchilega* (Pallas). Sedimentology 34: 49–66

Dauer DM, Tourtelotte GH, Ewing RM (1982) Oyster shells and artificial worm tubes: The role of refuges in structuring benthic communities of the Lower Chesapeake Bay. Int Rev ges Hydrobiol 67(5): 661–677

Eckman JE (1979) Small-scale patterns and processes in a soft-substratum, intertidal community. J Mar Res 37: 437–457

Eckman JE (1983) Hydrodynamic processes affecting benthic recruitment. Limnol Oceanogr 28(2): 241–257

Eckman JE, Nowell ARM, Jumars PA (1981) Sediment destabilisation by animal tubes. J Mar Res 39: 361–374

Friedrichs M (1996) Auswirkungen von Polychaetenröhren auf die Wasser-Sediment-Grenzschicht. Diplomarbeit Universität Kiel

Giere O (1993) Meiobenthology. The Microscopic Fauna in Aquatic Sediments. Springer, Berlin Heidelberg New York

Goerke H, Weber K (1990) Locality-dependent concentrations of bromophenols in *Lanice conchilega* (Polychaeta: Terebellidae). Comp Biochem Physiol 97B: 741–744

Hartmann-Schröder G (1996) Die Tierwelt Deutschlands, Polychaeta. Gustav Fischer Verlag, Jena

Pfeifer D, Bäumer H-P, Schleier U, de Valk V (1995) Recommendations on the use of statistics in benthos monitoring. Texte zur Statistischen Ökologie. Institut für Mathematische Stochastik der Universität Hamburg, Bericht Nr. 4/95

Pfeifer D, Bäumer H-P, Ortleb H, Sach G, Schleier U (1996) Modeling spatial distributional patterns of benthic meiofauna species by Thomas and related processes. Ecol Model 87: 285–294

Reise K (1985) Tidal Flat Ecology. Springer, Berlin Heidelberg New York

Sach G, van Bernem K-H (1996) Spatial patterns of harpacticoid copepods on tidal flats. Senckenbergiana marit 26: 97–106

Steward CC, Pinckney J, Piceno Y, Lovell CR (1992) Bacterial numbers and activity, microalgal biomass and productivity, and meiofauna distribution in sediments naturally contaminated with biogenic bromophenols. Mar Ecol Prog Ser 90: 61–71

Woodin SA (1978) Refuges, disturbance, and community structure: a marine soft-bottom example. Ecology 59(2): 274–284

Zühlke R, Blome D, van Bernem K-H, Dittmann S (1998) Effects of the tube building polychaete *Lanice conchilega* (Pallas) on benthic macrofauna and nematodes in an intertidal sandflat. Senckenbergiana marit 29: 131–138

5.6
Size Frequency, Distribution and Colour Variation of *Carcinus maenas* in the Spiekeroog Backbarrier System

Sabine Dittmann & Marlies Villbrandt

Abstract: Shore crabs (*Carcinus maenas*) were studied in the backbarrier system of Spiekeroog to assess the size of the (sub-)populations and colour variations in sub- and intertidal areas. Juvenile shore crabs occurred exclusively in the intertidal and preferentially within mussel beds. The record of relative abundances and size frequency distributions showed that small and medium-sized crabs migrate from the subtidal into the intertidal during the summer months. Thus the crabs occurring in the intertidal are not representative for the entire population. After the ice winter 1995/96 very few shore crabs appeared on the tidal flats and not until late summer. The majority of the shore crabs in the study area were of the green morphotype, especially in the intertidal. The number of male red crabs increased during the mating season. Female red crabs were frequent in spring and early summer. Implications of the colour morphotypes as a phenotypic plasticity of the shore crabs are discussed in relation to the variable living conditions in the Wadden Sea.

5.6.1
Introduction

The shore crab *Carcinus maenas* occurs in the coastal seas of the entire Northeast Atlantic and has been introduced to many areas of the world (Grosholz & Ruiz 1995; Cohen et al. 1995; Hayward & Ryland 1996). The crabs live mainly subtidally, but a part of the population occurs sporadically on tidal flats. Shore crabs settle in the tidal flats and migrate into the subtidal late in autumn (Thiel & Dernedde 1994) or remain in the intertidal over winter (Klein-Breteler 1976a). Scherer & Reise (1981) report a mean density of about 15 crabs 100 m^{-2} on sandflats of the North Frisian Wadden Sea and much higher values of juvenile crabs in seagrass beds and mussel beds (e.g. 500 O-group crabs m^{-2}). Migrations of adult shore crabs between the sub- and intertidal vary with crab size, but also with season, daytime and tides (Naylor 1962; Atkinsons & Parsons 1973; Klein-Breteler 1976b; Scherer & Reise 1981; Warman et al. 1993; Hunter & Naylor 1993; Thiel & Dernedde 1994). Over winter, the crabs are inactive in deeper areas of the subtidal. They don´t feed during this time and their metabolism is reduced (Crothers 1968). In spring, the first crabs appear in shallow subtidal areas and become more active with rising water temperatures. In the course of the summer increasing numbers of crabs occur in the shallow subtidal and intertidal (Beukema 1991). Towards the end of the year the crabs withdraw to the subtidal. *C. maenas* can theoretically reproduce all year round, but most pre- and in-copula pairs are recorded in late summer (van der Meeren 1994). The larvae hatch in the following spring.

C. maenas plays a central role in the food web of intertidal areas and is a prominent mobile predator in the Wadden Sea with a wide food spectrum. Their prey choice and predation pressure varies with the abundance and availability of prey as well as with age, size and sex of the crabs (Ropes 1968; Walne & Dean 1972; Scherer & Reise 1981; Jubb et al. 1983; Rangely & Thomas 1987). Shore crabs are also scavengers (Moore & Howarth 1996). Juvenile crabs feed on meiofauna (nematodes) and juvenile stages or small-sized macrofauna (larvae, barnacles, molluscs or polychaetes) (Klein-Breteler 1975; Reise 1985; Rangeley & Thomas 1987). Given a respective supply and optimal size, blue mussels (*Mytilus edulis*) can constitute the greatest part of the crabs' diet (Elner & Hughes 1978; Elner 1981; Jubb et al. 1983). *C. maenas* itself is preyed upon by birds (for example curlews, redshanks, and gulls; see e.g. Dernedde 1993, 1994), fish (Jensen & Jensen 1985), and seals (Crothers 1968).

Adult shore crabs occur in different colour forms, refering to their ventral carapace colour varying from green to red. Colour morphotypes are known from many crab species and are often, but not always, indicators for moulting stages (Reid et al. 1997; Türkay, pers. comm.). Crabs can regulate their moulting cycle to a certain degree and delay a forthcoming moult (Reid et al. 1997). This can be triggered by external stimuli on the hormonal system via the sensory organs of the crabs or by photo-denaturation of pigments in the carapace (Carlisle 1957; Reid et al. 1997). In the course of their life, shore crabs can undergo individually different moulting frequencies and thus alternate between a "green" and "red" colour form (Reid et al. 1997). These different coloured phenotypes go along with physiological differences as well as modified behaviour and activities (McGaw & Naylor 1992 a,b; McGaw et al. 1992; Warman et al. 1993; Aagaard et al. 1995). Red crabs are in a prolonged intermoult and thus have a thicker and stronger carapace and chelae, giving them a competitive advantage during aggressive interactions with conspecifics (Kaiser et al. 1990). In turn, red colour forms have to cope with physiological disadvantages (higher oxygen demand, reduced tolerance towards temperature and salinity variations) (Aldrich 1986; Reid & Aldrich 1989; Reid et al. 1997). Green crabs are more tolerant to salinity fluctuations and aerial exposure and are thus better adapted to living conditions in the intertidal (Reid et al. 1989, 1997).

As part of the ecosystem research project ELAWAT, the shore crab population in the backbarrier system of Spiekeroog Island was investigated to assess distribution, size frequency, sex distribution and colour forms of the crabs in the sub- and intertidal. This chapter summarizes these studies that were carried out by Wolf (1997, 1998 and unpublished).

5.6.2 Material and Methods

Several methods were used by Wolf (1997, 1998) for surveying the shore crab population. In the subtidal, crabs were caught using a 2 m beam trawl from the research vessel Senckenberg. Samples were taken at a water depth of 5 m in the main tidal channel of the backbarrier area of Spiekeroog Island, on the northern edge of the Janssand (Fig. 3.1.2), over a trawling distance of 1 nautical mile. The net had a mesh size of 10 mm which captured crabs > 15 mm carapace width (CW). These trawls were taken from late 1992 to 1994 with varying sampling frequencies between the years and varying replication of the monthly trawls. In total, more than 25 000 crabs were caught during this survey and their size and sex were recorded.

An epibenthos sledge was used to capture crabs < 15 mm in size. This sledge had a net opening of 100 x 17 cm and a mesh size of 200 μm. It was lowered to the seafloor around low tide with the net opening facing the current and retrieved after 10 minutes. These samples were taken monthly in 1995.

A mark-recapture experiment was carried out from 1992 until 1994 alongside the subtidal survey. Crabs caught were colour-coded by gluing a drop of paint on the dried carapace. Experiments in an aquarium had shown that this colour-code persisted for about 3 months and caused no harm to the crabs. During the experiment, as many crabs as possible were colour-coded and released in August 1992, September and October 1993 and June 1996. A different colour was applied at each date. Their recapture was recorded during the regular monthly surveys in the subtidal.

In the intertidal, crabs were caught with sediment traps. Buckets of about 500 cm^2 surface area were implanted level with the sediment surface. Five traps were implanted at each of four transect sites from a mussel bed into a sandflat on the Swinnplate (Fig. 3.1.3 and see Chap. 5.2). Monthly samples were carried out in 1995 and 1996. Therefore the traps were first emptied and cleared of sediment and the crabs collected in the traps within the following 24 h were recorded. In addition, samples were taken with a corer of about 500 cm^2 surface area to 5 cm sediment depth. These samples were sieved through a sieve with 2 mm mesh size to record the smallest shore crabs of < 15 mm in size. Three replicate samples were taken monthly in 1995 and 1996 on three of the transect sites, the mussel bed, near the mussel bed and in the sandflat.

5.6.3 Results

Population size

Mark-recapture experiments had been carried out to assess the population size of the shore crabs in the subtidal of the Spiekeroog backbarrier system. The rate of shore crabs retrieved in these experiments was extremely low, which indicates a large population size. Out of a total of 9368 colour-coded crabs only 68 individuals were retrieved (0.73 %) during the entire experiment. In 1992, a few crabs were retrieved in the same area four months after their release.

Size frequency and sex distribution of shore crabs in the subtidal

In 1994, about 15 000 shore crabs were caught in the subtidal using the dredge. These were three times as many crabs as in the previous years, which were not sampled as frequently as 1994. Over the year, the abundances of male and female crabs in the subtidal were highest in August/September (Fig. 5.6.1). Ovigerous females were more abundant early in the year from February to June.

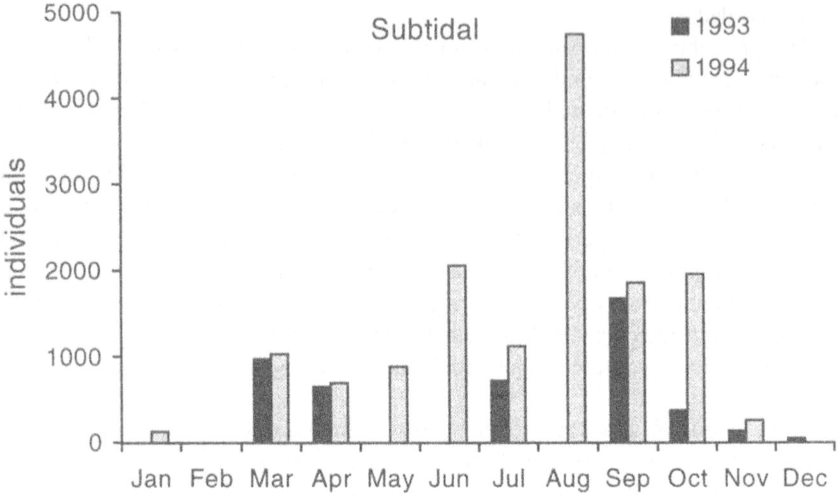

Fig. 5.6.1 Abundances and seasonal development of shore crabs in the subtidal of the backbarrier system of Spiekeroog in 1993 and 1994, caught with a beam trawl. (Data from Wolf 1997)

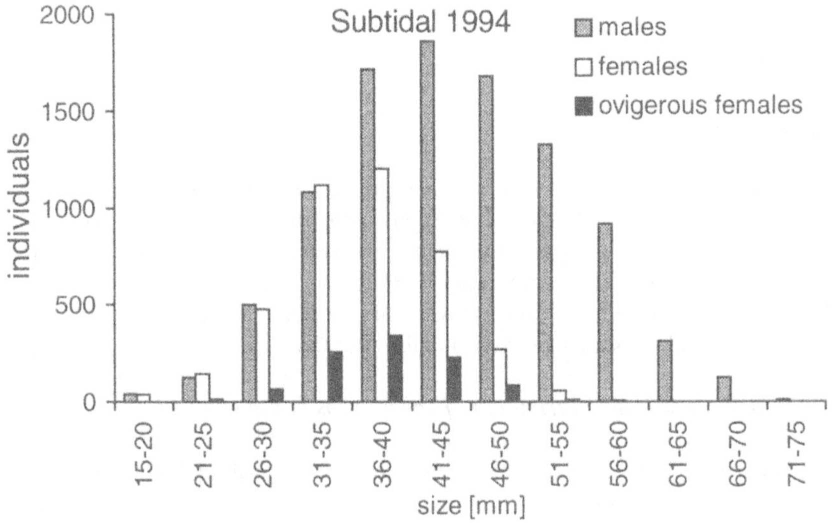

Fig. 5.6.2 Size frequency distributions of all shore crabs caught with a beam trawl in the subtidal in 1994. (Data from Wolf 1997)

The sex ratio of adult crabs was male-dominated (1 : 0.52 based on the survey 1994; 55–65 % of all crabs caught were male), but almost equal in the small crabs caught in the subtidal.

The average size was 47 mm carapace width (CW) for male and 37 mm CW for female crabs (Fig. 5.6.2). Not a single crab < 15 mm CW was caught in any of the samples taken with the epibenthos sledge in the subtidal.

Size frequency and sex distribution of shore crabs in the intertidal

In the intertidal, the sediment traps caught little over 2200 crabs in 1995. Male and female crabs were recorded with highest frequencies from May to July and hardly any shore crabs were caught after September (Fig. 5.6.3). Females were more frequent in the intertidal (sex ratio 1 : 1.35, 57 % of all crabs caught), but almost no ovigerous females were recorded here. Following the ice winter 1995/96, only very few crabs (75 individuals) were recorded with the traps in the tidal flats. These were mainly larger male crabs (sex ratio 1 : 0.17) which occurred late in the year in August/September 1996. Only 17 crabs < 15 mm were caught in the intertidal after the ice winter.

The average size of crabs in the intertidal was 41 mm CW for male and 31 mm CW for female crabs (Fig. 5.6.4). The size frequency distribution was shifted to smaller sizes and even crabs < 15 mm in size were recorded. These small shore crabs entered the tidal flats in two colonization waves in March and September.

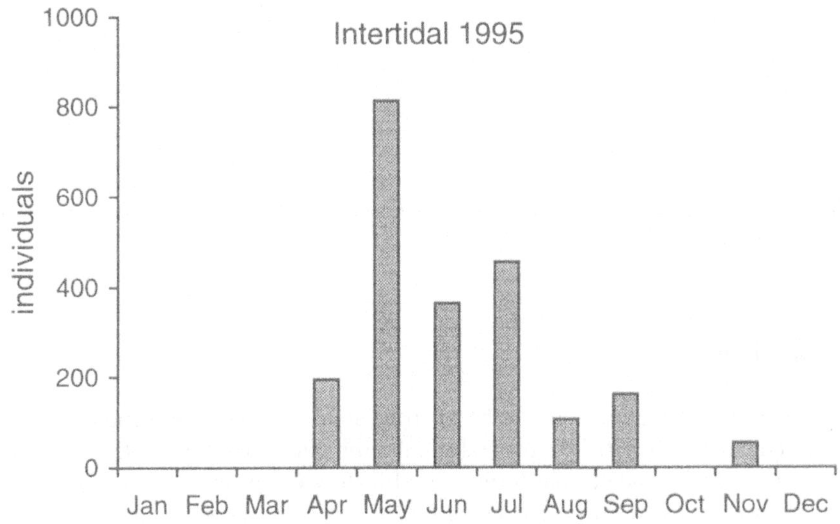

Fig. 5.6.3 Abundances and seasonal development of the shore crabs in the intertidal of the backbarrier system of Spiekeroog in 1995, caught with traps. (Data from Wolf 1997)

Their numbers were higher in the trap samples in April to June, whereas the sediment core samples recorded higher numbers in September.

Along the transect on the Swinnplate, a high number of small crabs (about 30 mm in size) was recorded with the traps in April and May at the sandflat sites. Numbers were lower in and around the mussel bed, only larger females occurred here in July and August. In 1995, very small shore crabs (< 15 mm in size) were more abundant in the mussel bed. Here, a total of 478 crabs was recorded with the sediment core samples compared to only 9 crabs of this size range on the sandflat.

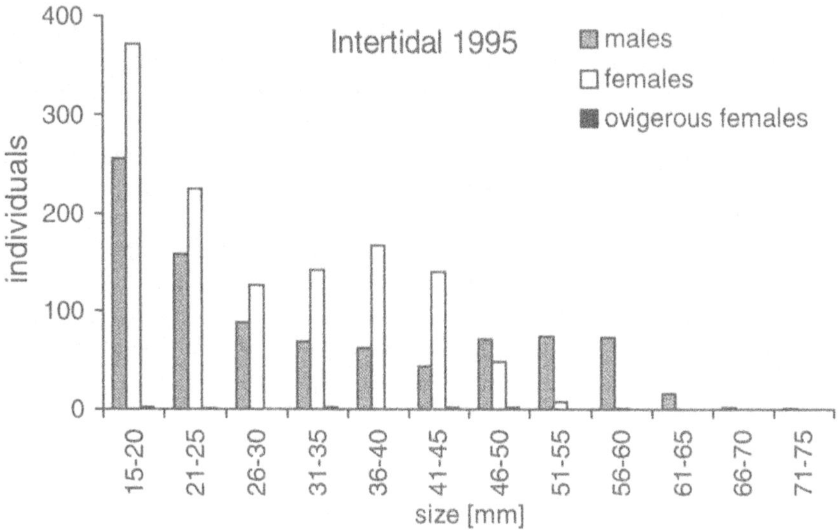

Fig. 5.6.4 Size frequency distributions of all shore crabs caught with traps in the intertidal in 1995. (Data from Wolf 1997)

Colour forms

The majority of the crabs in the backbarrier area of Spiekeroog occurred in the green colour form (> 70 % of the crabs in the sub- and intertidal, Table 5.6.1). The frequency of the single colour forms varied with size, sex and season. Comparing the average sizes of crabs, red crabs turned out to be larger than green crabs, which was the case both in the sub- and intertidal and for both sexes (Table 5.6.2). Size frequency distributions revealed that green forms of both sexes were more common among smaller sized crabs than the red form. Green crabs were especially frequent in the intertidal, where juvenile crabs prevailed. Ovigerous females, which occur almost exclusively in the subtidal, were mainly of a red colour. The few crabs recorded in 1996 after the ice winter were almost entirely green forms.

The share of red colour forms varied with season. In the subtidal, more large male crabs of the red form were found in late summer/autumn. The number of male red crabs was also increasing later in the year in the intertidal. Female red crabs were frequent from May to July/August, both in the sub- and intertidal.

Table 5.6.1 Number (n) of crabs (*Carcinus maenas*) caught in the backbarrier system of Spiekeroog over the respective year using a beam trawl in the subtidal (1994) and traps in the intertidal (1995). The frequency of colour forms is given in percent of male, female and ovigerous female crabs respectively. (Data from Wolf 1997)

	n	% red	% red/green	% green
Subtidal 1994				
males	9740	14	11	75
females	4091	26	13	61
ovigerous females	992	88	10	2
Intertidal 1995				
males	968	9	4	87
females	1297	26	3	71
ovigerous females	9	56	22	22

Table 5.6.2 Average, smallest and largest size (carapace width) of crabs of the different colour forms recorded in the sub- and intertidal of the backbarrier area of Spiekeroog Island. (Data from Wolf 1997).

	red crabs			green crabs		
	mean	min	max	mean	min	max
Subtidal 1994						
males	53	23	71	39	<15	73
females	38	21	51	36	<15	59
ovigerous females	38	24	57	36	29	46
Intertidal 1995						
males	52	32	65	29	15	76
females	40	26	58	23	15	53
ovigerous females	43	35	49	28	20	35

5.6.4 Discussion

Surveying mobile crabs is a tedious and difficult affair. The population size of the shore crabs in the backbarrier system could not be assessed, as too few of the marked crabs were retrieved (Munch-Petersen et al. 1982; Wolf unpublished).

However, the low recovery rate from the mark-recapture experiments indicates a large population of *C. maenas* in the Wadden Sea with a high degree of migration activity.

The seasonality in the abundance of shore crabs in the sub- and intertidal of the backbarrier system of Spiekeroog was in agreement with previous records (Klein-Breteler 1976b; Beukema 1991). A higher share of male shore crabs in the subtidal, as recorded in the backbarrier system of Spiekeroog, had not been observed before (Wolf 1997). The frequency distributions of crab sizes indicated that the average size of crabs was larger in the subtidal than in the intertidal, but this could be biased by the different sampling techniques which had to be used sub- and intertidally. A comparison of the population structure in the sub- and intertidal was also not possible as the two areas were sampled in different years and variations can occur between years (see Beukema 1991).

Following the ice winter, a severe decline was recorded in the intertidal population of the shore crabs. Low abundances and late appearance of crabs after cold winters was also reported by Beukema (1991). Up to the end of the field season for ELAWAT in autumn 1996, neither the abundances nor population structure had returned to the values of the preceding year. Since newly settled and juvenile crabs were missing, a prolonged effect of the ice winter on the crab population was predicted. Giving such a low recovery rate, the stability properties of this species can only be assessed with long-term surveys.

The part of the shore crab population occurring in the intertidal is not representative for the entire population (Wolf 1997). No crabs < 15 mm had been caught in the subtidal, confirming that the crabs settle exclusively in the intertidal (Klein-Breteler 1976a). The early benthic stages of shore crabs were almost exclusively recorded in intertidal mussel beds, which has been documented before and is seen as protection from predation (Klein-Breteler 1976b; Thiel & Dernedde 1994; Moksnes et al. 1998). The high number of small crabs caught with the traps in the sandflat was related to their foraging activity, as these traps catch those crabs active on the tidal flats at high tide (Wolf 1997). Thus the spatial distribution in the intertidal is related to the activity of the shore crabs and reflects the function of different intertidal habitats for the crabs of different sizes. The occurrence of the largest crabs was restricted to the subtidal. Differences in the population structure of sub- and intertidal shore crabs were also shown by Hunter & Naylor (1993). Although the intertidal hosts only a small part of the population, it is an important habitat for *C. maenas*. The crabs frequenting the tidal flats could be grouped as follows:

- early crab stages (< 15 mm in size) in the mussel beds,
- small juvenile crabs (1.5–2.5 cm in size) in high numbers, derived from the spatfall of the previous year,
- medium sized or few large crabs, possibly migrating into the tidal flats to avoid competition with larger crabs in the subtidal.

The size distribution, frequency and seasonality in the occurrence of colour forms of *C. maenas* recorded from the backbarrier system of Spiekeroog confirms records from other seas. The higher share of green colour forms in the intertidal results both from the abundance of juvenile and medium sized crabs in the tidal flats (see above) and the better adaptation of green crabs to living conditions in the intertidal (Reid et al. 1989, 1997). A higher proportion of the green morphotypes of *C. maenas* and smaller sized crabs closer to the shore has also been recorded by Kaiser et al. (1990); Hunter & Naylor (1993); Warman et al. (1993) and Abello et al. (1997).

In female shore crabs, the seasonal cycle of colour forms was related to their reproductive cycle. Ovigerous female crabs were red because they do not moult until the larvae are released in spring. Afterwards they have to complete a moulting cycle until the mating season in autumn. In male crabs, the red colour form was more frequent in the sub- and intertidal during the mating season in autumn. Together with the larger individual size of male red crabs, especially in the subtidal, this supports a relation of the red colour form in males with competitive dominance and reproductive success (Kaiser et al. 1990; Reid et al. 1997).

Wolf (1997) discusses the colour forms of the shore crabs as either growth or reproduction strategies. Given that the crabs change between the red and green colour form during their life cycle, they are not committed to a certain strategy once and forever. Instead, the colour morphotypes represent rather a "life stage strategy". The shore crabs can alternate their phenotype repeatedly according to their life cycle or environmental conditions, adding a further dimension to their phenotypic plasticity. For ELAWAT, it was discussed whether this phenotypic plasticity is a relevant mechanism for stability properties of the Wadden Sea ecosystem. The individual variability within a population due to phenotypically different ecotypes (in the case of *C. maenas* the colour forms with their associated life strategies) could substitute niche development and speciation. Different phenotypes allow a survival under varying environmental conditions, giving a species with a high phenotypic plasticity an evolutionary advantage over a species with a low phenotypic plasticity (Aldrich 1989; Hadfield & Strathmann 1996). In a constantly changing ecosystem like the Wadden Sea, phenotypic plasticity may be a better individual strategy to achieve a maximum reproductive output (Wolf 1997, Chap. 9). Furthermore, since every crab can change the phenotype depending on environmental or reproductive requirements, intraspecific competition is reduced. The generalistic character of the shore crab population is not attributable to the ecological width of single individuals, but to the phenotypic variability within the population (Wolf 1997).

Acknowledgements

This chapter is based on the studies in ELAWAT by Frank Wolf, Michael Türkay and their colleagues from the Senckenberg Institute. The results were published in the PhD thesis by Frank Wolf and as cited in the text, and further unpublished data were kindly provided by him. We thank the reviewer and Volker Grimm for comments and discussions on this chapter. The project was funded by the German Bundesministerium für Bildung, Wissenschaft, Forschung und Technologie (BMBF) under grant number 03F0112 A and B. The responsibility for the contents of the publication rests with the authors.

References

Aagaard A, Warman CG, Depledge MH (1995) Tidal and seasonal changes in the temporal and spatial distribution of foraging *Carcinus maenas* in the weakly tidal littoral zone of Kertemine Fjord, Denmark. Mar Ecol Prog Ser 122: 165–172

Abello P, Aagaard A, Warman CG, Depledge MH (1997) Spatial variability in the population structure of the shore crab *Carcinus maenas* (Crustacea: Brachyura) in a shallow-water, weakly tidal fjord. Mar Ecol Prog Ser 147: 97–103

Aldrich JC (1986) The influence of individual variations in metabolic rate and tidal conditions on the response to hypoxia in *Carcinus maenas* (L.). Comp Biochem Phys 83A: 53–60

Aldrich JC (1989) The world beyond the species, an argument for greater definition in experimental work. In: Aldrich JC (Ed) Phenotypic responses and individuality in aquatic ectoterms. Co. Wicklow, Ireland, Japaga. pp 3-8

Atkinson RJA, Parsons AJ (1973) Seasonal patterns of migration and locomotory rhythmicity in populations of *Carcinus*. Neth J Sea Res 7: 81–93

Beukema JJ (1991) The abundance of shore crabs *Carcinus maenas* (L.) on a tidal flat in the Wadden Sea after cold and mild winters. J Exp Mar Biol Ecol 153: 97–113

Carlisle DB (1957) On hormonal inhibition of moulting in decapod crustacea. II. The terminal anecdysis in crabs. J Mar Biol Ass UK 36: 291–307

Cohen, AN, Carlton JT & MC Fountain (1995) Introduction, dispersal and potential impacts of the green crab *Carcinus maenas* in San Francisco Bay, California. Mar Biol 122: 225–237

Crothers JH (1968) The biology of the shore crab *Carcinus maenas*. 2.) The life history of the adult crab. Field Studies 2: 579–614

Dernedde T (1993) Vergleichende Untersuchungen zur Nahrungszusammensetzung von Silbermöwe (*Larus argentatus*), Sturmmöwe (*L. canus*) und Lachmöwe (*L. ridibundus*) im Königshafen/Sylt. Corax 15: 222–240

Dernedde T (1994) Foraging overlap of three gull species (Larus spp.) on tidal flats in the Wadden Sea. Ophelia Suppl 6: 225–238

Elner RW (1981) Diet of green crab *Carcinus maenas* (L.) from Port Hebert, southwest Nova Scotia. J Shellfish Res 1: 89–94

Elner RW, Hughes RN (1978) Energy maximization in the diet of the shore crab *Carcinus maenas*. J Anim Ecol 47: 103–116

Grosholz ED, Ruiz GM (1995) Spread and potential impact of the recently introduced European green crab, *Carcinus maenas* in central California. Mar Biol 122: 239–247

Hadfield MG, Strathmann MF (1996) Variability, flexibility and plasticity in life histories of marine invertebrates. Oceanol-Acta 19: 323–334

Hayward PJ, Ryland JS (1996) Handbook of the Marine Fauna of North-West Europe. Oxford University Press, Oxford

Hunter E, Naylor E (1993) Intertidal migration by the shore crab *Carcinus maenas*. Mar Ecol Prog Ser 101: 131–138

Jensen KT, Jensen JN (1985) The importance of some epibenthic predators on the density of juvenile benthic macrofauna in the Danish Wadden Sea. J Exp Mar Biol Ecol 89: 157–174

Jubb CA, Hughes RN, Rheinalt T (1983) Behavioural mechanisms of size-selection by crabs, *Carcinus maenas*, feeding on mussels, *Mytilus edulis*. J Exp Mar Biol Ecol 66: 81–87

Kaiser MJ, Hughes RN, Reid DG (1990) Chelal morphometry, prey-size selection and aggressive competition in green and red forms of *Carcinus maenas* (L.). J Exp Mar Biol Ecol 140: 121–134

Klein Breteler WCM (1975) Food consumption, growth, and energy metabolism in the juvenile shore crab *Carcinus maenas*. Neth J Sea Res 9: 86–99

Klein Breteler WCM (1976a) Settlement, growth and production of the shore crab, *Carcinus maenas*, on tidal flats in the Dutch Wadden Sea. Neth J Sea Res 10: 354–376

Klein Breteler WCM (1976b) Migration of the shore crab, *Carcinus maenas*, in the Dutch Wadden Sea. Neth J Sea Res 10: 338–353

McGaw IJ, Kaiser MJ, Naylor E, Hughes RN (1992) Intraspecific morphological variation related to the moult-cycle in color forms of the shore crab *Carcinus maenas*. J Zool 228: 351–359

McGaw IJ, Naylor E (1992a) Distribution and rhythmic locomotor patterns of estuarine and open-shore populations of *Carcinus maenas*. J Mar Biol Ass UK 72: 599–609

McGaw IJ, Naylor E (1992b) The effect of shelter on salinity preference behaviour of the shore crab *Carcinus maenas*. Mar Behav Physiol 21: 145–152

Moore, PG & J Howarth (1996) Foraging by marine scavengers: effects of relatedness, bait damage and hunger. J. Sea Res 36: 267–273.

Moksnes PO, Pihl L, van Montfrans J (1998) Predation on postlarvae and juveniles of the shore crab *Carcinus maenas*: importance of shelter, size and cannibalism. Mar Ecol Prog Ser 166: 211–225

Munch-Petersen S, Sparre P, Hoffmann E (1982) Abundance of the shore crab, *Carcinus maenas* (L.), estimated from mark-recapture experiments. Dana 2: 97–121

Naylor E (1962) Seasonal changes in a population of *Carcinus maenas* in the littoral zone. J Animal Ecol 31: 601–609

Rangeley RW, Thomas MLH (1987) Predatory behaviour of the juvenile shore crab *Carcinus maenas*. J Exp Mar Biol Ecol 108: 191–197

Reid DG, Aldrich JC (1989) Variations in response to environmental hypoxia of different color forms of the shore crab, *Carcinus maenas*. Comp Biochem Pysiol 92: 535–539

Reid DG, Abello P, McGaw IJ, Naylor E (1989) Phenotypic variation in sympatric crab populations. In: Aldrich JC (Ed) Phenotypic responses and individuality in aquatic ectoterms. Co. Wicklow, Ireland, Japaga. pp 89–95

Reid DG, Abello P, Kaiser MJ, Warman CG (1997) Carapace colour, inter-moult duration and the behavioural and physiological ecology of the shore crab *Carcinus maenas*. Est Coast Shelf Sci 44: 203–211

Reise K (1985) Tidal flat ecology. Springer, Berlin Heidelberg New York

Ropes JW (1968) The feeding habits of the green crab *Carcinus maenas*. Fish Bull Fish Wildl Serv US 67: 183–203

Scherer B, Reise K (1981) Significant predation on micro- and macrobenthos of the Wadden Sea. Kieler Meeresforsch. Sonderheft 5: 490–500

Thiel M, Dernedde T (1994) Recruitment of shore crabs (*Carcinus maenas*) on tidal flats: mussel clumps as an important refuge for juveniles. Helgoländer Meeresunters 48: 321–332

Walne PR, Dean (1972) Experiments on predation by the shore crab, *Carcinus maenas*, on *Mytilus* and *Mercenaria*. J Cons Perm Int Expl Mer 34: 190–199

Warman CG, Reid DG, Naylor E (1993) Variations in the tidal migratory behaviour and rhythmic light-responsiveness in the shore crab, *Carcinus maenas*. J Mar Biol Ass UK 73: 355–364

Wolf F (1997) Untersuchungen zum Auftreten roter und grüner Farbvarianten bei der Strandkrabbe, *Carcinus maenas* (Linnaeus 1758) (Crustacea: Decapoda: Brachyura). Dissertation, Universität Frankfurt a. M.

Wolf F (1998) Red and green colour forms in the common shore crab *Carcinus maenas* (L.) (Crustacea: Brachyura: Portunidae): theoretical predictions and empirical data. Journal of Natural History 32: 1807–1812

van der Meeren Gro I. (1994) Sex- and size-dependent mating tactics in a natural population of shore crabs *Carcinus maenas*. Journal of Animal Ecology 63: 307–314.

6 Recolonization of Tidal Flats After Disturbance

Sabine Dittmann, Carmen-Pia Günther & Ulrike Schleier

Abstract: The benthic recolonization of experimentally disturbed sandflat sites by diatoms, nematodes and small macrofauna was studied with several treatments, differing in intensity and timing of disturbance. Furthermore, recolonization following a natural disturbance, the ice winter 1995/96, was studied and compared with the results of the experiments.

Diatoms were the fastest organisms recolonizing the disturbed sites. Within a few days, their species number, species composition and individual abundances were comparable to control situations. This colonization pattern depended on the diatoms available in the ambient sediment at the time of disturbance. The mode of recolonization was passive via sediment transport. The ice winter had little effect on the diatoms and abundances were comparable to preceding years. Yet, three species, which had not been recorded in the Wadden Sea before, occurred in the tidal flats in 1996.

Nematodes reached ambient densities in the experimentally disturbed plots within a few weeks or months, with a slower recolonization the longer the disturbance had lasted. After the ice winter, nematode abundances increased steadily and reached even higher values in summer 1996 than in the years before.

Smaller sized macrofauna recolonized disturbed experimental plots in spring faster than plots disturbed in autumn. The recolonization was also faster, the shorter the disturbance had lasted. The first stages of recolonization reflected the ambient situation of the species composition and abundance of benthic organisms. After several months, species numbers, abundance and community composition in the disturbed plots could not be distinguished from the reference plots. The first macrofaunal species colonizing the experimentally disturbed plots were the polychaete *Pygospio elegans*, Ostracoda and the amphipod *Urothoe poseidonis*. These were also dominant in the benthic community in 1996 after the ice winter. Juvenile polychaetes colonized the experimental sites to a higher degree than the ambient sediment. The course of the recolonization was mainly dependent on the availability of settling organisms, i.e. larvae as well as postlarvae. The study period was not long enough to record all successional stages.

6.1 Introduction

The analysis of recolonization after disturbances allows one to assess processes that regulate and structure communities and to detect successional stages and their underlying causes. Studies on the recolonization in soft sediment communities have resulted in a scheme where an early successional stage dominated by opportunistic species and r-strategists is followed by further stages with mainly k-selected species (Grassle & Grassle 1974; McCall 1977; Pearson & Rosenberg 1978; Rhoads et al. 1978). This classic succession model was further elaborated by investigations on the colonization mode of larvae, postlarvae or adults (Dauer & Simon 1976; Santos & Simon 1980; Smith & Brumsickle 1989; Savidge & Taghon 1988; Frid 1989), the dependence on the size of the disturbance (Probert 1984; Thrush et al. 1996) as well as the timing of the disturbance (Simon & Dauer 1977; Zajac & Whitlatch 1982 a; Bonsdorff & Österman 1985). However, this scheme was not found to be valid for all soft-sediment communities, as in some cases a recolonization occurred that was more dependent on the area and time instead of following clear successional stages (Zajac & Whitlatch 1982 b). This is especially the case when the benthic community is composed of opportunistic, stress-tolerant species and is naturally in a state resembling an early successional stage of the above mentioned scheme (Zajac & Whitlatch 1982 b; Gamenick et al. 1996).

In the ecosystem research project ELAWAT, the recolonization experiments were carried out under the central topic of the reaction of benthic organisms to disturbances and the search for possible stability properties of the Wadden Sea ecosystem (Chap. 2). This chapter summarizes unpublished results of experiments carried out by Tecklenborg, Keuker-Rüdiger, Blome and Langner. Their experiments targeted the question how timing, kind and intensity of a disturbance may affect a benthic community. The recolonization of small infauna and microphytobenthos were recorded. The mode of recolonization and relevant geochemical parameters were assessed in collaboration with other sub-projects. The experiments were carried out in the main study area of ELAWAT in a sandflat on the Gröninger Plate (Fig. 3.1.4), where the community was characterized by the lugworm *Arenicola marina* and the sand mason *Lanice conchilega*.

As a disturbance, the plots were defaunated to simulate the cover of tidal flats by extensive mats of filamentous green algae. Such events had occurred in the Wadden Sea in the 1980s, depriving the sediment underneath the algal layer of oxygen and finally causing the death of benthic organisms (Reise et al. 1994). The "black spots" occurring in the East Frisian Wadden Sea can be interpreted as a long-lasting effect of this event (Höpner & Michaelis 1994).

The ice winter 1995/96 was taken as an occasion to study the regeneration of the tidal flat community after this natural disturbance and to compare the recolonization pattern with the results of the recolonization experiments of the preceding years. In contrast to the more localized effects of the disturbance experiments, the ice winter affected the entire region (Beukema et al. 1993, Chap. 7). While the defaunation reduced all benthic organisms in the experimental plots, the ice winter selectively killed the cold sensitive species (Beukema 1979, 1989; Chaps. 5, 7). Thus regeneration of benthic communities can be different for the experimental and natural disturbances investigated here.

6.2
Material and Methods

As a disturbance, the macroalgal cover mentioned above was simulated by covering the sediment with a plastic tarpaulin. Thus, a defaunation was achieved. With different experiments the intensity and timing (start of recolonization) of the disturbance were modified. The recolonization by small macrofauna, nematodes and benthic diatoms was recorded. The experiments were set up on the Gröninger Plate in the backbarrier system of Spiekeroog (Fig. 3.1.4). The single treatment plots were interspersed.

The following experiments were carried out:

- Recolonization by macrofauna in dependence of season
  In 1994, a **spring experiment** and an **autumn experiment** were set up. For the former, the plots were covered with a tarpaulin in April; for the latter in September. The cover was removed after 4 weeks in May and October, respectively. For each treatment, three plots of 4 x 5 m in size were arranged. Three undisturbed sandflat plots of the same size each were marked in spring and autumn and served as controls. They were used as controls for all experiments. All of these plots were sampled until the end of 1995.
- Recolonization by macrofauna in dependence of disturbance intensity
  In 1995, three plots of 4 x 5 m in size each were covered for 4 weeks (**4-week disturbance**) or one week (**1-week disturbance**) and uncovered simultaneously in May. Immediately after removal of the tarpaulin, the plots were raked to counteract the compaction of the sediment which was caused by the cover. Three further plots of 4 x 5 m in size each were raked only (mimicking a **mechanical disturbance**) and all cockles (*Cerastoderma edule*) thus appearing on the surface were removed. As control plots, those of the previous experiments were used.
- Recolonization by benthic diatoms and meiofauna
  In 1994, a plot of 3 x 3 m in size was covered with a tarpaulin and uncovered after six weeks in August (**6-week disturbance**). To study the recolonization of nematodes in **azoic sediment**, sediment was frozen and washed and subsequently filled into 4 plastic containers (20 x 20 cm in size and perforated) which were implanted into the sandflat next to the other experimental plots. The containers were implanted simultaneously with the uncovering of the 6-week disturbance plot in August.
  In 1995, the 4-week and 1-week disturbance experiments mentioned above were sampled as well. To study the mode of recolonization by diatoms, traps were implanted in the experimental and control plots. These traps had an opening just above the sediment surface, allowing lateral access only.

Sampling frequencies were high (daily) in the beginning of the recolonization and slowly extended to monthly intervals. Small macrofauna was sampled with a corer of 38.5 cm² surface area to 15 cm sediment depth. The samples were sieved through 250 µm mesh size and sorted and identified alive. Juvenile and adult polychaetes were separated using the criteria given in Hartmann-Schröder (1971).

Nematode samples were taken with a corer of 2 cm^2 surface area to 2 cm sediment depth. The samples were pickled in 4 % buffered formalin and the nematodes were extracted using the McIntyre method (63 µm mesh size), followed by a microscopic classification of the species (see Blome 1983). For benthic diatoms, samples were taken with a multicorer of 5 x 5 cm (single corer size 1 cm^2). In addition, 25 subsamples of 1 cm^3 each were taken at random in an area of 0.25 m^2, divided into 625 fields. 10 of the latter samples were sorted alive and the remaining samples were stored in 4 % buffered formalin for later analysis.

Sulphide was measured in the defaunation experiments using both a pore water sampler and a diffusion sampler (Langner 1997).

Benthic recolonization after the ice winter was studied in a plot of 100 x 100 m on the Gröninger Plate. Sampling was carried out as in the preceding years.

6.3 Recolonization After Experimental Disturbances

In the experimental plots that were covered with a tarpaulin, the sediment was compacted by this cover and thus the sediment surface was lowered by about 0.5–2 cm compared to the surrounding ambient sediment. The tarpaulin had interrupted gas exchange and thus the anoxic sediment horizon reached to the sediment surface by the time the tarpaulin was removed. But within a few days after the removal, a thin (mm) oxidized sediment layer developed in the experimental plots and reached 1 cm thickness after 5 (experiments 1995) or 11 weeks (experiments 1994). Straight after removing the tarpaulin, sulphide concentrations of 3–3.8 mmol dm^{-3} were measured. In the course of the experiment, the sediment layer with the highest sulphide concentrations shifted further into greater sediment depths. No free sulphide was measured in the control plots (Langner, Oelschläger, unpubl. data).

All experimental plots were recolonized. The recolonization was faster in the spring than autumn and faster in the 1-week than the 4-week disturbance. Microphytobenthos and meiofauna reached background values within days, whereas the macrofauna required several weeks.

6.3.1 The Course of Recolonization

Microphytobenthos

Benthic diatoms appeared as quick colonizers (Keuker-Rüdiger, unpubl. data). Within a few days after the plots were opened for recolonization, species number reached background values (9 days after 6-week disturbance in 1994 and 7 days after 4-week disturbance in 1995, Fig. 6.3.1 a, c). Abundance increased 20-fold in the first week of recolonization after the 6-week disturbance, but did not reach background values. After the 4-week disturbance in 1995, the abundances of diatoms in the experimental plots were as high as control values after 7 days (Fig. 6.3.1 b, d). Some sessile diatoms survived the experimental disturbance;

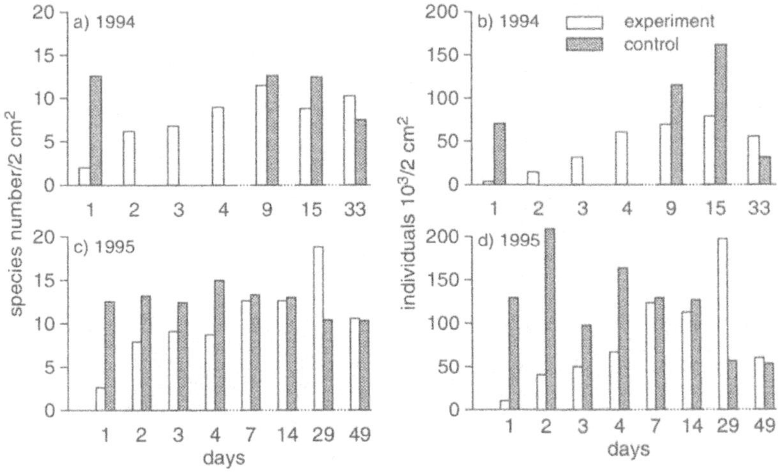

Fig. 6.3.1 Mean species and individual densities of diatoms in the recolonization experiments. In 1994 the experimental plot was defaunated for 6 weeks, in 1995 for 4 weeks (see text for details). (Keuker-Rüdiger, unpublished data)

were the same species which survived in laboratory experiments, where sediment from the field site was kept in dark, sealed containers for 5 to 8 weeks.

The relative proportion of dominant species and their composition classified by life modes (benthic mobile, benthic sessile, colony-forming, planktonic) resembled the control situation after 14 days in the 6-week disturbance experiment of 1994. In the 4-week disturbance experiment of 1995, it took 2 days to reach this situation. This difference was related to seasonal variations in the diatom composition of ambient sediments at the time the tarpaulins were removed (August 1994 and May 1995 respectively).

The diatoms caught in the traps implanted into the experimental and control plots in 1995 were mainly (> 90 %) mobile species, especially of the genus *Navicula*. Thus, the potential for colonization by mobile diatoms was the same for experimental and control plots. But a different colonization pattern was recorded. In the experimental plots, a higher share (60–80 %) of colony-forming and sessile diatoms were found during the first few days of recolonization. This indicates a passive transport of diatoms into the experimental plots with lateral sediment movement. Mobile diatom species may have avoided the experimental plots after removal of the tarpaulin, as long as living conditions were unfavourable.

Nematodes

Nematodes were also fast in recolonizing the disturbed plots, with a return time depending on the intensity of the disturbance (Blome, unpubl. data). Species numbers of control conditions were reached 1 month after the 6-week disturbance and 14 days after the 4-week disturbance (Fig. 6.3.2 a, c). In the latter experiment,

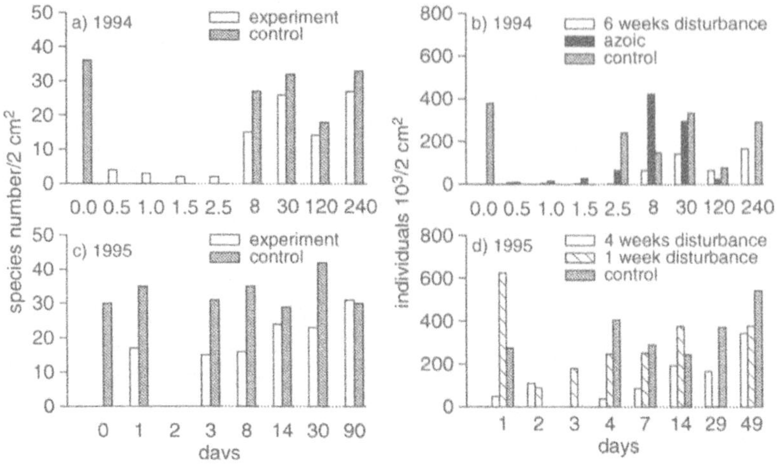

Fig. 6.3.2 Mean species and individual densities of nematodes in the recolonization experiments. In 1994 the experimental plot was defaunated for 6 weeks (experiment in Fig. 6.3.2 a) and azoic sediment was implanted in addition. In 1995 experimental plots had been defaunated for 4 weeks (experiment in Fig. 6.3.2 b) (see text for details). (Blome, unpublished data)

about 10 species had survived the cover. In each treatment, a species-specific recolonization sequence was recorded, but they differed between the two experiments. Mainly diatom-feeding nematodes were among the first settlers in the experiments of 1994. In the 6-week disturbance experiment, nematode abundances reached control levels after 4 months. The azoic sediment was colonized much faster, already after 8 days the abundances exceeded control values by three times. After the 1-week disturbance in 1995 it took 1 week, and 2 weeks after the 4-week disturbance until nematode abundances reached control values (Fig. 6.3.2 b, d).

Macrofauna

Depending on the timing of the disturbance, recolonization by macrofauna was much slower in the autumn experiment than in the spring experiment (Tecklenborg, unpubl. data). The variation of disturbance intensity had only a small influence on the recolonization. With the exception of the autumn experiment, similar species compositions were recorded between control and experimental plots after about 2 weeks. Especially the first stages of recolonization reflected the species composition and frequency of the ambient tidal flats. Overall, almost all species which occurred on the sites prior to the experimental disturbance, recolonized the plots. Exceptions were *Arenicola marina*, *Lanice conchilega* and its commensal *Malmgreniella lunulata* (Chap. 5.5), which did not establish again in the disturbed plots. Species recorded in the experimental plots within the first two weeks included the polychaetes *Pygospio elegans*, *Polydora* sp., *Nephtys hombergii*, *Capitella capitata*, *Heteromastus filiformis*, *Eteone longa*, *Nereis diversicolor* and

Scoloplos armiger, and the molluscs *Macoma balthica* and *Mytilus edulis*. Ostracoda and the amphipod *Urothoe poseidonis* were among the first settlers as well.

Several stages of recolonization were distinguished in the spring experiment of 1994. Until August, the species composition quickly approximated the control situation. Dominant species in the experimental plots were *P. elegans*, ostracods, *S. armiger*, *Polydora* sp. and *C. capitata*. Caused especially by the increase of juvenile polychaetes (mainly *P. elegans*), abundances were not significantly different between experimental and control plots after three months and all plots were similar with respect to species composition. Hardly any molluscs settled in the experimental plots in spring. A storm at the end of August caused the end of this stage and benthic abundances were reduced everywhere. In the following year, a similar seasonal development of species composition and abundances prevailed in both the experimental and control plots. But in June and August, recruitment by *P. elegans* and *Polydora* sp. was much higher in the experimental than in the control plots with abundances twice as high as control values (Fig. 6.3.3).

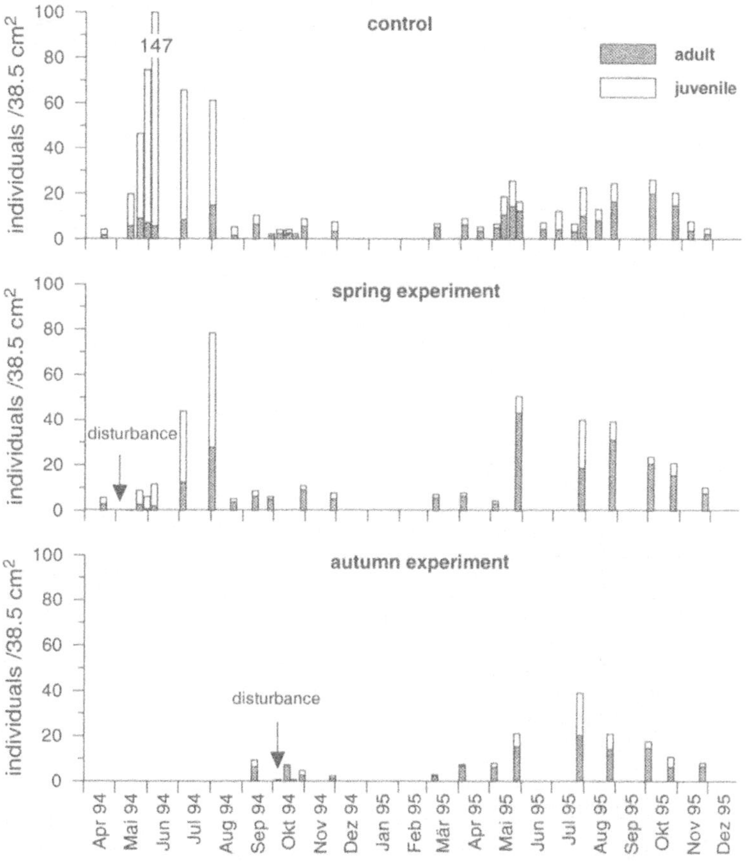

Fig. 6.3.3 Mean individual densities of small macrofauna in the course of the experiments opened for recolonization in spring and autumn 1994 (see text for details). The share of juvenile polychaetes is indicated by no shading. (Tecklenborg, unpublished data)

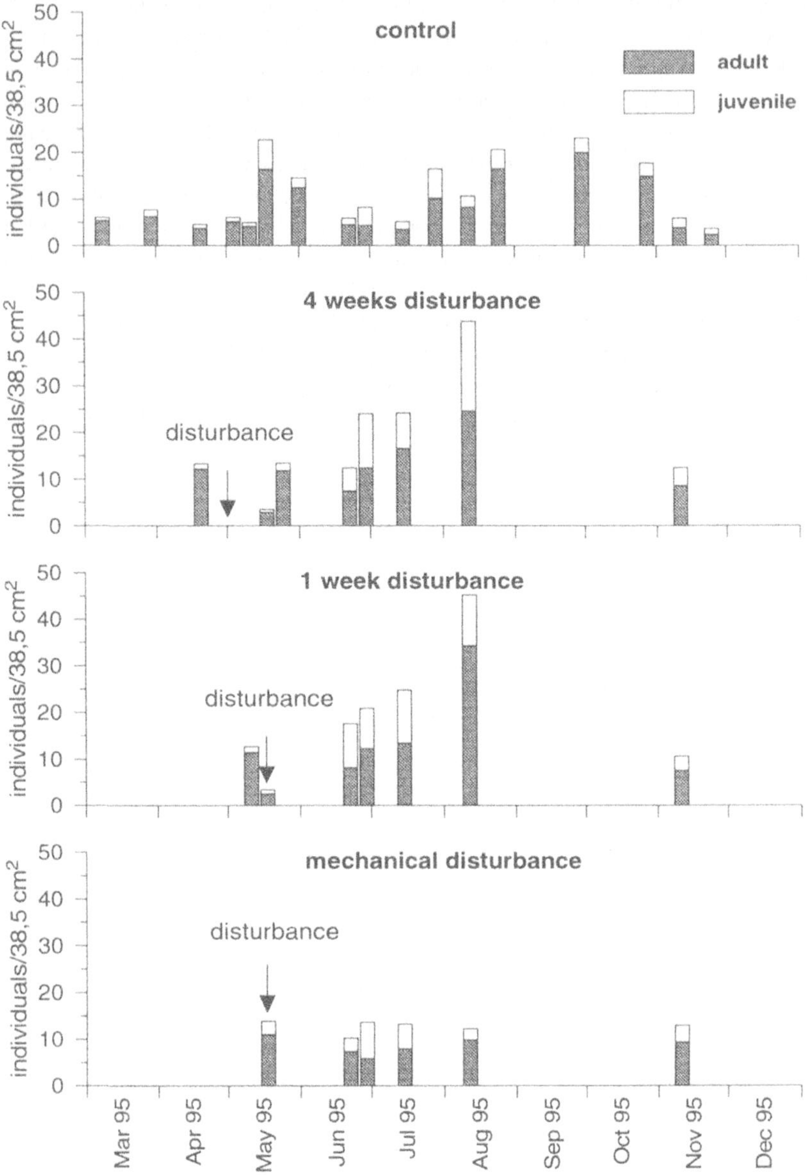

Fig. 6.3.4 Mean individual densities of small macrofauna in the course of the experiments in 1995, opened for recolonization after 4 weeks, 1 week and mechanical disturbance (see text for details). The share of juvenile polychaetes is indicated by no shading. (Tecklenborg, unpublished data)

Hardly any species settled in the autumn experiment, as the abundances in the adjacent sandflat were low and almost no larvae or postlarvae were available (Fig. 6.3.3). High abundances of individuals were not recorded until the following year. In spring and summer 1995, benthic colonization of the spring and autumn experiment were comparable to each other, only capitellid polychaetes were more abundant in the autumn experiment and molluscs were absent. As in the colonization of the spring experiment in the preceding year, seasonal events influenced the course of development. Juvenile polychaetes occurred in higher individual numbers in the experimental than in the control plots.

The experiments with varying intensities of disturbance carried out in 1995 showed that the weaker the disturbance, the faster the recovery of the plots (Fig. 6.3.4). The mechanical disturbance caused a temporary decline in abundances only. But no recruitment of juvenile spionid polychaetes, as recorded in all other experimental plots, was observed in this treatment. In the plots that were disturbed for 1 and 4 weeks, the recruitment of polychaetes was twice as high as in the control plots. Dominant species in these experimental plots were *P. elegans*, ostracods, *Polydora* sp., *H. filiformis* and *C. capitata.*

The mean species density reached background values fastest after the mechanical disturbance (4 days). This took 5–7 weeks after the spring experiment of 1994 and the 1- and 4-week disturbances in 1995. The autumn experiment still had a lower species density than control plots after half a year.

The abundance of the small macrofauna species were reduced depending on the type of disturbance and reached background values in dependence of the timing and intensity of disturbance. The recovery time ranged from a few days (the mechanical disturbance), over 2–3 months (spring experiment, 1- and 4-week disturbances) to more than 6 months (autumn experiment).

6.3.2 Mode of Recolonization

To assess the mode of transport and the age of the organisms colonizing the experimental plots, the macrofauna was divided into adults and juveniles and samples were grouped according to their sample location at the periphery or centre of the plots. A comparison of the colonization in the central and peripheral areas should allow to distinguish between a dispersal through the water column (mainly larvae and highly mobile postlarvae) or a dispersal with bedload transport (mainly less mobile postlarvae/adults). With a prevailing water column transport, larvae/postlarvae should show similar abundances all over the plots, with a prevailing bedload transport it was expected that abundances of adults were higher in the periphery. However, this analysis is biased, as four samples had been taken from the periphery and one from the centre of each site. An exemplary data analysis of the spring experiment 1994 and of the control plots gave no significant differences for the individual densities between central and peripheral areas. Juveniles were recorded in the experimental plots after one week, whereas the first adults occurred only after 3 weeks. Thereby it was not possible to separate whether these adults were derived from recruitment in the plots or from colonization from ambient sandflat sediments.

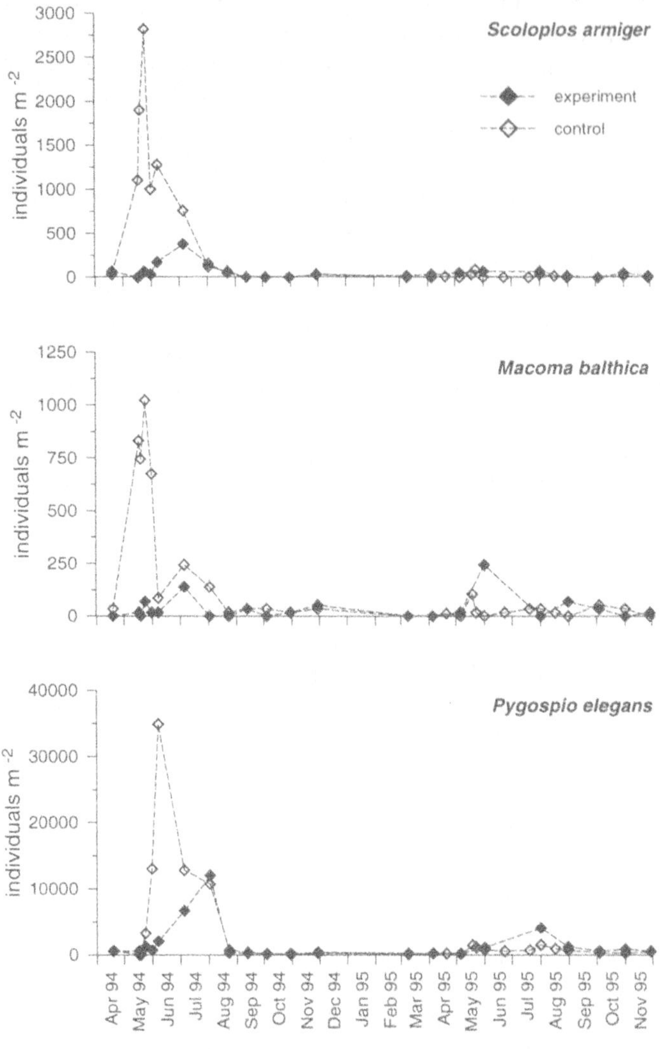

Fig. 6.3.5 Mean individual densities of juveniles of three benthos species in the spring experiment of 1994 (see text for details). (Tecklenborg, unpublished data)

The importance of initial settlement and immigration of postlarvae for the recolonization was demonstrated by the faster regeneration of the spring than the autumn experiment (Fig. 6.3.3). Juveniles were especially abundant during the first few weeks of recolonization in the spring of 1994 and 1995 and in the latter year they occurred with even high numbers in the experimental plots. Among the first colonizers were species who recolonized via postlarval or juvenile stages, e.g. *M. balthica* and *S. armiger* (Fig. 6.3.5). *P. elegans*, which dominated the recolonization in all treatments with high individual numbers, has probably settled in the experimental plots as both larvae and postlarvae (compare Chap. 5.5). In 1995, its abundances in the experimental plots exceeded those from control plots by 2–3 fold. But at that time *P. elegans* was very abundant throughout the study area of the Gröninger Plate, which, together with a high spatial variability of the benthos, requires careful interpretation.

6.4. Recolonization Following the Ice Winter 1995/96

Microphytobenthos

The individual abundances and species number of benthic diatoms were not affected by the ice winter (Keuker-Rüdiger, Ramm, unpubl.). Even immediately after the winter, mean abundances of 75 000–110 000 ind. cm^{-2} were recorded, which was comparable to values recorded in the previous summer. In spring 1996, species number and abundances were a little lower than values from the same time of 1995. The composition of the diatom assemblage according to the abundances of their life modes followed the same seasonal development as in previous years, with mobile diatoms prevailing in the spring period and a higher proportion of sessile and colony-forming diatoms residing during the summer period. As a consequence of the ice winter, three species (*Coscinodiscus concinnus*, *Achnanthes* sp., *Nitzschia* sp.) were recorded in the backbarrier system of Spiekeroog, which had not been found here before. In the summer months, cyanobacteria were frequently found in the sediment.

Nematodes

Some nematodes and other meiofauna organisms were found alive in the ice and thus survived the winter (Blome, unpubl.). But in February 1996 the lowest nematode abundances of the entire study period were recorded (160 ind. 2 cm^{-2}). The individual density doubled monthly and reached values of 1145 ind. 2 cm^{-2} in June, which was higher than abundances recorded in previous years. For single species, the time of year with their highest abundances had shifted. Two scavenging nematode species (*Oncholaimellus calvadosicus* and *Viscosia rustica*) dominated the assemblage in winter. The species number of nematodes recorded in 1996 was about the same as in preceding years.

Macrofauna

Immediately after the ice winter only 4 species 200 cm^{-2} were recorded (Tecklenborg, unpubl.). With increasing colonization of the tidal flats, this value increased to 10–12 species 200 cm^{-2} by June. This value was comparable to the control situation of the spring experiment of 1994.

The abundances of the small macrofauna were very low after the ice winter (about 52 ind. 38.5 cm^{-2} in spring). They increased only with the recruitment in early June. In the following two months, the abundances of juveniles exceeded those of adult benthic organisms by 3–5 times. Due to this recruitment, individual densities rose until the end of the study period in September.

Multivariate analysis showed distinct benthic assemblages before and after the major settlement event in early June. Until the end of May, ostracods and *P. elegans* accounted for 88 % of the individuals in the benthic assemblage, which was composed of 15 species that spring. *S. armiger* and *E. longa* were the next most abundant species. A higher diversity characterized the assemblage again in summer and autumn 1996, when overall 25 species were recorded. *P. elegans* remained the dominant organism (57 % of the individuals). Besides *S. armiger* and *E. longa* the molluscs *Mytilus edulis* and *Macoma balthica* were frequent members of the benthic assemblage.

6.5 Discussion

The sandflats in the backbarrier system of Spiekeroog were recolonized by benthic diatoms, meio- and small macrofauna after both experimentally simulated disturbances and an ice winter event. Neither the experimental nor the natural disturbance caused any long-lasting changes in the benthic assemblages. Given the size of the experimentally disturbed plots in ELAWAT, the study period was not long enough to follow all successional stages. The two prominent macrobenthic species found on the tidal flats, *A. marina* and *L. conchilega*, were recorded on the disturbed plots with relatively few individuals only. Also, few adults recolonized the disturbed plots within the study period.

The defaunation experiments in ELAWAT are the largest of such experiments (20 m^2 sized plots each) carried out in benthic ecology to date. Previous experiments using a similar methodological approach took place on plots of several cm^2 up to m^2 in size, whereby Thrush et al. (1996) could show a scale-dependent recolonization pattern. Also Smith & Brumsickle (1989) proved that the smaller the disturbed site, the faster the recolonization. Varying with the size of disturbed plots is the relative contribution of larval, postlarval or adult stages to the recolonization process (Smith & Brumsickle 1989; Günther 1992; Whitlatch et al. 1998).

To assess the recolonization in the intertidal zone after larger-scale defaunations than those experimentally simulated in ELAWAT would require natural or human-related disturbance events. Thus the disturbance of the ice winter 1995/96 was fortunate for the project (Chap. 7). A further large-scale disturbance occurred in June 1996, mainly in the backbarrier system of the island of Baltrum, and was induced by the bloom of *Coscinodiscus concinnus*. It was later inferred that alga-produced lipids had covered the sediment like an oilfilm and interrupted the water-sediment gas exchange, resulting in a mass mortality of benthic organisms on several km^2 of tidal flats (Höpner 1996). In retrospect, it was a missed opportunity that this event could not be studied during the course of the ELAWAT project.

The design of the large-scale experiments in ELAWAT had artefacts. The sediments covered by tarpaulins for several weeks in 1994, compacted and formed depressions in the sandflat, which were visible for several months. At each low tide, these depressions filled with residual water after the tide receded, which could have affected the recolonization process. A lateral input of sediment can occur into depressions. Furthermore, the water-filled depressions could have contained a concentration of pelagic or other mobile stages of benthic organisms. Juvenile shrimps (*Crangon crangon*) were also recorded frequently in the experimental plots, where they could have preyed upon the developing benthic assemblages. Thus in 1995 the plots were raked after the removal of the tarpaulins to reduce sediment compaction. In a further experiment the time of cover had been shortened to simulate a disturbance of reduced intensity.

The variety of treatments allowed comparisons of the recolonization process in dependence of the timing and intensity of disturbance. It was also possible to compare recolonization in the two consecutive years 1994 and 1995, as in both years plots which had been covered for 4 weeks were uncovered in May. The year 1995 differed from the preceding year by low benthic abundances in the studied tidal flats and an overall low recruitment after a mild winter 1994/95 (Chaps. 5.3; 5.5). These low abundances made it difficult to analyse the data from the experiments in 1995. The lower individual numbers recorded in the experimental plots in 1995 illustrated that recolonization also depends on year to year climatic variations (Chaps. 3.2; 5.1).

The recolonization experiments were of a central interest for ELAWAT (Chap. 2) and thus several sub-projects collaborated for this experiment to sample macro- and meiofauna, microphytobenthos and sediment chemistry. Still, experimental designs to jointly assess recolonization, dispersal and exchange rates of benthic organisms are recommended for the future.

6.5.1
Recolonization in Dependence of the Sediment Chemistry

In the sediments that were covered with a tarpaulin for several weeks, sulphide concentrations increased due to the remineralization activity of the microbes. Although the overall concentrations measured were not especially high, they reached toxic levels for benthic organisms temporarily and locally. Macrobenthic organisms vary in their oxygen demand and sulphide tolerance (Thierman et al. 1996). In similar defaunation experiments on the Baltic coast, sulphide-tolerant ostracods were the first colonizers. They were also among the first colonizers in the experiments in the backbarrier tidal flats of Spiekeroog. However, it remains unclear here, whether they had survived in the disturbed plots because of their sulphide tolerance or whether they had immigrated into these plots actively or by passive sediment transport. A sulphide tolerance is also discussed for opportunistic *Capitella* spp. (Grassle & Grassle 1974). Capitellids occurred in the experimental plots, but did not dominate the benthic assemblage at any stage.

Organic enrichment can also affect the benthic species composition (Pearson & Rosenberg 1978; Flothmann & Werner 1992). The decay of the benthic organisms who died underneath the tarpaulins has probably increased the content of organic matter in these plots. Unfortunately this was not measured and it remains unclear whether there was a related effect on the colonization.

6.5.2 Temporal Development of Recolonization

Diatoms were the first organisms to recolonize the disturbed plots. Individuals of some species survived the defaunation treatment. A similar rapid colonization within days was recorded by Davis & Lee (1983). Diatoms can be transported by resuspension in the water column (Baillie & Welsh 1980; Chap. 5.1). Although the sediment traps implanted at the study site on the Gröninger Plate contained mainly mobile diatoms, colony-forming and sessile species were first to colonize the disturbed plots. These diatoms were probably transported with lateral sediment movement. The composition of the diatom assemblage developing in the disturbed plots depended on the diatoms present in the ambient sediments at the moment of recolonization.

The rapid colonization of diatoms provided a food source for the subsequent recolonization by other benthic organisms (Davis & Lee 1983; Reise 1992). Among the nematodes, who also appeared in the disturbed plots within a few days and weeks, diatom-feeding species (e.g. *Atrochromadora microlaima*, *Innocuonema tentabundum*, *Paracanthonchus caecus*, *Chromadoropsis vivipara*, *Daptonema normandicum*, *Daptonema setosum*) made up the majority of the first colonizers. Meiofauna is known to react actively to diatom (Lee et al. 1977) and bacteria aggregations in the sediment (Thistle 1981; Rieper 1982). The availability of resources can be more decisive for recolonization than dispersal abilities (Fegley 1988). Both emergence and drift as well as passive lateral transport are important mechanisms for meiofauna recolonization (Bell & Sherman 1980; Sherman & Coull 1980; Hagerman & Rieger 1981; Palmer & Brandt 1981; Palmer 1988; Armonies 1994). The rapid colonization by ostracods could have been an indication for bedload transport in the area.

The first macrofauna organisms recorded in the disturbed plots belonged to several feeding modes. The most abundant colonizer, *Pygospio elegans*, has a versatile feeding strategy (Fauchald &Jumars 1979). Polychaetes proved to be early recolonizers in many investigations, as they are mobile and often enough opportunistic (Grassle & Grassle 1974; Simon & Dauer 1977; Smith & Brumsickle 1989; Thrush et al. 1996). The settlement of suspension-feeding polychaetes can be facilitated by the mucus production of microbes and meiofauna (Probert 1984). Furthermore, the developing tube-lawns of spionids can solidify the sediment and thus affect the further recolonization (Probert 1984; Thrush et al. 1996). Thus, the recruitment success of *P. elegans* in the experimental plots could have affected the course of recolonization by other infauna, yet its density did not reach that of tube-lawns reported in the literature cited above. *P. elegans* was also the most dominant small macrobenthic organism in ambient sediments and its abundance in the experimental treatments caused the quick similarity between abundances of disturbed and control plots.

The amphipod *Urothoe poseidonis* was also among the first colonizers in the disturbed plots. This species is very mobile and was often recorded in driftnets on the ambient sandflats (Steuwer & Heuers, pers. comm). Next to polychaetes, amphipods have been recorded as early colonizers in other investigations (Bess & Devlin 1993; Thrush et al. 1996). Their colonizing potential depends on their dispersal ability (Grant 1981), tolerance towards disturbance induced sediment modifications (Gamenick et al. 1996), season and reproductive mode (Simon & Dauer 1977).

The dependence of macrobenthic recolonization on the season and thus the availability of larvae and postlarvae became apparent by the spring and autumn disturbance experiments. Disturbances in spring and summer can be recolonized quickly due to larvae, the high ambient density of juveniles and the high mobility of the organisms (Simon & Dauer 1977; Zajac & Whitlatch 1982 a, b; Bonsdorff & Österman 1985). Adult macrobenthic organisms can also recolonize areas through the water column or by sediment transport (Dauer & Simon 1976; Santos & Simon 1980; Bell & Devlin 1983; Frid 1989). The larger the disturbed site, the more important is the colonization by larvae and postlarvae (Probert 1984; Smith & Brumsickle 1989; Günther 1992; Thrush et al. 1996; Whitlatch et al. 1998). The experiments on the Gröninger Plate were homogeneously colonized in their peripheral and central areas. As the size classes of benthic organisms were not recorded, it was impossible to distinguish between recruitment in those plots and continuous immigration of postlarval and adult stages. Thus the dependence of colonization on the mode of dispersal cannot be solved for these experiments.

Several models exist to explain the succession of soft-bottom communities, taking resource availability, life history strategies, reduced competition and other species interactions into account (Connell & Slayter 1976; Grassle & Sanders 1973; Thistle 1981; Ambrose 1984; Chesney 1985). Zajac & Whitlatch (1984) combine these explanations in a hierarchical approach of succession. During the experiments on the Gröninger Plate only seasonally influenced colonization stages were recorded, but no clear successional stages. The recolonization depended on the availability of settlers and food resources. In some experiments, abundances of diatoms and small macrofauna exceeded background values (cf. "peak of opportunists" Pearson & Rosenberg 1978). The plots of the spring and autumn experiments arranged in 1994 had higher abundances of small macrofauna in the following year than the control plots, which could indicate a reduced competition, as larger macrobenthic species were still missing. Successional stages can be absent, when the benthic community naturally occurring in an area is characterized by opportunistic species (Zajac & Whitlatch 1982b; Gamenick et al 1996). All these findings support the approach by Zajac & Whitlatch (1984) of the combined relevance of environmental factors, life history strategies and biotic interactions for succession in benthic communities.

6.5.3 Comparison of Recolonization After the Ice Winter and After Experimental Disturbances

The rapid increase of species numbers after the ice winter compared to the recolonization in the experiments was due to the successful recruitment of many species in the summer of 1996 and the selective reduction of cold-sensitive species by the winter (Beukema 1989, Chap. 5.6). Timing of recruitment as well as year class strength of juvenile macrobenthos (mainly polychaetes and molluscs) was comparable with the recolonization in the spring experiment 1994. Thus, recolonization after both the natural ice winter disturbance and the experimental disturbances depended on the availability of settling stages. This demonstrates that for the Wadden Sea ecosystem, source populations are essential for a recolonization after ice winters or other disturbances (Heiber 1985). The comparable species numbers and abundances in the spring of 1994 and 1996 are mainly a result of better recruitment of many benthic organisms after cold winters (Chaps. 5.3; 7), as the winter 1993/94 had been cold as well (Chap. 3.2). Beukema (1990) points out that ice winter effects on the species composition are no longer apparent after 1–2 years in the tidal flats of the Wadden Sea. Those species which were frequent in the benthos after the ice winter had also been among the first colonizers in the experiments (except for the autumn experiment). The most prominent species in the recolonization after experimental disturbances and the ice winter was the spionid polychaete *P. elegans*.

Acknowledgements

This chapter is based on the studies in ELAWAT by Anita Tecklenborg, Reinhard Wilhelm, Carlo van Bernem, Dietrich Blome, Roswitha Keuker-Rüdiger, Georg Ramm, Jörn Reichert, Günther Sach, Ruth Zühlke, Ingo Langner, Bärbel Oelschläger and Thomas Höpner. Unpublished data were kindly provided by Anita Tecklenborg, Dietrich Blome, Roswitha Keuker-Rüdiger and Ingo Langner. I thank the reviewer for helpful comments on the manuscript. The project was funded by the German Bundesministerium für Bildung, Wissenschaft, Forschung und Technologie (BMBF) under grant number 03F0112 A and B. The responsibility for the contents of the publication rests with the authors.

References

Ambrose WG (1984) Influence of residents on the development of a marine soft-bottom community. J Mar Res 42: 633–654

Armonies W (1994) Drifting meio- and macrobenthic invertebrates on tidal flats in Königshafen: a review. Helgoländer Meeresunters 48: 299–320

Baillie Welsh (1980) The effect of tidal resuspension on the distribution of intertidal epibenthic algae in an estuary. Estuar Coast Mar Sci 10: 165–180

Bell SS, Devlin DJ (1983) Short-term macrofaunal recolonization of sediment and epibenthic habitats in Tampa Bay, Florida. Bulletin of Marine Science 33 (1): 102–108

Bell SS, Sherman KM (1980) A field investigation of meiofaunal dispersal: tidal resuspension and implications. Mar Ecol Prog Ser 3: 245–249

Beukema JJ (1979) Biomass and species richness of the macrobenthic animals living on a tidal flat area in the Dutch Wadden Sea: effects of a severe winter. Neth J Sea Res 13: 203–223

Beukema JJ (1989) Long-term changes in macrozoobenthic abundance on the tidal flats of the western part of the Dutch Wadden Sea. Helgoländer Meeresunters 43: 405–415

Beukema JJ (1990) Expected effects of changes in winter temperatures on benthic animals living in soft sediments in coastal North Sea areas. In: Beukema JJ, Wolff WJ, Brouns JJWM (Eds) Expected Effects of Climate Change on Marine Coastal Ecosystems. Kluwer Academic Publ, Dordrecht, pp 83–92

Beukema JJ, Essink K, Michaelis H, Zwarts L (1993) Year-to-year variability in the biomass of macrobenthic animals on tidal flats of the Wadden Sea: How predictable is this food source for birds? Neth J Sea Res 31: 319–330

Blome D (1983) Ökologie der Nematoden eines Sandstrandes der Nordseeinsel Sylt. Mikrofauna Meeresboden 88: 1–76

Bonsdorff E, Österman CS (1985) The establishment, succession and dynamics of a zoobenthic community - an experimental study. In: Gibbs P-E (Ed) Proceedings of the 19th EMBS. Cambridge University Press, pp 287–297

Chesney EJ jr (1985) Succession on soft-bottom benthic environments: are pioneering species really outcompeted? In: Gibbs PE (Ed) Proceedings of the 19th EMBS 1984. Cambridge University Press, pp 277–286

Connell JH, Slayter RO (1976) Mechanisms of succession in natural communities and their roles in community stability and organisation. Am Nat 111: 1119–1144

Dauer DM, Simon JL (1976a) Repopulation of the polychaete fauna of an intertidal habitat following natural defaunation: species equilibrium. Oecologia (Berlin) 22: 99–117

Davis MW, Lee H II (1983) Recolonization of sediment-associated microalgae and effects of estuarine infauna on microalgal production. Mar Ecol Prog Ser, Vol. 11: 227–232

Fauchald K, Jumars PA (1979) The diet of worms: a study on polychaete feeding guilds. Oceanogr Mar Biol Ann Rev 17: 193–284

Fegley SR (1988) A comparison of meiofaunal settlement onto the sediment surface and recolonization of defaunated sandy sediment. J Exp Mar Biol Ecol 123: 97–113

Flothmann S, Werner I (1992) Experimental eutrophication on an intertidal sandflat: effects on microphytobenthos, meio- and macrofauna. In: Colombo G, Ferrari I, Ceccherelli VV, Rossi R (Eds) Marine Eutrophication and Population Dynamics, Proceedings of the 25th EMBS. Tolsen & Olsen, Fredensborg, pp 93–100

Frid CLJ (1989) The role of recolonization processes in benthic communities, with special reference to the interpretation of predator-induced effects. J Exp Mar Biol Ecol, Vol. 126:: 163–171

Gamenick I, Jahn A, Vopel K, Giere O (1996) Hypoxia and sulphide as structuring factors in a macrozoobenthic community on the Baltic Sea shore: colonisation studies and tolerance experiments. Mar Ecol Prog Ser 144: 73–85

Grant J (1981) Sediment Transport and Disturbance on an Intertidal Sandflat: Infaunal Distribution and Recolonization. Mar Ecol Prog Ser 6: 249–255

Grassle JF, Grassle JP (1974) Opportunistic life histories and genetic systems in marine benthic polychaetes. J Mar Res 32: 254–289

Grassle JF, Sanders HL (1973) Life histories and the role of disturbance. Deep Sea Research 20: 643–659

Günther C-P (1992) Dispersal of intertidal invertebrates: a strategy to react to disturbances on different scales? Neth J Sea Res 30: 45–56

Hagerman GM, Rieger RM (1981) Dispersal of benthic meiofauna by wave and current action in Bogue Sound, North Carolina, USA. Mar Ecol 2: 245–270

Hartmann-Schröder G (1971) Annelida, Borstenwürmer, Polychaeta. Die Tierwelt Deutschlands 58: 1–594

Heiber W (1985) Möglichkeiten der Wiederbesiedlung von Wattflächen nach "Umweltkatastrophen". Seevögel 6: 89–97

Höpner T (1996) Schwarze Tage im Nationalpark. Spektrum der Wissenschaft, August, pp 16–21

Höpner T, Michaelis H (1994) Sogenannte "Schwarze Flecken" - ein Eutrophierungssymptom des Wattenmeeres. In: Lozàn JL, Rachor E, Reise K, v. Westernhagen H, Lenz W (Eds) Warnsignale aus dem Wattenmeer. Blackwell Berlin, pp 153–159

Langner I (1997) Statistische Klassifizierung der räumlichen und zeitlichen Heterogenität von Nährstoffkonzentrationen im Porenwasser von Wattsedimenten: Identifizierung von räumlichen Mustern und zeitlichen Entwicklungen. Berichte Forschungszentrum Terramare 6

Lee JJ, Tietjen JH, Mastropaolo C, Rubin H (1977) Food quality and the heterogeneous spatial distribution of meiofauna. Helgoländer wiss Meeresunters 30: 272–282
McCall PC (1977) Community patterns and adaptive strategies of the infaunal benthos of Long Island Sound. J Mar Res 35: 221–266
Palmer MA (1988) Dispersal of marine meiofauna: a review and conceptual model explaining passive transport and active emergence with implications for recruitment. Mar Ecol Prog Ser 48: 81–91
Palmer MA, Brandt, RR (1981) Tidal variation in sediment densities of marine benthic copepods. Mar Ecol Prog Ser 4: 207–212
Pearson TH, Rosenberg R (1978) Macrobenthic succession in relation to organic enrichment and pollution of the marine environment. Oceanogr Mar Biol Ann Rev 16: 229–311
Probert PK (1984) Disturbance, sediment stability, and trophic structure of soft-bottom communities. J Mar Res 42: 893–921
Reise K (1992) Grazing on sediment shores. In: Plant-Animal Interactions in the Marine Benthos. DM John, SJ Hawkins & JH Price (Eds). Systematics Association Special Volume No 46, Clarendon Press, Oxford: 133–145
Reise K, Kolbe K, de Jonge V (1994) Makroalgen und Seegrasbestände im Wattenmeer. In: Lozàn JJ, Rachor E, Reise K, v. Westernhagen H, Lenz W (Eds) Warnsignale aus dem Wattenmeer. Blackwell Berlin, pp 90–100
Rhoads DC, McCall PL, Yingst JY (1978) Disturbance and production on the estuarine seafloor. Am J Sci 66: 577–586
Rieper M, (1982) Feeding preferences of marine harpacticoid copepods for various species of bacteria. Mar Ecol Prog Ser 7: 303–307
Santos SL, Simon JL (1980) Marine soft-bottom community establishment following annual defaunation: larval or adult recruitment? Mar Ecol Prog Ser 2: 235–241
Savidge WB, Taghon GL (1988) Passive and active components of colonization following two types of disturbance on intertidal sandflat. J Exp Mar Biol Ecol 115: 137–155
Sherman KM, Coull BC (1980) The response of meiofauna to sediment disturbance. J Exp Mar Biol Ecol 46: 59–71
Simon JL, Dauer DM (1977) Reestablishment of a benthic community following natural defaunation. In: Coull BC (Ed) Ecology of marine benthos. Belle W. Baruch Library in Marine Science, No. 6. Columbia University of South Carolina Press, pp 139–154
Smith CR, Brumsickle SJ (1989) The effects of patch size and substrate isolation on colonization modes and rates in an intertidal sediment. Limnol Oceanogr 34 (7), pp 1263–1277
Thiermann F, Niemeyer A-S, Giere O (1996) Variations in the sulphide regime and the distribution of macrofauna in an intertidal flat in the North Sea. Helgoländer Meeresunters 50: 87–104
Thistle D (1981) Natural physical disturbances and communites of marine soft bottoms. Mar Ecol Prog Ser 6: 223–228
Thrush SF, Whitlatch RB, Pridmore RD, Hewitt JE, Cummings VJ, Wilkinson MR (1996) Scale-dependent recolonization: The role of sediment stability in a dynamic sandflat habitat. Ecology 77 (8): 2472–2487
Whitlatch RB, Lohrer AM, Thrush SF, Pridmore RD, Hewitt JE, Cummings VJ, Zajac RN (1998) Scale-dependent benthic recolonization dynamics: life-stage-based dispersal and demographic consequences. Hydrobiologia 375/376: 217–226
Zajac RN, Whitlatch RB (1982a) Responses of Estuarine Infauna to Disturbance. I. Spatial and Temporal Variation of Initial Recolonization. Mar Ecol Prog Ser 10: 1–14
Zajac RN, Whitlatch RB (1982b) Responses of Estuarine Infauna to Disturbance. II. Spatial and Temporal Variation of Succession. Mar Ecol Prog Ser 10: 15–27
Zajac RN, Whitlatch RB (1984) A hierarchical approach to modelling soft-bottom successional dynamics. In: Gibbs PE (Ed) Proceedings of the 19th EMBS 1984. Cambridge University Press, pp 265–276

7 Effects of the Ice Winter 1995/96

Carmen-Pia Günther & Verena Niesel

Abstract: Investigations in the framework of ELAWAT showed that cold winters with ice formation lead to mass mortalities of cold sensitive species in all parts of the Wadden Sea ecosystem and to an increased mobility of organisms. While in the phytoplankton cold resistant species became dominant due to their increased growth, changes in dominance in the benthos were due to mortality of cold-sensitive species. Inspite of the massive die-off of large individuals of the macrofauna, no strong increase in TOC (total organic carbon) was measured in the sediment.

In the following regeneration, the seasonal development in the water column differed markedly from the preceding years, which could be explained by extraordinarily calm weather, low water temperatures and competition between algal species. In the zooplankton the meroplankton showed severe changes in the temporal sequence of appearance and density of species when compared to 1994 and 1995.

During a "black area event" (end of May/beginning of June 1996) a short term reduction in depth of the oxidized sediment layer was measured together with a small increase in TOC and bacterial activity. Besides this event, regeneration of the benthic community by successful recruitment of several species was undisturbed. Due to the fact that the "black area event" and recruitment (larvae were already in the water column, settlement occurred after redevelopment of an oxidized layer) were temporally disconnected, a fast recovery of even strongly affected areas was possible.

Among the macrozoobenthos the trophic groups of the suspension-feeders and epibenthic predators were most negatively affected by the ice winter. The disappearance of suspension-feeding species such as *Mytilus edulis*, *Cerastoderma edule* or *Lanice conchilega* was partly outbalanced by very good recruitment of single species (*M. edulis*, *Mya arenaria*). Predation pressure by epibenthic predators was low and thus not a limiting factor for the recruitment of prey species.

Most parts of the macrobenthic association recovered fast from the effects of the ice winter (< 1 year), with the exception of populations of the more longliving cold sensitive species which presumably take 3–4 years to recover completely.

Comparing the ice winter effects observed in the backbarrier system of Spiekeroog (East Frisian Wadden Sea) and Sylt (North Frisian Wadden Sea) the duration of the ice cover was longer in the Sylt area, but had in principle the same effects on the pelagic and benthic system (modification of the species composition towards cold adapted/cold tolerant species). The largest differences between the areas were observed in the temporal development of the pelagic system (dominance of different species, different temporal course in the meroplankton). There are indications that additional differences occurred subtidally. As the source populations for the recolonization of the intertidal by cold sensitive species lie subtidally, future studies on recruitment after ice winters should include both the intertidal and subtidal benthic systems.

7.1 Introduction

In the three years of investigation meteorological parameters varied strongly, resulting in a pronounced variability of seasonal processes in the Spiekeroog backbarrier system (Chap. 3.5). Especially in the winter 1995/96 extreme conditions were observed, including a cold period of about three months with ice cover for 48 days in the East Frisian Wadden Sea. Because the effects of disturbances were of major interest to ELAWAT (Chap. 2), this cold period was an opportunity to study its effects on the chemical and biological processes in the Wadden Sea. The direct effects of the ice winter as well as the following regeneration were observed until September 1996. The results give an indication of mechanisms taking place after large-scale climatic perturbations in the Wadden Sea ecosystem.

7.2 The "Disturbing" Effects of the Ice Winter 1995/96

The effects of a cold winter vary with the duration of the ice cover, the air temperature and the location within the intertidal. Ice-scour mainly affects exposed parts of the intertidal such as fringes of tidal channels or mussel beds. The co-occurring long term periods of easterly winds result in a prolonged emergence of areas close to the high water level (HWL), which are subsequently exposed to the extreme temperature regime of the air. Accordingly, the effects on the biota may be ordered as follows.

7.2.1 Mechanical Effects

Already at the end of December 1995 an ice layer of several cm up to over one meter thickness had developed on the sandflats of the Gröninger Plate. At this time, the surface temperatures of the sediment reached about -2° C; the mean temperature over the three cold months (middle of December 1995 until the middle of March 1996) was -0.5° C.

By movements of the ice back and forth with the tides epibenthic structures such as *Lanice conchilega* tubes and mussel beds (*Mytilus edulis*) disappeared from the studied tidal flats Gröninger Plate and Swinnplate (Figs. 3.1.3. and 3.1.4). The biogenic mud layer on the Swinnplate was removed by the ice as well as by the loss of the *Mytilus* cover (Chap. 3.4). Ice-scour was also responsible for the appearance of mud in formerly pure sandy areas of the Swinnplate and Gröninger Plate and already in spring *Heteromastus filiformis* inhabited these areas (Chap. 3.5).

Besides the mechanical destruction of epibenthic structures by the ice movement, they were also frozen onto the ice and subsequently transported with the ice and tides. Freezing onto the ice was also seen as a reason for the density reduction of mussels observed after ice winters (Strasser pers. comm.). Both effects may occur simultaneously, depending on the different locations of mussel beds within the intertidal.

Not only epibenthic structures were frozen onto the ice at low tide, but also entire sediment layers, containing, according to ELAWAT results, living nematodes and other meiofaunal taxa.

7.2.2
Direct Effects of Low Temperatures

The ice winter of 1995/96 caused changes in the species composition not only in the benthic realm as already observed by Beukema (1990, 1992), but also in the pelagic realm of the Spiekeroog backbarrier system.

In the pelagic system of the German Bight a *Coscinodiscus wailesii* bloom in February and March was followed by a bloom of the closely related cold adapted species *C. concinnus* (Henke 1997). In February no cells of *Coscinodiscus* spp. were observed in the coastal zone between the island of Spiekeroog and the mainland. Here, the *C. concinnus* bloom started at the beginning of March (Niesel 1997).

Measurements of the bacterial enzyme activity showed that microorganisms were active over winter. This activity was most likely the reason that the TOC of the sediment did not increase during and after the ice winter although a mass mortality of some macrofauna species was recorded in the sediment.

The benthic fauna showed pronounced changes after the ice winter. Locally some cold sensitive nematode species disappeared, whereas other species found refuge in deeper sediment layers. In the intertidal and nearby subtidal apparently all *L. conchilega* died off, as well as all cockles (*Cerastoderma edule* only intertidal observations). Reichert & Dörjes (1980) assumed that the low abundance of more mobile species such as *Nephtys hombergii*, which was also observed after the ice winter 1995/96, may be explained by mortality as well as migration into the subtidal

Marine organisms living in the higher tidal flats were affected by long periods of low temperatures in combination with the lack of inundation caused by easterly winds. Here, cold sensitive species disappeared earlier than at periodically flooded sites. As a long-term effect an increased mortality was observed even in cold tolerant species (*Scoloplos armiger*, *Mya arenaria*), presumably resulting from oxygen deficiency (Reichert & Dörjes 1980; Beukema 1989). Thus, mortality in cold winters may be site-specific (compare also Beukema 1985, 1990).

In February 1996 the total amount of resting waders in the entire East Frisian Wadden Sea area was reduced by 35% compared to mean values for the same months of the preceding years (Exo, Ketzenberg, unpubl.). But exchange rates of local bird stocks had increased, as many birds left the area while others, e.g. oystercatchers, wigeons and common gulls came in from even more unfavourable regions. In March/April 1996 no differences to the preceding years could be detected concerning total bird numbers and species composition.

According to Beukema et al. (1993), migration will not take the birds to totally ice free areas, because cold temperatures normally affect the entire Wadden Sea, although to a different degree. The duration of ice cover varies within the Wadden Sea and the birds can migrate to an area of shorter ice cover duration.

7.3 The Seasonal Development 1996 (Regeneration)

Due to the effects of the ice winter, the seasonal development in 1996 was strongly modified. In the following section results from ELAWAT will be presented supplemented by data from nearby areas.

7.3.1 The Pelagic System

After the ice melt, the salinity was unusually high (32–35 PPT) until May, which was also due to an unusually low freshwater input (Henke 1997). Additionally an upwelling front situated off the East Frisian Island may have caused an import of water of higher salinity into the Wadden Sea (Henke 1997).

In spring, the turbidity of the coastal water was extremely low and at the same time sunny weather prevailed. While in previous years the euphotic zone had a "normal" extension of 1–2 m water depth (compare Colijn 1981; Postma 1984), it reached down to 6–8 m water depth in spring 1996. Thus, the light conditions for the growth of phytoplankton and microphytobenthos were extremely favourable, even in the subtidal.

Already in February the nitrate concentration dropped below the long term average. From April onwards ammonia was also strongly reduced. In 1996 the silicate concentrations declined earlier as usual, while from March onwards the concentrations of dissolved phosphate were higher than the long term average (compare Chap. 5.1).

Caused by the ice winter, the species composition of the phytoplankton was clearly modified. From March onwards, a bloom of the large, cold adapted diatom *Coscinodiscus concinnus* was observed in the Wadden Sea as well as in the German Bight. This species has the ability to produce intracellular oil. When the bloom died down, this fat was released, producing a 640 km^2 sized lipid film on the water surface, which remained stationary off the East Frisian Islands for about 3 weeks. The same phenomenon was observed in 1947 in the northern part of the North Sea, when – also after an extremely cold winter – a bloom of *C. concinnus* had occurred (Grøntved 1952). Diatom species such as *Asterinellopsis glacialis* or *Guinardia delicatula*, which are representative for the coastal spring phytoplankton bloom in normal years, were observed in low densities only (compare Chap. 5.1).

Instead of *Phaeocystis globosa*, the cold adapted species *Phaeocystis pouchetii* was observed after the diatom bloom in 1996, most likely resulting from the very low water temperatures (Chap. 5.1).

In 1996, the first meroplanktonic larvae were recorded later in the year compared to years with mild winters. The larval density reflected to some extent the sensitivity of the adults to cold water temperatures, e.g. larvae of *L. conchilega* and *Magelona papillicornis* were found in low densities only.

In contrast to preceding years, larvae of all bivalve species appeared with maximum densities at the same time (June 1996) in the plankton. The larval densities of all species, except for *Macoma balthica* and *M. edulis*, were comparable to results from 1994 and 1995. For *M. balthica*, the lowest larval densities of all three years of investigation were observed in 1996. This is not in accordance with the mechanism described by Honkoop & van der Meer (1997) and must probably be related to the unusual late spawning of this species in 1995 (Günther et al. 1998). The density of *M. edulis* larvae was increased after the ice winter 1995/1996.

Some copepod species apparently took advantage from the intracellular oil produced by *C. concinnus*. Already at the beginning of the algal bloom they had incorporated small oil droplets, which is normally observed at the beginning of autumn. Highest densities of copepodid stages were recorded as early as April, indicating that reproduction had already taken place early in the year.

7.3.2 The Benthic System

Directly after the ice melt, the oxidized layer of the sediment had the same depth as in the previous years. With the temperature related rise of remineralization in April/May the RPD-layer reached closer to the sediment surface. In summer the oxidized sediment layer of the sandflat at the Gröninger Plate (GP3, Fig. 3.1.4) had a depth of 0.5–1 cm, which was lower than in the previous years (> 1 cm). The lowest depth was measured in June and coincided with the appearance of black areas at the sediment surface (see below).

With the loss of mussel beds on the tidal flat area Swinnplate, biodeposition of fine sediment ceased. This caused a situation in early 1996 in which the morphodynamics of the area was characterized by erosion instead of sediment accumulation. In the sediment, the protein and fatty acid contents were lower than in the preceding years, indicating that the organic material was of lower nutritional quality (Chap. 5.2.1). Only in the few mud patches left, higher concentrations of these compounds were measured. In the sediments of the site were mussel beds existed in previous years, maximum numbers of bacteria were only half as high, but the enzyme activities (glucosidase and phosphatase) were higher.

At the Gröninger Plate, the seasonal development of the total amount of bacteria resembled the normal pattern. Bacterial density strongly increased from 29th of May until 11th of June, caused by an increase in sediment temperature and by the input of organic matter derived from the sedimentation of the blooms of *C. concinnus* and *P. pouchetti*. This was also followed by the occurrence of bacterial methanococci, *Merismopedia* and other cyanobacteria in the sediment.

During the course of the summer 1996 the Chl.-*a* concentration in the sediment was 3–4 times lower when compared to 1994 and 1995. In the period of the "black area event", a short-term decline in the density of benthic diatoms was observed. The number of microphytobenthic species was comparable to that of preceding years, but included three new species not observed in the benthos until then. From March onwards high numbers of the pelagic diatom *C. concinnus* were found in the intertidal sediments, later on *P. pouchetii* was observed also. This points to a high sedimentation rate of the bloom of both species.

The total abundance of meiofauna reached maximum values in June with nematodes being numerically dominant. For nematodes, similar to the macrofauna, temporal shifts in maximum abundances of some species as well as variations in dominance structure could be observed in 1996 (compare Chap. 6.4). In spring species of the nematode family Oncholaimidae, which are potentially scavengers, dominated. Later in the year, diatom feeding species took over. Total abundances of nematodes in spring and summer were generally higher compared to preceding years.

The effects of cold winters on the intertidal and subtidal macrofauna are well documented (Ziegelmeier 1964; Beukema 1979, 1982, 1985, 1989, 1990; Beukema & Cadée 1996; Beukema et al. 1978, 1988, 1993; Buhr, 1981; Dörjes 1992 a, b; Dörjes et al. 1986; Reichert & Dörjes 1980). In both habitats temporary changes in the macrobenthic association have been observed. Cold sensitive species were either drastically reduced in abundance or extinguished locally. The following regeneration varies with the potential of single species to reproduce and disperse and to the distance of source populations. Thus, depending on the species, regeneration may take up to several years (compare Chaps. 5.3.6 and 6.4).

For example, at the Gröninger Plate and Swinnplate the intertidal part of the *L. conchilega* population did not recover during the period of field observations. Measurements of the thorax diameter indicate that some of those specimens present in autumn 1996 may reproduce in the next summer. This implies that the age/size structure would be faster re-established as the abundance, which is estimated to take 2–4 years (Hertweck 1995; Chap. 5.3.6), provided no further cold winter occurs.

In summer 1996 mussel beds re-established at the Swinnplate by good recruitment. Similar to *L. conchilega* patches, the mussel beds consisted only of individuals of the 0-group. The development of a mature bed with several age classes takes several years. A similar regeneration pattern is assumed to take place in *C. edule* (Dörjes 1992).

Depending on the late availability of meroplanktonic larvae, settlement and recruitment of nearly all macrobenthic species occurred later in the year when compared to the previous two years of investigation. At the Gröninger Plate juveniles of the soft shell clam, *M. arenaria*, reached the extreme abundance of up to 200 000 ind m^{-2}. Other species, such as *H. filiformis* or *Ensis americanus* recruited in higher abundances as in 1994 and 1995.

The extraordinarily good recruitment of macrofauna, especially bivalves, after cold winters was attributed to several non exclusive causes, being probably all relevant to a different degree:

1. In a cold winter, gametes produced in the preceding summer are not resorbed for the nutrition of the adults. Thus, the number of eggs spawned is higher and subsequently the number of recruits is enlarged as well (Honkoop & van der Meer 1997; compare Chap. 5.1).
2. It is assumed that a reduction of suspension-feeders increases survival of planktonic larvae, thus positively affecting recruitment (Reichert & Dörjes 1980). This effect was often postulated, but has so far only been supported by

experimental data (e.g. André & Rosenberg 1991). The quantitative importance of suspension-feeders on recruitment cannot be estimated as the populations are usually not quantified on a system level.

3. The decline of epibenthic predators esp. *Crangon crangon* and *Carcinus maenas* due to the cold winter, positively affects the survival rate of newly settled macrofauna. According to Kruse (pers. comm.) young of the year 1996 of cockles outgrew the prey size range of their epibenthic predators. Due to the combination of a high growth rate with a low mortality rate, recruitment after the winter was good. In the frame of ELAWAT length frequency measurements of the species *Arenicola marina*, *C. edule*, *M. balthica* and *M. arenaria* also indicated good growth of juveniles after the cold winter. As for the predator, the intertidal population of the shore crab *C. maenas* was drastically reduced. Until autumn 1996 neither mean abundance nor age structure of their population had recovered. This effect probably persisted into the following years as the recruitment of shore crabs was low in 1996.

The three factors limiting recruitment mentioned above work after each cold winter. But they do not explain why different species show good recruitment after different cold winters or why recruitment of *M. balthica* was poor in 1996.

It seems that in addition to the abundance of adults the seasonal development in the preceding year has to be considered. For example in years of bad feeding conditions adults probably produce only small amounts of gametes. In case of a successive cold winter, even a high survival rate of larvae and juveniles may not result in good recruitment. The temporal development is of special importance if a series of cold winters occurs as in the period of 1985–1987 (Beukema 1992). First results of a model on effects of cold winters on the population dynamics of *L. conchilega* indicate long term disappearance of the intertidal population in case of successive cold winters.

At both study areas, the Swinnplate and Gröninger Plate, highest numbers of species were recorded in July 1996, as a consequence of the temporal delay of recruitment.

A potential indirect effect of the ice winter on the bird fauna might be the lack of food due to the mortality of some preferred prey species such as *L. conchilega*. The example of the grey plover showed that the range of potential prey organisms is wide enough for the birds to switch to those which are still available (Ketzenberg, Exo, unpubl. data).

7.4 The "Black Area Event" 1996

During the regeneration phase after the ice winter 1995/96 the occurrence of extended intertidal areas with a black sediment surface were observed. The macrofauna from these areas protruded from the sediment and died. The occurrence of such black areas and the reaction of the macrofauna varied within the different backbarrier systems. Most affected was the backbarrier system behind the island of Baltrum.

Parallel to the appearance of the black areas an extended lipid film was observed on the water surface north of the East Frisian Islands. Investigations carried out by the BSH (Bundesamt für Seeschiffahrt und Hydrographie) showed that this film was of natural origin produced by lipids derived from the diatom *Coscinodiscus concinnus* (compare Delafontaine & Flemming 1997). Cells of this species were also found in the sediments of the East Frisian Wadden Sea. This indicates that algae and parts of the lipid film were transported by tidal currents into the Wadden Sea where sedimentation took place followed by the development of anoxic sediment areas. At the same time, an increase in the density of bacteria, a white surface film covering the sediment and an increase of lipid concentrations in the sediment were observed. The additional input of organic matter derived from the algal bloom as well as the reduced gas exchange between the sediment and the water column due to a sealing effect of the lipid film may have caused the change in sediment chemistry from aerobic to anaerobic conditions (Höpner 1996).

Aerial surveys from the 28/30th of May and 12th of June 1996 over the entire East Frisian Wadden Sea showed that anoxic sediment surfaces occurred in the Spiekeroog backbarrier tidal flats as well. The oxidized sediment layers of the Gröninger Plate and the Swinnplate were extremely thin, indicating that the main study area of ELAWAT was also affected by this event. Increases of bacterial counts and TOC concentrations were measured in the sediments. At the same time, the density of diatoms dropped in the microphytobenthos. According to experimental results (Villbrandt, unpubl. data) the regeneration of the oxic layer depends on temperature, the concentration of organic matter, current speed and bioturbation. Under stagnant conditions reoxidation and regeneration of anoxic sediment surfaces was either strongly slowed down or stopped completely.

Neither at the Gröninger Plate nor at the Swinnplate an increased macrofauna mortality could be observed. As larval settlement in most of the species took place after the "black area event", effects on recruitment were not detectable. Possibly recruitment of the polychaete *H. filiformis*, feeding on bacteria, was positively affected by the increase in bacterial density (compare Chap. 5.3).

In 1996 the *C. concinnus* bloom and its sedimentation in the Wadden Sea were an indirect effect of the ice winter and its hydrographic peculiarities. But *C. concinnus* blooms may occasionally also develop after mild winters and their sedimentation may cause – as sedimentation of blooms of other algal species – an overload of organic matter in the sediment and an increased mortality of macrofauna (Cadée 1996; Michaelis, pers. comm.).

Combined effects of the ice winter and the "black area event" could not be detected at the study sites of ELAWAT. They were more likely to have happened in areas such as the Baltrum tidal flats. But also in these areas good recruitment of macrofauna took place (Günther, pers. obs.) and in 1997, after a period of one month of high summer temperatures, no return to anoxic conditions was observed (Höpner 1997).

7.5 Comparison of Ice Winter Effects in Different Parts of the Wadden Sea

The comparison of results obtained in the East and North Frisian part of the Wadden Sea may elucidate effects of cold winters and subsequent regeneration on a geographical scale.

With 66 days of ice cover the cold period prevailed longer in the North Frisian Wadden Sea around the island of Sylt when compared to 48 days on the East Frisian coast. In both systems salinity had increased, either caused by a lack of precipitation or by an upwelling front. In the pelagic of both areas a bloom of cold adapted agal species was observed: close to Sylt island it was produced by *Odontella aurita*, close to Spiekeroog at both sites by *C. concinnus*. Simultaneously low concentrations of silicate were recorded, indicating that this nutrient became limiting soon, the same applied for nitrate. The intracellular oil production of *C. concinnus* was not observed in the North Frisian part of the Wadden Sea.

Reactions of cold sensitive species varied between the two regions (Table 7.5.1). While *L. conchilega* disappeared in the entire Wadden Sea the intertidal population of the cockle (*C. edule*) was reduced by 69 % in the northern Wadden Sea. The local differences may be due to reinvasion of adult cockles surviving in the subtidal. Another site-specific response was observed for *N. hombergii*. After the winter 1995/96 this species disappeared almost completely from the intertidal flats at the island of Sylt, while it was regularly recorded in the backbarrier tidal flats of Spiekeroog Island, although in very low abundances. According to Beukema (1985) the likelihood for survival of *N. hombergii* is higher in areas close to the low water level than in areas with higher tidal elevation. Thus, the differences may result from different tidal elevations of the sampling sites in the two areas. Following the ice winter, an extremely good recruitment of *Nereis diversicolor* took place in the Spiekeroog backbarrier tidal area, but not of *Phyllodoce mucosa* or *Capitella capitata*, as observed in tidal flats at Sylt.

In both areas, the recruitment of bivalves was of special interest, as the existing literature indicated a link between cold winters and recruitment success in intertidal bivalve species (e.g. Beukema 1979, 1982). With the exception of *M. balthica* whose larval densities were low in the backbarrier area of Spiekeroog, but dominated the meroplankton in the waters around Sylt, the rank order of species by their larval densities was similar in both regions: *E. americanus*, followed by a cluster containing *M. arenaria*, *M. edulis* and *C. edule*. But the temporal pattern of occurrence differed clearly. In the Wadden Sea near Sylt, several maxima of bivalve larvae were observed in the plankton, with each one clearly dominated by a specific species. In the backbarrier tidal flats of Spiekeroog, nearly all species reached maximum larval densities at the same time. In both areas the proportion of larval densities was not reflected in the proportion of abundances in recruits (Table 7.5.1, comparison meroplankton/recruitment strength).

Table 7.5.1 Comparison of ice winter effects. Results were obtained at different sites in the East Frisian Wadden Sea (ELAWAT) and in the North Frisian Wadden Sea close to the island of Sylt (data of the Biologische Anstalt Helgoland, Sylt; Strasser pers. comm.). Meroplankton was not sampled at Norderney and Dornumer Nacken.

	East Frisian coast		**North Frisian coast**
Norderney	Dornumer Nacken	Spiekeroog	Sylt
Meroplankton (Bivalves)			
Rank order of abundance			
		1. *Ensis americanus* 2. *Mya arenaria* *Mytilus edulis* *Cerastoderma edule* 3. *Macoma balthica*	1. *Macoma balthica* 2. *Ensis americanus* 3. *Mya arenaria* *Mytilus edulis* *Cerastoderma edule*
Temporal sequence of occurrence			
		simultaneously	1. *Macoma balthica* 2. *Ensis americanus* *Mya arenaria* *Mytilus edulis* *Cerastoderma edule* 3. *Ensis americanus*
Macrofauna			
Localized extinction of population			
Lanice conchilega *Cerastoderma edule* *Nephtys hombergii* *Phyllodoce maculata* *Harmothoe sarsi* *Crangon crangon*	*Lanice conchilega* *Cerastoderma edule* *Nephtys hombergii* *Malmgreniella lunulata*	*Lanice conchilega* *Cerastoderma edule* *Mytilus edulis*	*Lanice conchilega* *Nephyts hombergii* *Lepidochiton cinerea*
Decline in intertidal abundance			
Aphelochaeta marioni *Capitella capitata* *Heteromastus filiformis* *Hydrobia ulvae* *Macoma balthica* *Mya arenaria* *Scoloplos armiger* *Tubificoides benedeni* *Urothoe poseidonis*	*Nephtys cirrosa* *Phyllodoce mucosa*	*Capitella capitata* *Macoma balthica* *Nephtys hombergii* *Phyllodoce maculata*	*Cerastoderma edule* *Corophium arenarium* *Crangon crangon* *Mya arenaria*

Table 7.5.1 (continued)

Norderney	East Frisian coast Dornumer Nacken	Spiekeroog	North Frisian coast Sylt
Unaffected species			
Arenicola marina *Corophium arenarium* *Eteone longa* *Nereis diversicolor* *Pygospio elegans*	*Arenicola marina* *Macoma balthica*	*Arenicola marina* *Pygospio elegans* *Scoloplos armiger*	
Extremely good recruitment			
Cerastoderma edule *Hydrobia ulvae* *Macoma balthica* *Mya arenaria* *Nereis diversicolor* *Pygospio elegans*	*Cerastoderma edule* *Macoma balthica* *Mya arenaria*	*Heteromastus filiformis* *Hydrobia ulvae* *Mytilus edulis* *Nereis diversicolor* *Pygospio elegans* highest: *Mya arenaria*	*Macoma balthica* *Mya arenaria* *Mytilus edulis* highest: *Capitella capitata* *Cerastoderma edule* *Nereis diversicolor* *Phyllodoce mucosa*
Poor recruitment			
Lanice conchilega *Nepthys hombergii*		*Carcinus maenas* *Lanice conchilega* *Malmgeniella lunulata*	*Carcinus maenas* *Lanice conchilega* *Nephtys hombergii*

It appears that the subtidal parts of the macrofauna populations were not strongly affected by the ice winter in the North Frisian Wadden Sea near Sylt (Strasser, pers. comm.). According to results from Beukema et al. (1988), which showed stronger subtidal effects of ice winters the more north-westerly the area of investigation was situated, one should have assumed that effects of low temperatures should extend far into the subtidal especially at the North Frisian coast. Adversely, at the East Frisian part of the Wadden Sea, the subtidal parts of the population of at least *L. conchilega* were severely affected, eventually caused by upwelling of cold water.

The comparison of both regions elucidates that the main differences in ice winter effects take place in the subsequent seasonal development in the pelagic system (dominating algal species, temporal dynamics of meroplankton). In the benthos of both regions, the expected effects on cold sensitive species were observed, followed by a localized good recruitment of some macrofauna species.

Acknowledgements

This chapter is based on the studies of all sub-projects of ELAWAT. They were published as cited in the text, and additionally unpublished data were kindly provided by the different working groups. Many thanks to the reviewer for the valuable comments to the text. We also thank the Biologische Anstalt Helgoland on Sylt, especially M. Strasser for providing information on ice winter effects in the Sylt area. ELAWAT was funded by the German Bundesministerium für Bildung, Wissenschaft, Forschung und Technologie (BMBF) under grant number 03F0112 A and B. The responsibility for the contents of the publication rests with the authors.

References

André C, Rosenberg R (1991) Adult-larval interactions in the suspension-feeding bivalves *Cerastoderma edule* and *Mya arenaria*. Mar Ecol Prog Ser 71: 227–234

Beukema JJ (1979) Biomass and species richness of the macrobenthic animals living on a tidal flat area in the Dutch Wadden Sea: Effects of a severe winter. Neth J Sea Res 13: 203–223

Beukema JJ (1982) Annual variation in reproductive success and biomass of the major macrozoobenthic species living in a tidal flat area of the Wadden Sea. Neth J Sea Res 16: 37–45

Beukema JJ (1985) Zoobenthos survival during severe winters on high and low tidal flats in the Dutch Wadden Sea. In: Gray JS, Christiansen ME. Marine Biology of Polar Regions and Effects of Stress on Marine Organisms. John Wiley & Sons Ltd, pp 351–361

Beukema JJ (1989) Long-term changes in macrozoobenthic abundance on the tidal flats of the western part of the Dutch Wadden Sea. Helgoländer Meeresunters 43: 405–415

Beukema JJ (1990) Expected effects of changes in winter temperatures on benthic animals living in soft sediments in coastal North Sea areas. In: Beukema JJ, Wolff WJ, Brouns JJWM (Eds) Expected effects of climatic change on marine coastal ecosystems. Kluwer Academic Publishers, Dordrecht, Boston, London, pp 83–92

Beukema JJ (1992) Expected changes in the Wadden Sea benthos in a warmer world: lessons from periods with mild winters. Neth J Sea Res 30: 73–79

Beukema JJ, Cadée GC (1996) Consequences of the sudden removal of nearly all mussels and cockles from the Dutch Wadden Sea. P.S.Z.N.I. Mar Ecol 17: 279-289

Beukema JJ, de Bruin W, Jansen JJM (1978) Biomass and species richness of the macrobenthic animals living on the tidal flats of the Dutch Wadden Sea: Long-term changes during a period with mild winters. Neth J Sea Res, 12: 58–77

Beukema JJ, Dörjes J, Essink K (1988) Latitudinal differences in survival during a severe winter in macrobenthic species sensitive to low temperatures. Senckenbergiana marit 20: 19–30

Beukema JJ, Essink K, Michaelis H, Zwarts L (1993) Year-to-year variability in the biomass of macrobenthic animals on tidal flats of the Wadden Sea: How predictable is this food source for birds? Neth J Sea Res 31: 319–330

Buhr K-J (1981) Auswirkungen des kalten Winters 1978/79 auf das Makrobenthos der *Lanice*-Siedlung im Weser-Ästuar. Veröff Inst Meeresforsch Bremerh 19: 115–131

Cadée GC (1996) Accumulation and sedimentation of *Phaeocystis globosa* in the Dutch Wadden Sea. J Sea Res 36: 321–327

Colijn F (1981) Influence of turbidity on primary production of phytoplankton. In: Biological Research Ems-Dollart-Estuary. Some aspects of an estuarine ecosystem in a series of ten poster summaries. Kieler Meeresforsch Sonderheft 5: 278–283

Delafontaine MT, Flemming BW (1997) Large-scale sedimentary anoxia and faunal mortality in the German Wadden Sea (Southern North Sea) in June 1996: a man-made catastrophe or a natural black tide? Dt Hydrogr Zeitschr, Suppl 7: 21–27

Dörjes J (1992a) Langfristige Veränderungen des Artenbestandes der Makroendofauna im Vorstrand der Düneninsel Norderney in der Zeit von 1976 bis 1988 (Nordsee). Senckenbergiana marit 22: 11–19

Dörjes J (1992b) Zur Populationsdynamik von *Cerastoderma edule* (L.) nach dem Eiswinter 1978/79 am Beispiel zweier Stationen der Jadewatten (Nordsee) in der Zeit von 1979–1988. Senckenbergiana marit 22: 21–28.

Dörjes J, Michaelis H, Rhode B (1986) Long-term studies of macrozoobenthos in intertidal and shallow subtidal habitats near the island of Norderney (East Frisian coast, Germany). Hydrobiologia 142: 217–232

Grøntved J (1952) Investigations on the phytoplankton in the southern North Sea in May 1947. Medd Komm Danm Fisk-og Havunders Ser Plankton, 5: 1–49

Günther C-P, Boysen-Ennen E, Niesel V, Hasemann C, Heuers J, Bittkai A, Fetzer I, Nacken M, Schlüter M & Jaklin S (1998) Observations of a mass occurrence of *Macoma balthica* larvae in midsummer. J Sea Res 40: 347–351

Hertweck G (1995) Verteilung charakteristischer Sedimentkörper und Benthosbesiedlungen im Rückseitenwatt der Insel Spiekeroog, südliche Nordsee. I. Ergebnisse der Wattkartierung 1988–92. Senckenbergiana marit 26: 81–94

Höpner T (1996) Schwarze Tage im Nationalpark Wattenmeer. Spektrum der Wissenschaft 8/96: 16–21

Höpner T (1997) Black days in the German Wadden Sea. The Ocean Challenger 2: 36–39

Honkoop PJC, van der Meer J (1997) Reproductive output of *Macoma balthica* populations in relation to winter-temperature and intertidal-height mediated changes of body mass. Mar Ecol Prog Ser 149: 155–162

Niesel V (1997) Populationsdynamische und ökophysiologische Konsequenzen des Wattaufenthaltes für Phytoplankter der Nordsee. Forschungszentrum Terramare Berichte 7

Postma H (1984) Introduction to the symposium on organic matter in the Wadden Sea. Neth J Sea Res Publ Ser 10: 15–22

Reichert A, Dörjes J (1980) Die Bodenfauna des Crildumersieler Watts (Jade, Nordsee) und ihre Veränderungen nach dem Eiswinter 1978/79. Senckenbergiana marit 12: 213–245

Henke S (1997) Black spots in the Wadden Sea – Causes, Effects, Ecological Implications. Umweltbundesamt Texte 3/97

Ziegelmeier E (1964) Einwirkungen des kalten Winters 1962/63 auf das Makrobenthos im Ostteil der Deutschen Bucht. Helgoländer Wiss Meeresunters 10: 276–282

8 Grid-Based Modelling of Macrozoobenthos in the Intertidal of the Wadden Sea: Potentials and Limitations

Volker Grimm, Carmen-Pia Günther, Sabine Dittmann, Hanno Hildenbrandt

Abstract: Grid-based ecological models are a valuable tool in terrestrial ecology which have considerably improved the understanding of many terrestrial systems. Therefore an attempt is made here to apply the grid-based approach to marine benthic soft bottom communities. The concept of patch-dynamics and the theory of mosaic-cycles are the theoretical background which motivated this attempt. Both these concepts are based on the notion of a spatially structured landscape whose spatial and temporal dynamics is ecologically significant. Using two grid-based models we try to find out whether in the intertidal of the Wadden Sea the spatial and temporal dynamics of dominant macrozoobenthic species meet the criteria of the patch-dynamics concepts and the mosaic-cycle theory. The first model takes into account the processes settlement, succession, disturbance events (storms, ice winters) and the dispersal of mussels (*Mytilus edulis*) into adjacent areas. This first model demonstrates that the frequency of disturbance events and dispersal of *M. edulis* are the major factors which determine the spatial distribution of the model species. However, the first model has four main methodological deficiencies: (1) the results of the model indicate that the spatial scale to which the model may apply must be rather small ($<100\text{x}100\text{m}^2$). (2) For the same reason the model does not reproduce large-scale zonation patterns of benthic communities. (3) Abiotic factors apart from disturbance events which are known to be decisive in the Wadden Sea, e.g. topography, are not taken into account. (4) Interactions between neighbouring spatial units ("cells") are all but neglected in the model. Thus, based on the topography of a selected part of the Wadden Sea (sandflat "Swinnplate" and surroundings), an advanced second model is developed (TOPOGRID) which only considers the two species *Lanice conchilega* and *M. edulis*. This model demonstrates elements of a future model of macrozoobenthos in the intertidal in the Wadden Sea which would be supported by much more expert knowledge than TOPOGRID, more field studies which are designed parallel to the model, and oceanographic models of the current regime. With respect to the theoretical concept which was to be tested with the two models presented here, we conclude that neither the concept of patch-dynamics nor the theory of mosaic cycles are applicable to the Wadden Sea in their original, full meaning.

8.1 Introduction

It is rarely possible in ecology to work empirically at the scales which are relevant to the patterns and processes studied. Instead of decades and square kilometres, field studies generally cover only square metres over a period of one to three years (Kareiva & Anderson 1988). This is also the case for studies of macrozoobenthic species in tidal flats. Additional problems arise in tidal flats: most organisms living here are hard to investigate because they live in the sediment and because of their high mobility. Moreover, frequently occurring disturbance events strongly affect distribution and abundance, and therefore the significance of short-term studies, which are no more than snapshots of spatial distribution, is doubtful. The problem of how to extrapolate small-scale studies to larger scales is thus decisive for advancing the benthic ecology of soft-bottom communities (Hall et al. 1994).

A tool to this end might be ecological models which explicitly take space into account – especially "grid-based" models, which are powerful tools (Grimm & Jeltsch 1996). The basic idea underlying grid-based models is to divide space into many small grid "cells" and to ignore spatial effects *within* the grid cells, but not *among* them. Empirical ecological knowledge usually only exists at the local scale within the grid cells: some idea exists of how the ecological state of the cell will change under normal conditions, how it will react to disturbance events, and how it interacts with the neighbouring cells. This local empirical knowledge can easily be taken into account in simulation programmes by using simple "if–then" rules. The programme then calculates how the whole region modelled will change. Chance events on both the local and regional scale are taken into account by using random numbers or random distributions to determine their occurrence and impact.

In terrestrial ecology, the grid-based approach has been used for different scientific and applied studies (e.g., Turner et al. 1995; Wennergren et al. 1995). In marine ecology, however, the grid-based approach has so far only been occasionally applied. In particular, as far as questions which address the spatial and temporal dynamics of macrozoobenthos in tidal flats are concerned, to our knowledge no grid-based models exist.

In this chapter we shall try to apply the grid-based approach to the regional distribution of macrozoobenthos in the intertidal of the Wadden Sea. We address the question of whether and how succession, extreme disturbance events and mutualistic interactions determine the distribution of macrozoobenthic species. This question – that of spatial and temporal patterns and their causes (Chap. 2) – was central to the ELAWAT project. We will also use the grid-based model to test ecological concepts which in a spatially explicit manner refer to succession and disturbance events, and which were considered as possible approaches to explain the distribution of the macrozoobenthos. These are the concept of "patch-dynamics" (Pickett & White 1985, Jax et al. 1993) and the theory of "mosaic-cycles" (Remmert 1991; Reise 1991).

8.1.1
Theoretical Background

In descriptions of macrozoobenthic communities, reference is often made to "distribution patterns" and "patches". The former term refers to the spatial distribution of species and communities, and does not necessarily imply what is usually considered as a "pattern", i.e. any structure in space or time that is beyond random variation. A "patch" may be defined as "an area which, on a specified scale of observation, has some level of consistency in the density of individuals" (Hall et al. 1994, p. 333). According to this definition, the spatial (and temporal) scale of observation has to be specified in any reference to patches or "patchy" distributions.

The notion behind the concept of "patch-dynamics" (Picket & White 1985; Paine & Levin 1981) is that natural disturbance events create patches in the landscape. Within these patches, the ecological "cards" are reshuffled. For instance, the disturbance may cause dominant species (e.g. mussels on rocky shores, certain tree species in forests) to disappear and thereby create space for earlier successional stages. Depending on the type, frequency, extent and intensity of natural disturbances, different kinds of landscape and ecological systems may emerge, e.g. a patch-work of different successional stages.

The significance of the concept of patch-dynamics is that it explains the natural structural diversity (or the lack thereof) of ecosystems. This structural diversity may be a precondition for species richness or stability properties. Heterogeneous systems are not uniformly affected by catastrophic disturbance events which otherwise would have the potential to affect the entire system (Kolasa & Picket 1991). The concept of patch-dynamics has frequently been discussed as an approach to study the Wadden Sea (Jax et al. 1993) or, in general, benthic species in soft-bottom communities (Hall et al. 1994).

Remmert (1991) formulated a concept related to patch-dynamics: the theory of mosaic-cycles (Jax 1994). He assumes that in many systems local succession is cyclic. After a disturbance event which clears the local community, a pioneer community is first established, followed by further successional stages up to the (local) climax community (e.g. dominant canopy trees; Wissel 1992). If the climax community breaks down due to senescence or disturbance events, the pioneer community follows, etc. Locally, the successional cycles may be out of phase, but regionally an equilibrium distribution of successional stages may emerge.

Based on our experience with the grid-based model presented in this chapter, we will at the end of this chapter discuss whether the concept of patch-dynamics and the theory of mosaic-cycles are suitable for application to the macrozoobenthos of the Wadden Sea.

8.1.2
Empirical Background

Besides the general knowledge about macrozoobenthos in the Wadden Sea, two findings obtained by ELAWAT were used as the biological background to the grid-based models developed in this chapter. Firstly, studies of *Lanice conchilega* (Chap. 5.3) show that after an ice winter in which almost all adult *Lanice* were killed, *Lanice* quickly recolonized areas where it had occurred previously. And secondly, relief casts (Chap. 3.5) suggest that over the course of years and decades, at a certain location different dominant communities replace each other (e.g., communities dominated or characterized by *Mytilus edulis*, *Arenicola marina*, *Lanice conchilega*).

Besides these results of ELAWAT, the formulation of the grid-based model TOPOGRID presented below would not have been possible without the survey of the benthos in the entire backbarrier tidal flat of Spiekeroog performed by Hertweck (1995; Chap. 3.5; Fig. 3.5.1). Hertweck's map gives an impression of the patchiness of benthic communities on the scale of the entire backbarrier tidal flat. Moreover, the maps allow to draw up hypotheses about local characteristics (e.g., topographic height, proximity to tidal channels) that determine the local type of community (e.g., areas with low or high density of *Lanice* tubes, mussel beds).

8.2 The First Model: A Demonstration

The first model presented is – in the positive meaning of the word – a "naive" attempt to transfer an extremely simple grid-based approach, which has nevertheless been successfully applied to some terrestrial systems (Wissel 1992, Jeltsch & Wissel 1994), to the macrozoobenthos of the Wadden Sea. The model is presented here because its deficiencies highlight the extensions and modification the grid-based approach needs if it is to be applied to the Wadden Sea. However, the model also has some value of its own because it allows some basic relationships to be *demonstrated*. Very often models are nothing but "demonstrations" of possible mechanisms (Crick 1988).

In the model, the area under consideration is divided into 32x32 grid "cells" which may be in one of four different states: empty (i.e. not occupied) because of a disturbance (indicated by "0" in the parameter names); dominated by lugworms (*Arenicola marina*, "A"), sand masons (*Lanice conchilega*, "L"), or blue mussels (*Mytilus edulis*, "M").

Every year, the following processes occur one after the other in the model: settlement, succession, the dispersal of *Mytilus*, and disturbance events. Each process is characterized by a certain set of transition rules, i.e. rules which determine the transitions between different states of a certain grid cell. Thus, time is not described continuously but in four discrete steps. Each step summarizes the effects the corresponding processes have within one year. Running the simulation then means applying the four sets of transition rules representing the processes in the model one after the other to each grid cell.

The model rules are probabilistic rules. The probabilities of transitions between states (Table 8.2.1) of a cell do not reflect hard field data but are estimated based on general knowledge and assumptions. This underpins the "if-then" character of ecological models emphasized by Wissel (1989). Ecological models do not claim to map reality with all its details into the computer or a mathematical formula. Instead, they are designed to explore the consequences of our assumptions (see also Starfield 1997).

Table 8.2.1 Meaning, name, and default of parameters used in the first model.

Settlement:			**Sensitivity to storms:**		
Empty cell → *Arenicola*	*p0A*	0.45	*Arenicola* → empty cell	*pStA*	0.02
Empty cell → *Lanice*	*p0L*	0.4	*Lanice* → empty cell	*pStL*	0.04
Empty cell → *Mytilus*	*p0M*	0.01	*Mytilus* → empty cell	*pStM*	0.4
Succession:			**Sensitivity to ice winters:**		
Lanice → *Mytilus*	*pLM*	0.1	*Arenicola* → empty cell	*pEisA*	0.02
Arenicola → *Mytilus*	*pAM*	0.02	*Lanice* → empty cell	*pEisL*	0.99
Lanice → *Arenicola*	*pLA*	0.02	*Mytilus* → empty cell	*pEisM*	0.9
Arenicola → *Lanice*	*pAL*	0.05			
			Dispersal of *Mytilus*:		
Disturbance events:			non-*Mytilus* → *Mytilus*[a]	*p8*	0.2
One storm in a year	*p1St*	0.33			
Two storms in a year	*p2St*	0.3			
Ice winter in a year	*pEis*	0.1			

All parameters are probabilities of a transition between cell states or of the occurrence of a disturbance event, → denotes a transition between two states, [a] if all neighbours are in state "Mytilus"

Settlement

An empty cell will be colonized by one of the three species in the model in accordance with the probabilities *p0A*, *p0L* and *p0M* (Table 8.2.1). In the simulation programme, this is implemented in the following way: a random number is drawn from a uniform distribution, i.e. the number is between zero and one. If this number is smaller than *p0L*, the state of the cell changes from "0" to "L", i.e. the cell is colonized and in turn dominated by *L. conchilega*. If the random number is greater than *p0L* but smaller than *p0L+p0A*, "A" will be the new state. And if the random number is even greater but still smaller than *p0L+p0A+p0M*, "*M*" will be the new state. The choice of the three probabilities is constrained in that the sum of the three probabilities must not exceed one.

Succession

The presence of a certain species in a grid cell may facilitate the settlement of another species in the cell, for example the attachment of the larvae of *M. edulis* to the protruding tubes of *L. conchilega*. The estimated probabilities of local succession are listed in Table 8.2.1.

Dispersal of Mytilus edulis

The only spatially explicit rule of the model concerns the dispersal or expansion of *M. edulis* into neighbouring cells: the more neighbouring cells of a cell are dominated by *Mytilus* (i.e., are in the state "M"), the higher the probability that this cell will be overgrown by *Mytilus* and thus also change to state "M" in the next time step, irrespective of the current state of the cell.

The parameter quantifying the dispersal probability of *Mytilus*, *p8*, is the probability that a cell which is not in state "M" but whose eight neighbouring cells are in state "M" will itself change to "M" in the next time step. Accordingly, the probability that a cell with n neighbours, n=1..7, in the state "M", will itself change to "M", is $pn=1-(1-p8)^{n/8}$.

Disturbance events

Storms and ice winters are taken into account in the model. As far as storms are concerned, we distinguish between years without storm, with one storm, and those with two storms. Irrespective of the occurrence of storms, ice winters may also occur. The corresponding probabilities of these extreme disturbance events are listed in Table 8.2.1.

The simulation programme determines the occurrence of storms by drawing a uniformly distributed random number each year. If this number is smaller than *p1St*, one storm will occur. If it is greater than *p1St* but smaller than *p1St+p2St*, two storms occur. Otherwise, there is no storm. The occurrence of ice winters is determined in the same way using another random number.

Storms and ice winters affect individual grid cells, i.e. whether the cell will change to state "0" is determined randomly for each grid cell in accordance with the transition probabilities specified in Table 8.2.1. In the case of two storms, the cell is disturbed twice with the corresponding transition probability. This way of modelling the effects of a disturbance event means that we do not take into account the fact that disturbance events might alter entire areas, i.e. clusters of grid cells, at the same time.

Simulation

At the beginning of a simulation, the grid is randomly initialized, i.e. each grid is assigned to one of the four possible states with a probability of 1 in 4. The simulation is run for fifty years.

8.2.1
Results of the First Model

Fig. 8.2.1 presents the typical results of the first model for four different parameter combinations. It shows time series of the abundances of the three species (quantified by the number of occupied grid cells), time series of the disturbance events, and snapshots of the spatial distribution of the species on the grid. Note that within each year, the abundance of grid cells may take four different values, depending on the processes settlement, succession, dispersal and disturbance events. In Fig. 8.2.1, abundances are presented at the point of time in a year immediately before disturbance events might occur. Likewise, the spatial snapshots show the situation for a certain point of time. Figs. 8.2.1a and d show situations immediately after a disturbance has occurred (note the "white", i.e. empty cells), whereas the snapshots in Figs. 8.2.1b and c are taken immediately before a disturbance event might occur.

Fig. 8.2.1a shows the situation for the reference parameter set of Table 8.2.1. This set of parameters is characterized by the following assumptions: *Mytilus* has a very small probability of settling on empty cells (*p0M*) and is very sensitive to

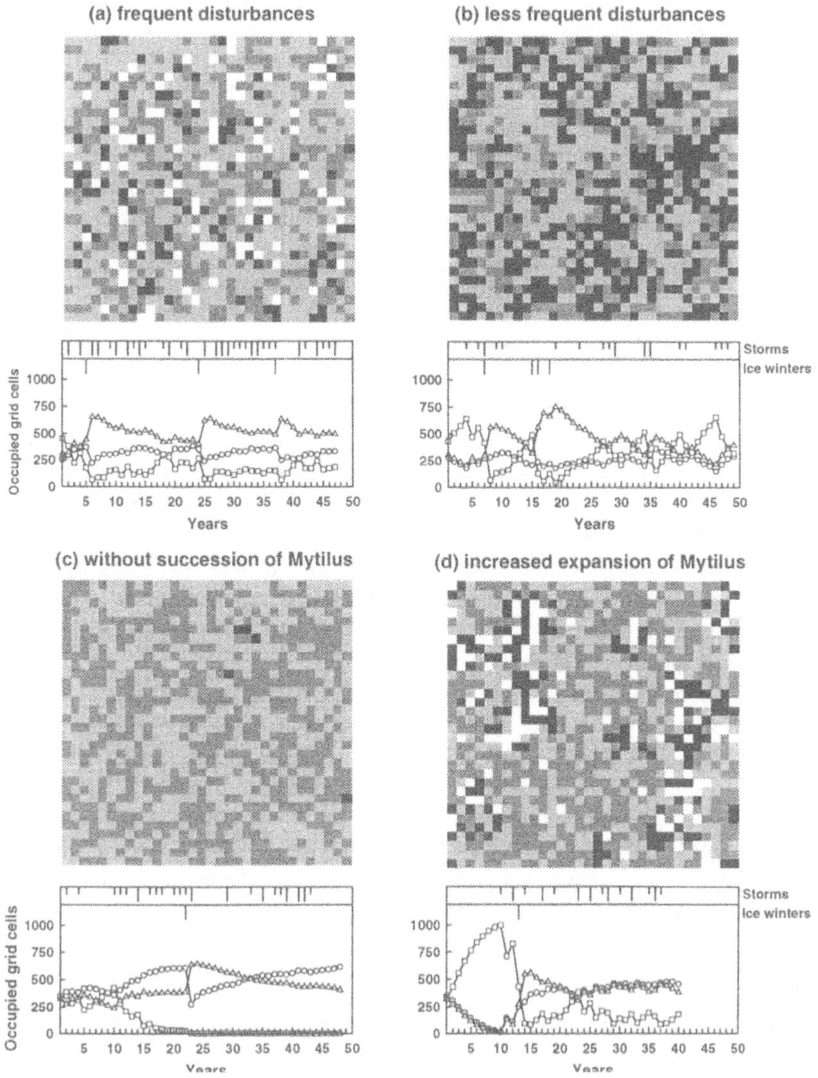

Fig. 8.2.1 Exemplary results of the first model showing the time series of abundances (circle: *Lanice*; square: *Mytilus*; triangle: *Arenicola*) and of the disturbance events (length of bars indicates whether one or two storms occurred) and a spatial image of the model grid taken in the year the time series stops (white: empty cell; light grey: *Arenicola*; dark grey: *Lanice*; black: *Mytilus*). Parameters: (a) see Table 8.2.1; (b) same as (a) except *p2St*=0.15 and *pIce*=0.05; (c) same as (b) except *pLM*=*pAM*=0; (d) same as (c) except *p8*=0.5.

disturbance events. *Lanice* is sensitive to ice winters but has a good ability to settle on empty cells. Finally, *Arenicola* is not very sensitive to disturbances and has little difficulty colonizing empty space. The frequency of disturbance events is assumed to be rather high. On average, one or two storms occur every two of three years, and an ice winter occurs every ten years.

With these assumptions, *Mytilus* only has two mechanisms to persist in the system: firstly, via succession by replacing *Lanice* or *Arenicola*, and secondly via dispersal into neighbouring cells. Nevertheless, the time series of Fig. 8.2.1a shows that the abundance of *Mytilus* is held at a constantly low level by the frequent disturbance events. Only in some rarely occurring periods with few or no disturbances is *Mytilus* able to build up higher abundance.

This effect of disturbances on *Mytilus* is confirmed and demonstrated by Fig. 8.2.1b, where both the probability that two storms occur within a year (*p2St*) and the probability of ice winters have been divided by two. *Mytilus* now shows markedly higher abundances but is still prevented by the disturbances from taking over more cells. Note that without disturbances *Mytilus* would take over the whole grid sooner or later because of the dispersal mechanism described above.

But what is the significance of succession for *Mytilus* in this model system? Fig. 8.2.1c shows the results for parameters which are the same as for Fig. 8.2.1b, except for parameters describing succession, *pAM* and *pLM*, which are set to zero. Under these conditions *Mytilus* almost becomes extinct rather quickly because the very low recruitment (*p0M*) can not compensate for losses due to disturbances. Interestingly, further simulations showed that even small values of *pAM* and *pLM* are sufficient to keep *Mytilus* in the system, albeit at a low level of abundance. This result demonstrates the potential significance of mutualistic interactions during the settlement of macrozoobenthic species (e.g., the settlement of larval *Mytilus* on tubes of *Lanice*), even if these interactions are rather weak, i.e. do not occur all the time and everywhere.

Another mechanism to keep *Mytilus* in the system is its dispersal into neighbouring cells. In Fig. 8.2.1d, the same parameters are used as in Fig. 8.2.1c, except for the probability *p8* that a non-*Mytilus* cell surrounded by eight *Mytilus* cells will change to state "M" in the next time step: this probability is now 0.5 instead of 0.2 in Fig. 8.2.1c. Despite the lack of succession of *Mytilus* in this scenario, *Mytilus* is well established in the system now because it spreads rather quickly during periods free of disturbance. Note that with this mechanism of persistence, the spatial distribution of *Mytilus* is characterized by clusters of cells in state "M".

8.3 Problems when Applying the Grid-Based Approach in the Wadden Sea

The first model is useful because it helps demonstrate the potential significance of disturbance events and succession in a hypothetical Wadden Sea. As a "conceptual" model (Wissel 1989, 1992) it helps analyse general relationships. The first model thus belongs to the same class of models as the well-know logistic equation or the Lotka-Volterra models of competition or predator-prey systems. However, without any way of testing the model, i.e. testing some of the model predictions, all explanations of natural phenomena deduced from the model are only *possible* explanations; it is impossible to determine which of a set of competing possible explanations is correct (Grimm 1994).

For some terrestrial systems (e.g., forests, the spread of rabies among red foxes), the approach used in the first model did indeed result in models producing testable predictions (Wissel 1992, Jeltsch & Wissel 1994, Jeltsch et al. 1997). Why did this not work with our first model? To answer this question, the deficiencies of the first model are discussed below.

Abiotic factors

In many ecological systems the local environment is determined by the biota to such an extent that it seems reasonable to ignore the spatial heterogeneities of the abiotic environment as approximation. For example, most grid-based models of terrestrial systems ignore not only topography but also heterogeneities of edaphic factors, precipitation, etc.

Obviously, with the Wadden Sea this kind of simplification does not suffice. Here abiotic factors determine to a large degree the processes which in turn directly or indirectly determine the spatial distribution pattern (settlement, secondary dispersal, properties of the sediment, etc.). Thus, a grid-based model tailored to the Wadden Sea *has* to take into account essential abiotic factors, for example topography.

Pattern

More or less clear patterns in the distribution of the macrozoobenthos can only be perceived at the spatial scale of entire sandflats (Hertweck 1995; Chap. 3.5, Fig. 3.5.1). This pattern is partly a simple zonation of more or less clearly distinguished typical associations of species (Dörjes 1978) and is partly determined by the complex topography of the Wadden Sea and the complexity of the current regime interacting with topography (Chaps. 3.3; 3.4; 9). Therefore, in contrast to rocky shores, there is no simple zonation pattern following some simple gradient in the intertidal of the Wadden Sea.

An additional problem is that pattern recognition strongly depends on the criteria to define certain associations of macrozoobenthic species (Michaelis & Böhme 1994). The criteria used by Hertweck (1995; Chap. 3.5) seem to be most suitable for comparative studies performed by different research projects, because in addition to the dominant species, characteristics of the sediment are also taken into account (Michaelis & Böhme 1994).

The scale we aimed at with our model was either the scale of an entire backbarrier tidal flat or at least of an entire sandflat whose boundaries are defined by tidal channels. But after inspecting the spatial distributions produced by the first model (Fig. 8.2.1), it became apparent that the entire model area must be rather small (100 x 100 m^2 or smaller), because at larger spatial scales of observation the distribution of dominant macrozoobenthic species is not that scattered as in Fig. 8.2.1, but there are usually larger patches or zones dominated by one species.

It also follows from Hertweck's (1995) map of the biofacies in the backbarrier tidal flat of Spiekeroog that, to model the spatial and temporal dynamics of the macrozoobenthos in a spatially explicit manner, the modelled area has at least to embrace the area of an entire sandflat, for example the Swinnplate or Gröninger Plate (cf. Chap. 3.1).

Size of the grid cells

The basic elements of grid-based models are natural spatial units ("cells") which allow space to be divided into a grid of these units. With forests, for example, the space an adult canopy tree requires would be a natural unit, or with the spread of rabies among red foxes the mean size of a territory (Grimm et al. 1996; Jeltsch et al. 1997). Depending on the system studied and on the question, the size of a grid cell may vary between one square centimetre and several square kilometres (Grimm & Jeltsch 1996).

For the Wadden Sea, there is no clear indicator what size grid cell would be appropriate for the benthos. Using the cells as building blocks, the size of the grid cells has to be small enough to reproduce the typical features of the patches which can be observed. This ought to be possible with grid cells measuring $50x50m^2$, but seems more reliable with $10x10m^2$ cells. Unfortunately, spatial autocorrelations were not available to help define the model resolution.

Neighbourhood interactions

The most important advantage of the grid-based approach is that it allows for an easy investigation of the regional consequences of local interactions between neighbouring cells. However, this requires a minimum degree of spatial continuity for the components which interact locally, achieved for example by sessile life forms. With the macrozoobenthos, however, only the adults of *L. conchilega* are sessile. Not even *M. edulis* is sessile in the strict sense because most individuals are only attached to each other and not to the sediment.

In fact, our model takes into account only one single local interaction between neighbouring cells: the dispersal or expansion of *Mytilus* into adjacent cells. Without this interaction our model would not need to be spatially explicit but could be replaced by a simple non-spatial model.

There are of course local interactions between the organisms of the macrozoobenthos, but they mainly seem to be significant on smaller spatial scales. The high mobility of most benthic organisms indicates the ecological and evolutionary significance of these small-scale interactions (Günther 1990, 1992), for example with respect to the ability to react to disturbances (Chap. 9).

But in contrast to many other ecological systems, it seems difficult, if not impossible, to find a simple, direct link between local interactions and large-scale distribution patterns in the intertidal of the Wadden Sea. Evidently, factors which do not play a major role in other systems and may therefore be neglected in models are decisive in the Wadden Sea, i.e. abiotic factors.

8.4 A Grid-Based Approach Tailored to the Wadden Sea: TopoGrid

Based on the lessons learned from our first, "naive" model, a grid-based model (TopoGrid) is developed in the following which is tailored to the specific characteristics of the Wadden Sea, or of tidal flat ecosystems in general. The main difference between the first model and TopoGrid is that with TopoGrid the main deficiencies of the first models are avoided. However, it must be emphasized that our focus is on the general concept of the model. Still, in the first version of TopoGrid which is presented here, extremely simplified assumptions are deliberately made. TopoGrid does therefore not claim to be realistic or to make testable predictions. The aim of TopoGrid is to show what a grid-based model of the macrozoobenthos in the Wadden Sea would in principle have to look like.

Grid cell size and time steps

To make the simulation programme run fast enough, not the entire backbarrier tidal flat of Spiekeroog is modelled, but only the Swinnplate and adjacent areas (Fig. 8.4.1). This model area is divided into 500x250 cells measuring 10x10 m^2 each. One time step in the model corresponds to one year. Within one year, the following three processes occur one after the other: disturbance events, settlement, and the dispersal of *Mytilus*. A simulation year starts at the end of the summer, before the storms of autumn or winter might occur.

For each process in the model, the abundance of the species is calculated for every single cell. In the case of the dispersal of *Mytilus*, the abundances in the neighbour cells may also change.

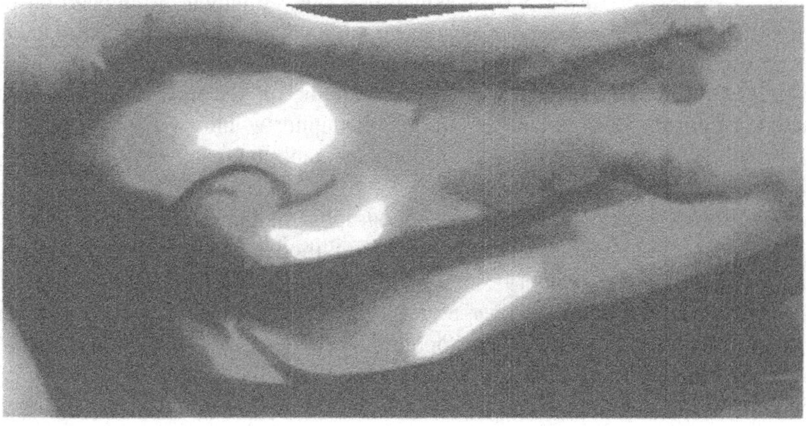

Fig. 8.4.1 Part of the Spiekeroog backbarrier tidal flat modelled by TopoGrid (Swinnplate and adjacent areas; cf. Fig. 3.1.3). This area is divided into 500x250 cells measuring 10x10 m^2. The maximum difference of topographic height in this area was mapped to the interval [0 , 255] (relative units) and represented here by continuously varying grey scales (0: white; 255: black)

State variables

Only two species, *Mytilus edulis* and *Lanice conchilega*, are considered in TopoGrid. The state of a cell is characterized by the abundances of these two species. Thus, in contrast to the first model, the state variable of a cell is no longer binary (occupied or not) and both species may be present within one cell. Abundances are specified in relative units which may take values between zero and 255. For reasons of convenience, *Arenicola marina* is not considered explicitly; it is thus assumed that *Arenicola* does not closely interact with the other two species.

In cells which contain *Mytilus*, an additional state variable is considered: the time since the last major spatfall of *Mytilus* occurred. This variable is recorded because *Mytilus* banks at the higher parts of a sandflat are often dominated by just one cohort, i.e. they go back to one spatfall (Obert & Michaelis 1991; Ruth 1991). Accordingly, if the mussels start to grow old, the entire mussel bank will disappear sooner or later. In the model, the *Mytilus* population of a cell will die with a probability of *pMold* (=0.7) if the last spatfall occurred seven or more years ago.

Abiotic factors: topography

The spatial distribution of the macrozoobenthos is mainly affected by larval settlement and disturbance events, which themselves depend on a host of abiotic factors: the direction and velocity of tidal currents, swell, duration of emergence, sediment properties, etc. How all these factors affect settlement and sensitivity to disturbance is not yet known in detail. In particular, little is known about the relative effects of these abiotic factors. For both practical and principal reasons it is not possible to take into account in a model *all* factors which are potentially important. But what are the most important factors, i.e. the *key factors*?

Since there is still no clear, unequivocal answer to this question, we decided to allow all parameters in the model, which describe settlement, to depend on topographic height (an analogous dependence of local sensitivity to disturbance on topographic height is – for reasons of simplicity – not considered in the model). "Topographic height" is thus used as a metaphor for the most important abiotic factors. In an advanced model version, this metaphor would have to be replaced by empirical information about real mechanisms and by results of hydrodynamic models describing the current regime.

Nevertheless, taking into account topographic height is a major achievement with respect to the "flat" topography of the first model. The assumptions which are used in the following model rules are mainly based on the conjectures of Hertweck (1995) that, at least in the backbarrier tidal flat of Spiekeroog, *L. conchilega* settles with low densities at intermediate topographic heights, and with high densities in higher regions. With *Mytilus* it is assumed that they only become established at the top parts of the sandflats.

Settlement of Lanice

After ice winters *Lanice* settles at a certain, "spontaneous" rate *Lspontan*. Epibenthic substrate for settlement include tubes of adult worms that died during winter, macroalgal mats which are partly buried in the sediment, eroded tube mats of other tube-building polychaetes, or other, more or less ephemeral epibenthic structures.

The population growth of a local population of *Lanice* after an ice winter ought, at the beginning, to be exponential with a growth rate r. On the other hand, there is obviously a local "carrying capacity", K_L, as an upper limit of local abundance (cf. Hertweck 1995). The most simple way to combine exponential growth and density dependence in a model is the "logistical equation" (e.g., Wissel 1989) which, in its time-discrete form, is:

$$L_A = L_B + rL_B\,(1 - L_B\,/\,K_L).$$

The indices "B" and "A" indicate the abundance before ("B") and after ("A") settlement. According to the map of Hertweck (1995) we assume that *Lspontan* and K_L depend on topographic height (Fig. 8.4.2). The growth rate r is then chosen to mimic the observation that it takes three to four years before abundance approaches local carrying capacity. If, however, a cell is very densely populated by *Mytilus* (80 % of the maximum local density of *Mytilus*), we assume that *Lanice* will not settle in this cell.

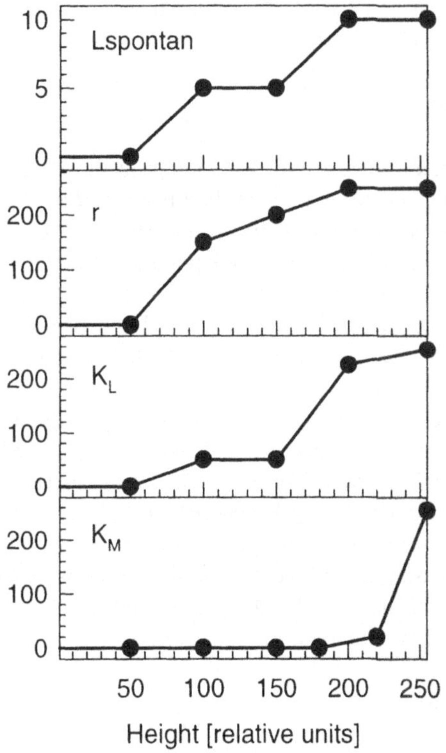

Fig. 8.4.2 Dependence of the parameters describing the settlement and local carrying capacity of *Lanice* (*Lspontan*, r [%], K_L) and *Mytilus* (K_M) on topographic height. Only the values indicated by points in the diagram are specified in the simulation programme; all other values are determined by linear interpolation

Settlement of Mytilus

In contrast to *Lanice*, the settlement of *Mytilus* is characterized by episodic spatfalls. Although larvae of *Mytilus* occur each year, this "normal" supply of larvae does not seem to be able to establish new mussel beds or to sustain existing mussel beds (Zens et al. 1997). Therefore we neglect this "normal" recruitment. Major spatfalls of *Mytilus* often, but not exclusively, occur after ice winters. Thus, we assume that after an ice winter a spatfall will occur with a probability of *pIceSpat*=0.9, whereas in years without ice winter the probability is *pSpat*=0.05.

Similar to *Lanice*, the abundance of *Mytilus* within a cell cannot exceed the capacity K_M. By specifying how K_M depends on topographic height (Fig. 8.4.2), we define the parts of a sandflat where *Mytilus* may in principle become established. The outcome of a spatfall depends on the situation in a cell. If only *Arenicola* is present (which we assume to be present everywhere implicitly), every second spatfall will be successful (*pSpatSucc*=0.5), i.e. the abundance of *Mytilus* will be at its maximum value K_M in the next year.

If *Lanice* is present in a cell, but not *Mytilus*, a spatfall will be successful with a probability of *pSuccession*=0.8. In TopoGrid, this is the *only* rule to take succession into account! Finally, if *Mytilus* is already present in a cell, a spatfall will irrespective of the actual abundance of *Mytilus* turn abundance to K_M in the next year. In addition, the "age" of the local mussel bed in the cell is reset to one.

Disturbance events

Two kinds of disturbance events are modelled: storms or extreme wind events, and ice winters. Ice winters occur with a probability of *pIce*=0.1 per year. As for *Lanice*, the strength of an ice winter is irrelevant because even rather short periods of temperatures near 0°C will kill the entire population of *Lanice* in the intertidal (*LIceMort* = 0.95). As for *Mytilus*, we distinguish between "normal", "severe", and "extremely severe" ice winters, which reduce the abundance of *Mytilus* in each cell by *MyIceMort* = 10, 30, or 80 % respectively. The three kinds of ice winters occur with a relative frequency of 2:1:1.

Storms occur on average every second year. Again, we distinguish between different degrees, i.e. "normal", "severe", and "very severe", which occur with a relative frequency of 4:2:1. These storms reduce *Lanice* by *LStormMort* = 10, 25, or 50 % respectively, and *Mytilus* by *MyStormMort* = 15, 25, or 65 %. It should be noted that all these mortalities and frequencies only reflect intuitive assumptions and are not based on hard, quantitative data. If the model were to be fully analysed (which will not be done in this chapter), these model parameters in particular would have to be varied systematically.

Dispersal of Mytilus

If sustained by successful spatfalls, mussel beds increase in size. To model this, we use a filter algorithm in TopoGrid in which the abundance of *Mytilus* in a certain cell is calculated as a weighted mean of the abundances in that cell and its eight neighbouring cells. The abundance of the focal cell is weighted with the factor *weight* (=200), and the sum of the abundances is divided by *divi* (=1000). Without

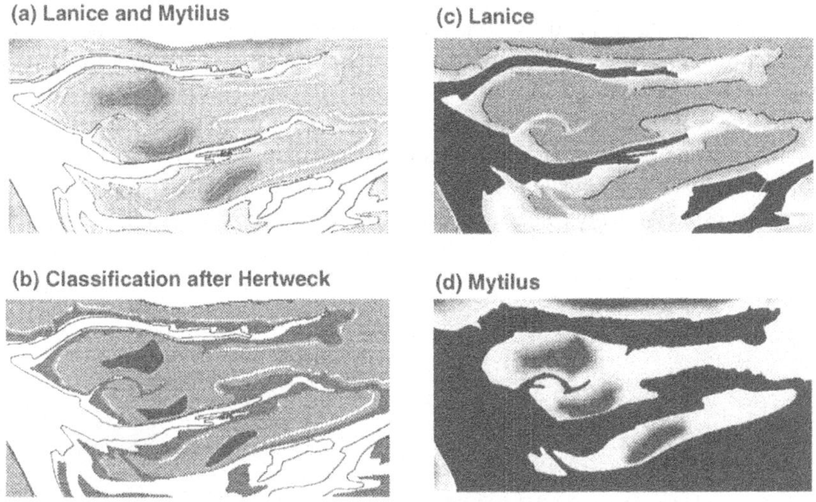

Fig. 8.4.3 Four alternatives for presenting the spatial output of TopoGrid. In the corresponding bitmaps, one pixel corresponds to one grid cell (as in Fig. 8.4.1): (a) Combined presentation of *Lanice* (light grey) and *Mytilus* (dark grey). (b) Classification of abundances analogous to Hertweck (1995) with low-density (medium grey scale) and high density (light grey) patches of *Lanice*, and with *Mytilus* (dark grey). (c) *Lanice* and (d) *Mytilus*: density is represented by grey scales (or colour intensity on the computer screen)

going into a detailed explanation of this approach here, the effect of the filter algorithm is that in years with no disturbances dense stands of *Mytilus* will expand but maintain a rather sharp boundary between low and high density regions.

Simulation procedure

Simulations start with an empty Swinnplate or, to be more precise, a Swinnplate without *Lanice* and *Mytilus*. Other species, in particular *Arenicola*, are assumed to be present implicitly. Simulations are run for 30 to 100 years. The output of the model takes the form of coloured (bit)maps of the spatial distribution of *Lanice* and *Mytilus,* where abundance is coded by different intensities of the colours (Fig. 8.4.3). Alternatively, a map is shown where the types of biofacies used by Hertweck are used, in particular the distinction between high and low density patches of *Lanice* (Fig. 8.4.3b). Additionally, time series of the abundance of *Lanice* and *Mytilus* as well as of the disturbance events are displayed (Fig. 8.4.4).

8.4.1 Typical Results of TopoGrid

The spatial distribution of the species considered is mainly determined by our assumptions about how capacity, recruitment etc. depend on topographic height (Fig. 8.4.2). Thus, in principle, we specified the distribution pattern with our assumptions. Indeed, if storms and ice winters are deactivated in the model, the distribution pattern determined by both the topography (Fig. 8.4.1) and the assumptions of Fig. 8.4.2 emerge: *Mytilus* occurs on the top parts of the sandflat, *Lanice* builds up high density patches at higher parts of the sandflat and low density patches at lower parts (see, for example, Fig. 8.4.3b). But despite this more or less obvious outcome of the model, the actual distribution pattern often differs considerably from this deterministic ideal due to the effect of disturbance events and the interaction between the two species (succession, dispersal of *Mytilus*).

Fig. 8.4.4 shows time series of the overall abundance (in relative units) of *Mytilus* and *Lanice* as well as the time series of the random events that shape the

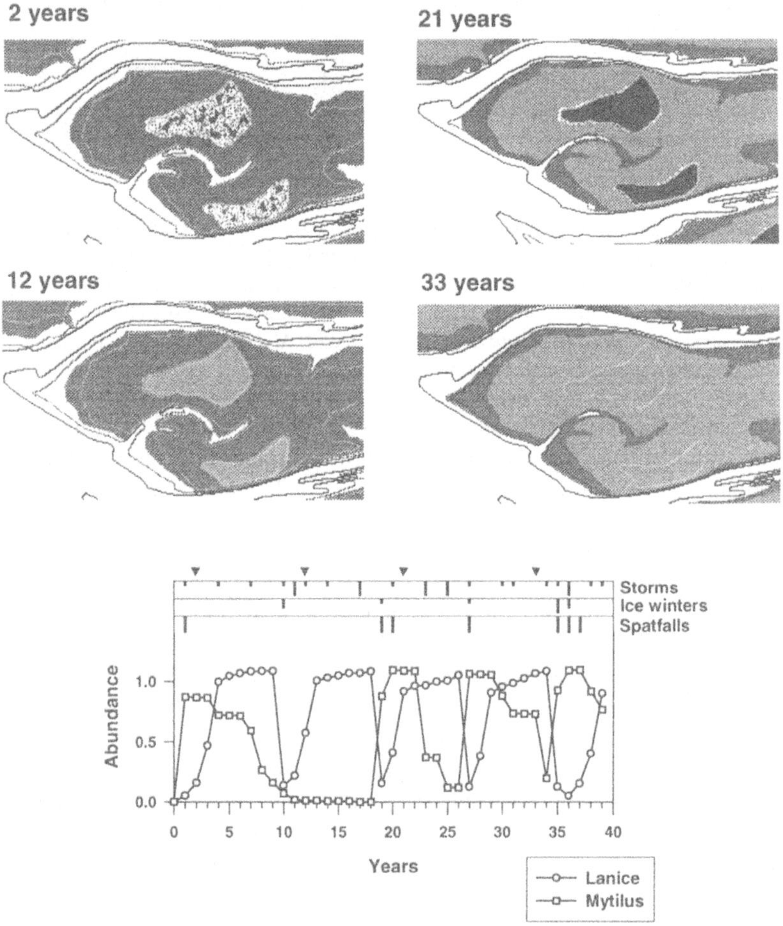

Fig. 8.4.4 Exemplary results of TopoGrid for the parameter values specified in the text. The significance of the different grey scales is as in Fig. 8.4.3b. Arrows on top of the time series indicate the years in which the spatial images are recorded

system: storms, ice winters and spatfalls. Although with our assumptions we determined a "normal" distribution pattern, surveys carried out in the years indicated and presented in Fig. 8.4.4 would lead to completely different results. This emphasizes the major problem of empirical benthic research of how to infer underlying mechanisms generating distribution from snapshots of the spatial distribution. Even in our model, where the mechanisms are, in contrast to reality, very simple and always the same, the distribution patterns differ enormously from year to year because they reflect not only more or less deterministic, continuous processes but also the *history of events* which determine distributions.

This contingency of actual distribution patterns would also be reflected by a virtual biofacies survey on the central parts of the hypothetical Swinnplate. Similar to the real relief casts of Hertweck (1995), a sequence of different benthic communities would show up (i.e., *Lanice*, *Mytilus*, or *Arenicola*).

The time series in Fig. 8.4.4 show that the dynamics of *Lanice* is determined by ice winters. After ice winters the population needs a few years to restore the predisturbance abundances. *Mytilus*, on the other hand, is affected by all three kinds of random events. Spatfalls initiate the dynamics. Storms reduce abundance and after eight years mussel beds begin to grow old and eventually die out. Likewise, ice winters reduce the abundance of *Mytilus* but at the same time may induce spatfalls.

Note again that all these results reflect model assumptions (as is always the case with models). The aim with TopoGrid was to run an initial model and to demonstrate what the assumptions of a more realistic, grid-based model would have to look like, and how these would lead to certain spatial patterns.

8.5 Methodological Conclusions

Applying the grid-based approach of ecological modelling to the macrozoobenthos of the Wadden Sea proved to be much more difficult than anticipated in view of the potential of this approach in terrestrial systems. The reasons for these difficulties became clear with the first model. TopoGrid shows how a grid-based model which is adapted to the Wadden Sea would have to look. TopoGrid is, however, still a demonstration mainly because of our *ad hoc* assumption about the dependence of major model parameters on topographic height.

Nevertheless, demonstrations offer the chance to go beyond the empirical restrictions of small spatial and temporal scales and to at least *think* on larger scales. Still, the ultimate goal of modelling is the production of models that are testable. However, this goal cannot be achieved in one step.

Our knowledge about local processes is still far from sufficient. At this scale, i.e. at the scale of tens to hundreds of metres, grid-based models similar to our first model may help gain a better understanding of local spatial and temporal dynamics. The model of *Lanice* (Chap. 5.4) is another example of models of this kind. Models and empirical studies would, however, have to be designed in conjunction with one another. The results of small-scale field studies, experiments and models could thus be aggregated to stochastic rules and be fed into large-scale models like TopoGrid. To compare model results and reality, large-scale surveys and long-term studies of abundance and occurrence of benthic species are needed.

Moreover, the models presented in this chapter underscore the plea of Michaelis & Böhme (1994) for more detailed studies of individual, dominant macrozoobenthic species, for example *L. conchilega* and *M. edulis*. In any case, emphasis must be focused even more strongly on abiotic factors than was the case in the ELAWAT project, in particular on the hydrodynamic regime. This would mean both empirical studies of tidal currents and swell, and oceanographic models. It would in principle be possible to adapt the large-scale oceanographic model, which was parameterized for the Wadden Sea at the end of the ELAWAT project (Chap. 3.3), to the sandflat shown in Fig. 8.4.1. The same spatial resolution could be used as in TopoGrid. We see such a combination of grid-based ecological modelling, an oceanographic model, measurements of the current regime at certain points, and experimental studies of the macrozoobenthic species, as a promising direction of future research of macrozoobenthos in tidal flat ecosystems.

Models are not ends in themselves but a means to an end. At the present state of knowledge of benthic research, the purpose of modelling is more to focus on the most important gaps in our knowledge and to explore effective ways of closing these gaps (see above) than to give definite answers. Moreover, the attempt to make a group of experts formulate a model like TopoGrid, for example at an informal, interdisciplinary workshop, would be an ideal means for an efficient communication beyond one's own discipline.

With these conclusions we tried to outline the actual potentials and limitations of spatially explicit, grid-based ecological models in the Wadden Sea. But since models are still often misinterpreted as "prediction machines", we would like to emphasize here an additional principle limitation of ecological models – namely, that deterministic predictions of the distribution pattern are impossible (cf. Fig. 8.4.4). Ecological models can, at best, reveal principle mechanisms and predict certain probabilities of certain outcomes.

8.6 Ecological Conclusions

Concerning the ecological concepts mentioned in 8.1.1, we are uncertain whether the concept of patch-dynamics or the theory of mosaic-cycles are directly applicable to the Wadden Sea. In the past, the concept of patch-dynamics was too uncritically transferred from marine hard bottoms to soft bottoms. Important basic differences between these two habitat types were not adequately considered. Frid and Townsend (1989) stress that patch-dynamics may only be invoked as an underlying mechanism which structures ecological systems if the organisms in question are sessile, e.g. plants or sessile benthic organisms living on hard substrate (barnacles, molluscs, etc.). In such systems, space is evidently a limiting resource. In addition, the local consequences of disturbance events remain visible and ecologically effective for a longer period of time. In contrast, on the soft bottoms of the intertidal of the Wadden Sea the limiting resource is more difficult to identify. Most organisms are so mobile that they may escape local competition. This mobility and the "great equalizer", the tidal currents, ensure that the local effects of disturbance events are only short-lived.

Our first model confirms these considerations. It was difficult to identify a defined "natural" spatial unit which could serve as a building block for a grid-based model. "Patch-dynamics" should therefore only be considered as a general framework for explaining the distribution and abundance of macrozoobenthic species in the intertidal of the Wadden Sea. This framework "acknowledge[s] the existence of significant variation in community structure, often over relatively small spatial scales" (Downes 1990, p. 411).

Similarly, the theory of mosaic-cycles is not applicable to the Wadden Sea and its soft bottom communities (cf. Reise 1991). There are no local "cycles" of communities. At best, one could speak of "erratic" cycles (Reise 1991), i.e. the sequence of communities in a "cycle" is not predictable. Moreover, there is no large-scale equilibrium of the distribution and abundance of macrozoobenthic species in the Wadden Sea. Therefore, concepts and theories which, like the theory of mosaic-cycles, assume or aim at equilibria are not suitable for exploring benthic communities of the Wadden Sea.

The two models presented in this chapter demonstrate that any such exploration has to start with the concept of a zonation pattern which follows an abiotically determined gradient. In contrast to rocky shores, however, "zonations" in the Wadden Sea are more variable in space and time because of the three-dimensional abiotic gradients, the disturbance regime and the high mobility of the organism (Raffaelli & Hawkins 1996). Interactions between macrozoobenthic species are probably not very strong – if compared to the key role some predators or competitors play on rocky shores. Nevertheless they may, as has been suggested by our models, have considerable influence on the distribution pattern.

Acknowledgements

We would like to thank Thomas Leu for his contribution to the first model, and Michael Reetz and Udo Hübner for the topographic map of the Spiekeroog back-barrier tidal flat in computer-readable form. We thank the reviewer for helpful comments on the manuscript. The project was funded by the German Bundesministerium für Bildung, Wissenschaft, Forschung und Technologie (BMBF) under grant number 03F0112 A and B. The responsibility for the contents of the publication rests with the authors.

References

Crick F (1988) What made pursuit. Basic Books, New York.

Downes BJ (1990) Patch dynamics and mobility of fauna in streams and other habitats. Oikos 59: 411–413.

Dörjes J (1978) Das Watt als Lebensraum. In: Reineck H-E (Ed) Das Watt. Ablagerungs- und Lebensraum, Waldemar Kramer, Frankfurt, pp 107–143

Frid CLJ, Townsend CR (1989) An appraisal of the patch dynamics concept in stream and marine benthic communities whose members are highly mobile. Oikos 56: 137–141

Grimm V (1994) Mathematical models and understanding in ecology. Ecol Model 75/76: 641–651

Grimm V, Frank K, Jeltsch F, Brandl R, Uchmanski J, Wissel C (1996) Pattern-oriented modelling in population ecology. Sci Total Environm 183: 151–166

Grimm V, Jeltsch F (1996) Ökologisches Modellieren am UFZ Leipzig-Halle. In: Mathes K, Breckling B, Ekschmitt K (Eds) Entwicklung und aktuelle Bedeutung der Systemtheorie in der Ökologie, Ecomed, Landsberg, pp 87–93

Günther C-P (1990) Zur Ökologie der Muschelbrut im Wattenmeer. Dissertation, Universität Bremen

Günter C-P (1992) Dispersal of intertidal invertebrates: a strategy to react to disturbances of different scales? Neth J Sea Res 30: 45–56

Hall SJ, Raffaelli D, Thrush SF (1994) Patchiness and disturbance in shallow water benthic assemblages. In: Giller PS, Hildrew AG, Raffaelli D (Eds) Aquatic Ecology. Scale, Pattern and Process. Blackwell Science, Cork, pp 333–375

Hertweck G (1995) Verteilung charakteristischer Sedimentkörper und Benthossiedlungen im Rückseitenwatt der Insel Spiekeroog, südliche Nordsee. I. Ergebnisse der Wattkartierung 1988-92. Senckenbergiana marit 26: 81–94

Jax K (1994) Mosaik-Zyklus und Patch-dynamics: Synonyme oder verschiedene Konzepte? Eine Einladung zur Diskussion. Z. Ökologie und Naturschutz 3: 107–112

Jax K, Vareschi E, Zauke G-P (1993) Entwicklung eines theoretischen Konzepts zur Ökosystemforschung Wattenmeer. Umweltbundesamt, Berlin. Texte 47/93

Jeltsch F, Wissel C (1994) Modelling dieback phenomena in natural forests. Ecol Model 75/76: 111–121

Jeltsch F, Müller MS, Grimm V, Wissel C, Brandl R (1997) Pattern formation triggered by rare events: lessons from the spread of rabies. Proc R Soc Lond B 264: 495–503

Kareiva PM, Anderson M (1988) Spatial aspects of species interaction: the wedding of models and experiments. In: Hastings A (Ed) Community Ecology, Springer, Berlin Heidelberg New York, pp 35–50

Kolasa J, Pickett STA (Eds), (1991) Ecological heterogeneity. Springer, Berlin Heidelberg New York

Michaelis H, Böhme B (1994) Benthosforschung im Ostfriesischen Wattenmeer. Umweltbundesamt, Berlin. Texte 24/94

Obert B, Michaelis H (1991) History and ecology of the mussel beds in the catchment area of a Wadden Sea tidal inlet. In: Elliott M, Ducrotoy J-P (Eds) Estuaries and coasts: spatial and temporal intercomparisons. Olson & Olson,Fredensborg, pp 185–194

Paine RT, Levin SA (1981) Intertidal landscapes: Disturbance and the dynamics of pattern. Ecol Monogr 51: 145–178

Pickett STA, White PS (Eds), (1985) The ecology of natural disturbance and patch dynamics. Academic Press, Orlando

Raffaelli D & Hawkins S (Eds) (1996) Intertidal Ecology. Chapman & Hall, London

Reise K (1991) Mosaic-cycles in the marine benthos. In: Remmert H (Ed) The Mosaic-Cycle Concept of Ecosystems, Springer, Berlin Heidelberg New York, pp 61–82

Remmert H (Ed), (1991) The mosaic-cycle concept of ecosystems. Springer, Berlin Heidelberg New York

Ruth M (1991) Miesmuschelfischerei im schleswig-holsteinischen Wattenmeer. Ein Beispiel für die Problematik einer Fischerei im Nationalpark. In: Probleme der Muschelfischerei im Wattenmeer, Schutzgemeinschaft Deutsche Nordseeküste e.V., Wilhelmshaven, pp 26–46

Starfield AM (1997) A pragmatic approach to modeling for wildlife managment. J Wildlife Manage 61: 261–270

Turner MG, Arthaud GJ, Engstrom RT, Hejl S, Liu J, Loeb S, McKelvey KS (1995) Usefulness of spatially explicit population models in land management. Ecol Appl 5: 12–16

Wennergren U, Ruckelshaus M, Kareiva P (1995) The promise and limitations of spatial models in conservation biology. Oikos 74: 349–356

Wissel C (1989) Theoretische Ökologie - Eine Einführung. Springer, Berlin Heidelberg New York

Wissel C (1992) Modelling the mosaic-cycle of a Middle European beech forest. Ecol Model 63: 29–43

Zens M, Michaelis H, Herlyn M, Reetz M (1997) Die Miesmuschelbestände der niedersächsischen Watten im Frühjahr 1994. Ber Forsch-Stelle Küste 41: 141–155

9 Stability Properties in the Wadden Sea

Volker Grimm, Hauke Bietz, Carmen-Pia Günther, Andreas Hild,

Marlies Villbrandt, Verena Niesel, Ulrike Schleier, Sabine Dittmann

Abstract: To achieve an understanding of the processes and mechanisms responsible for the reaction of the components of the Wadden Sea to disturbances, an attempt is made to assess the stability properties of the components studied by ELAWAT. It is discussed how this assessment depends on the choices of the variable of interest, the scale of observation, the type of disturbance, and the identification of the reference dynamics. The assessment shows that the components of the Wadden Sea possess certain stability properties: both biotic and abiotic components exhibit low constancy or high variability at small spatial and temporal scales, constancy (as well as resilience in the case of disturbances) at intermediate time scales of years to several decades and spatial scales of entire tidal flats, and persistence under natural conditions at temporal scales of many decades to centuries and for the entire Wadden Sea (or large parts thereof). Among the most important stability mechanisms of the organisms living in the Wadden Sea are their high taxonomic or functional diversity, their mobility and their high potential population growth rates. Benthic organisms may react to disturbances via "resilience *in-situ*" or "resilience by migration". However, the natural, tide-driven web of abiotic processes must continue to persist if the stability mechanisms of the organisms in the Wadden Sea are to unfurl.

9.1 Introduction

The goal of the ELAWAT project was to gain a better understanding of the properties and processes which enable the Wadden Sea to react flexibly to events such as storms, ice winters and anthropogenic influences. Therefore natural structures were studied, as were structure-generating processes and the manner in which the generation of structures is affected by disturbances (Chap. 2).

Although the reaction of ecological systems to disturbances can of course be studied without referring to stability concepts (e.g., Sousa 1980), the pivotal concept of ELAWAT was *resilience*, i.e. the ability of a system to return to reference dynamics following a temporary disturbance. The motivation behind studying resilience was to *understand* the processes in the Wadden Sea to a degree which would allow a general assessment of the stability properties emerging from these processes. Consequently, this chapter comprises a detailed discussion of the components of the Wadden Sea with respect to their stability properties (i.e. constancy, resilience, resistance, and persistence; cf. Chap. 2).

The same questions will be posed in connection with each component: How does the assessment of stability properties depend on the scale of observation (both spatial and temporal), on the type of disturbance, on the variable of interest being considered, and on the definition of the reference dynamics? This will grant insights which extend beyond individual components and which refer to the processes in the Wadden Sea which are generally essential for stability properties. However, deterministic predictions are impossible because of the openness and complexity of ecological systems, and in particular due to the unpredictability of individual events which may shape ecosystems to a large degree.

In the past, attempts to assess stability properties of ecological systems were often linked or even equated to the notion that ecological systems are entirely regulated, with the "natural" state being an equilibrium (cf. Reise 1990b). Here we use stability concepts in a different way, without connotations such as "regulation" or "equilibrium". We consider assessments of stability properties as a diagnostic tool which helps organize our thinking about properties and processes of the Wadden Sea (Chaps. 2 and 10). A diagnostic tool must not be confused with an entirely objective device. Diagnoses are necessarily subjective to some degree: they are an attempt to integrate and interpret measurements and observations. Interpretations, however, are not only influenced by experience and knowledge, but possibly also by prejudices. The enumeration of stability properties presented in this chapter is thus not "objective" in the way that, for example, hard data are; nevertheless it is still "scientific" in the sense of a scientific discussion.

Interpreting stability concepts as diagnostic tools also means that we consider general assessments of the "stability" or resilience of an entire ecosystem to be impossible (Grimm 1996; Gigon & Grimm 1997; Grimm & Wissel 1997). "Stability", like the concepts "energy" and "information", represents the hope of many ecologists for "a global explanatory power ... which makes tedious work with details more or less superfluous. This hope failed as a rule." (Schwegler 1985, p. 263).

In the following we will first define and explain the conceptual framework that we used for our synthesis of the ELAWAT project. Then we will discuss stability properties and their underlying mechanisms for all the components of the Wadden Sea studied by ELAWAT. Finally, we will summarize these mechanisms for the species living in the Wadden Sea.

9.2 The Conceptual Framework: A Glossary

The conceptual framework of ELAWAT and its development are explained in Chap. 2. Below, we compile a list of the concepts which are most important for the synthesis of the ELAWAT project and briefly explain them in more detail, as a clear synthesis is only possible with clearly defined concepts.

Stability

"Stability" as an individual, isolated term has become completely useless in ecology because it is impossible to assign a single, precise meaning to this term

(Grimm & Wissel 1997). Therefore, we only use "stability" here as a generic term for six certain properties of ecological systems (cf Table 2.1.1). These six properties, which we refer to in the following as "stability properties", are the essence of the extreme terminological and conceptual diversity of the debate about "stability" and related issues in ecology (Grimm et al. 1992; Grimm & Wissel 1997; cf. Connell & Sousa 1983).

Constancy and variability

"Constancy" refers to the property of a dynamics to remain essentially unchanged. "Variability", in turn, is the complementary concept (Pimm 1991). "Variability", however, also often refers to some statistical measure of variability, i.e. standard deviation, variance, or a coefficient of variation. In the following we will directly refer to these statistical measures whenever variability is to be *quantified*.

Constancy is not, as is often assumed, a simple and merely descriptive concept. Instead, it requires an initial, basic understanding of the system in question because what is "essential" must be decided (Jax et al. 1998). This may lead to the (seemingly) paradoxical situation that cyclical or even extremely irregular dynamics are considered "constant" if the characteristics of the dynamics remain unchanged over longer periods of time.

Resilience

"Resilience" refers to the property of a dynamics to return to the reference dynamics after a temporary disturbance. Resilience is the most important stability concept to understand ecological systems because "understanding" implies the ability to predict reactions to imposed changes. Resilience has two aspects: firstly, the speed with which the dynamics returns to the reference dynamics (mostly referred to as "elasticity") and secondly, the whole of the states from which the reference dynamics can be reached again after a disturbance (referred to as "domain of attraction").

Resistance

"Resistance" refers to the property of a dynamics to remain essentially unchanged despite the presence of disturbances which have the potential to affect changes. Applying this concept is difficult because it is hardly ever possible to prove that a certain impact really has the potential to induce (strong) changes, although in the case considered no (or only weak) changes are observed. We will use the concept of resistance only in cases where we know of mechanisms at the level of individuals which prevent the individual from reacting (strongly) to certain impacts (e.g., by changing the feeding mode).

Persistence

"Persistence" refers to the property of ecological systems to persist over time, i.e. to be an identifiable unit over a longer period of time (Shrader-Frechette & McCoy 1993; Grimm 1996; Jax et al. 1998; Grimm 1998). Persistence is a concept similar to constancy because neither refers to disturbances. Instead, they take ecological systems and disturbances affecting the system as a unity. Persistence, however, is – in contrast to constancy – a holistic and qualitative concept: holistic, because it refers to entire systems by definition, whereas constancy refers to the dynamics of certain variables of interest; and qualitative, because the focus is no longer on dynamics with its quantitative details, but on the qualitative question of whether the whole set of variables of interest used to characterize the system remains within certain boundaries. In the following, we will also extend the notion of persistence to the abiotic components of the Wadden Sea.

Stability mechanisms

"Stability mechanisms" are processes in ecological systems which are responsible for certain stability properties. Thus the ultimate goal of any attempt to assess stability properties is to understand the underlying processes.

Stability measures

To make stability concepts operational, i.e. to make them applicable for practical purposes, stability properties have to be quantified. To this end, depending on the particular ecological situation studied, certain "stability measures" are defined (Connell & Sousa 1983; Williamson 1987; Pimm 1991; Grimm 1996; Gigon & Grimm 1997). However, quantifying stability properties is subject to certain restrictions. One important condition is that different ecological situations are compared by means of stability measures, because stability measures are necessarily comparative. There is no theoretical or practical basis for assigning absolute values of stability measures to certain ecological situations. Thus, stability measures are used in combination with comparative approaches (Cole et al. 1991), comparative experiments or ecological modelling. Consequently, in the following we will mainly discuss qualitative rather than quantitative assessments of stability properties, because comparative approaches were not applied in ELAWAT (with the exception of recolonization experiments, Chap. 6).

It should be emphasized here that talking about the "assessment" of a certain stability property does not necessarily imply that this property indeed exists or even is very pronounced. Instead, "assessment" primarily means that a system or a component of a system is checked for the property.

Ecological checklist

Most statements about stability properties in the ecological literature are extremely vague and therefore utterly useless because they refer to entire ecological systems. But this is, except for the property of persistence, impossible as a rule because the stability properties of ecological systems may vary enormously depending on the point of view chosen. Although a number of species of a certain system may be resistant to a certain disturbance (e.g., ice winter), this finding does not allow any conclusions about the resistance of, say, biomass to the same disturbance. Similarly, assessments of the constancy (or variability) of the abundance of macrobenthic species clearly depend on the area to which the assessment refers, e.g. one, ten or ten thousand square metres.

To make the assessment of a stability property unambiguous, the following six different aspects have to be considered and communicated. These six aspects were compiled by Grimm et al. (1992) in an *ecological checklist*: variable of interest (e.g., abundance, number of species, biomass), level of description (e.g., individual, population, community), reference dynamics (e.g., equilibrium, cycles, certain range of variation), type of disturbance considered, and temporal and spatial scale considered. It would, of course, be too time-consuming to specify the whole checklist each time a statement about a stability property is made. Therefore in the following we will only specify the checklist in those cases where the aspects are not clear from the context.

In ELAWAT, an impressively broad spectrum of variables of interest has been studied in the various abiotic and biotic components of the Wadden Sea on a broad spectrum of temporal and spatial scales. However, these rather broad spectra, which are characteristic of a joint research project, are merely a condition which is necessary but not sufficient to develop an integrated view of the Wadden Sea, because to this end we would have to understand the interactions between the different components of the Wadden Sea. But not even a joint research project like ELAWAT can investigate more than a few of these interactions. Therefore, in the following we will often have to rely on *assumptions* about the interactions. This means that the integrated view of the Wadden Sea and, in particular, of its stability properties, will be partly based on assumptions and cannot be completely founded on empirical investigations.

Natural disturbances and disturbance regime

A "disturbance" is "any relatively discrete event in time that disrupts an ecosystem, community, or population structure and changes resources, substrate availability, or the physical environment" (White & Pickett 1985, p. 7). This definition combines the negative aspect of disturbances with possible ecological significance. In addition, the character of being an event is emphasized, thus distinguishing disturbances from continuous impacts (e.g., stress; Jax et al. 1993). The distinction between event and continuous impact is not absolute but depends on the time scale considered.

The ecological significance of disturbances, i.e. the way in which they shape ecological systems, usually becomes evident if entire "disturbance regimes" are considered instead of individual disturbance events. "Disturbance regimes" encompass all temporal and spatial attributes of disturbances in certain systems (White & Pickett 1985).

One decisive question in this context is whether a disturbance regime is constant, i.e. stationary in all its attributes, or whether it changes slowly, i.e. by following a trend in some or all of its attributes. If the regime is constant, it makes sense to speak of "reference dynamics" which may then serve as a baseline for a reference dynamics of the biotic components of the Wadden Sea. Reference dynamics are needed if the resilience of populations or communities are to be assessed with respect to certain disturbance events. However, if the disturbance regime follows a trend, any attempt to define reference dynamics is questionable.

(Un)predictability

In conjunction with "disturbance", reference is often made to "unpredictability" (Costanza et al. 1993), which refers to a broad range of abiotic conditions, i.e. it is not predictable how these conditions will change on small areas over short time intervals. The complementary term "predictability" refers to a narrow range of conditions and is therefore different from deterministic predictability. "Predictability" is a comparative concept which allows the abiotic conditions of different ecological systems to be compared. For example, according to Costanza et al. (1993), unpredictability is the key property of tidal zones and estuaries, whereas forests and coral reefs are characterized by predictable conditions.

Emergent properties

"Emergent properties" are properties which only exist on certain levels of description, but not below this level. Resilience, for instance, can only be observed at the level of populations or higher, since population resilience has no significance for a single individual. The concept of emergent properties helps system properties be explicitly defined and assessed. All too often in ecology people talk about "systems" without defining them, i.e. without stating explicitly the properties and interactions which are considered as essential to constitute the "system" (Jax et al. 1998; Grimm 1998). It should be noted that by "emergent properties" we do not refer to unexplainable, "mystical" properties which cannot be studied scientifically (cf. Wiegleb & Bröring 1996; Müller et al. 1997).

Community and ecosystem

By "ecosystem" we will refer in the following to a landscape with all its abiotic and biotic components, as well as the interactions between these components (Chap. 2). This means in particular that we do not necessarily infer any emergent property from the word "system". The same holds for "communities", which simply refers to different populations which occur in the same area at the same time.

9.3 A Canon of Questions

In the following we will discuss stability properties of those components of the Waddens Sea which were studied by ELAWAT. Instead of components, it might be more appropriate to talk of "ingredients": the ecosystem "Wadden Sea" is composed of different biotic and abiotic ingredients. As with a recipe, the properties of each individual ingredient influence the properties of the system as a whole. Referring to "ingredients" is very apt because we take into account not just material components and energy but also everything else which may more or less determine the properties of the system as a whole. Therefore, not only nutrients and primary production are considered, but also the presence or absence of certain species (Jones et al. 1994), properties of the landscape (morphology, hydrography), of the processes affecting the landscape (tides), and the entire disturbance regime.

In order to give a homogeneous, unifying structure to the discussion of the ingredients of the Wadden Sea, the following canon of questions will be addressed wherever possible:

(1) At what level of description and with what variables of interest is the ingredient considered?

In principle, an unlimited number of levels of description and, in particular, variables of interest could be used to characterize a certain system (Hall & DeAngelis 1985). The level of description and variables of interest actually used are often determined by methodological limits, as well as by the scale of human perception. Bacteria, for example, are described in much less detail than, for example, macrozoobenthic species (Fenchel 1992).

(2) What is the significance of the ingredient with respect to the dynamics of the Wadden Sea?

This question is designed to identify the elements of an integrated view of the Wadden Sea; it targets interactions between the ingredient and the Wadden Sea or other ingredients. Aspects which are only relevant at the level of the ingredient will not be discussed.

(3) What are the results of the assessments of constancy, persistence, resilience, and (if possible) resistance?

First of all, stability properties are considered which can be assessed without reference to disturbances, i.e. constancy and persistence. Resilience can only be assessed if there is some idea about the reference dynamics (see "ecological checklist" above). The task is then to characterize the reaction to certain disturbances. Resistance will only be discussed in exceptional cases.

(4) What mechanisms, processes and structures are relevant to the stability properties?

Understanding stability mechanisms is the main concern during the discussion of the stability properties of the ingredients. Therefore we will primarily concentrate our efforts on answering this question.

(5) Extreme scenarios: What would the Wadden Sea look like if the ingredient lost its stability properties?

Like question (2), the aim behind this question is to understand the interactions between the ingredient and the whole system. Once the extreme scenarios have been discussed, we will discuss the mechanisms which prevented these scenarios in the past and try to assess the probability that they will continue to do so in the future. However, this does not mean that we will define any precise tolerance limits as this would present a machine-like and therefore inappropriate view of the Wadden Sea.

9.4 The Abiotic Ingredients of the Wadden Sea

The meteorological, hydrographical, morphological, sedimentological and biogeochemical conditions determine the abiotic framework for life and abiotic processes within the Wadden Sea.

9.4.1 Meteorological Boundary Conditions and Disturbance Regime

Variable of interest

The most important meteorological variables of interest are wind speed and direction, temperature and, indirectly, water temperature. With respect to disturbance regimes, the most important variable of interest is the mean frequency of disturbance events.

Significance of the ingredients for the Wadden Sea

Temperature is an important autecological parameter. In the Wadden Sea, the winter temperature is decisive. Is the winter mild, i.e. without any frost? Or is the winter severe, featuring longer periods of temperatures below 0 °C and hence the formation of ice (an ice winter)? Ice winters have physiological and physical consequences. Some species exhibit reduced reproduction (e.g., *Carcinus maenas*, Chap. 5.6); others may be hit by extremely high mortality (*Lanice conchilega*, Chap. 5.3). On the other hand, species which are adapted to the cold may show blooms after ice winters (e.g., *Coscinodiscus concinnus*, Chaps. 5.1 and 7). For some species, ice winters are advantageous. Their lowered metabolic rates mean they do not have to use their gametes as energy reserves during periods of starvation. Consequently, they can produce unusually high numbers of larvae after ice winters (Chap. 7).

The physical effects of ice winters are mainly linked to the occurrence of ice. Blue mussels (*Mytilus edulis*), for example, may be frozen to ice-floes and thus transported to other regions during periods of drift ice. Besides organisms, sediment is also transported on the ice-floes (Chap. 3.4), leading to changes in morphology and sediment properties. Thus, ice winters are events which influence the Wadden Sea on many different levels. Extremely hot summers are not known to have similarly strong effects on the Wadden Sea.

Another major type of disturbance event linked to meteorological conditions are storms and storm tides. The amount of sediments and nutrients transported during these events may exceed that transported at all other times with no storm events by several orders of magnitude (Reise et al. 1998). In addition, storms may strongly affect the distribution of macrobenthic species, for example *Mytilus edulis* (Nehls & Thiel 1993). However, during the four years ELAWAT lasted, no major storm or storm tide event occurred.

Wind direction may affect the Wadden Sea if winds from the East predominate over longer periods of time, because both the area which falls dry at low tide and the time span of emergence may increase considerably.

Stability property: constancy

The data presented in Chap. 3.2 document the enormous variation of the meteorological conditions. Each study year had its own peculiarities, with the ice winter 1995/1996 being most prominent.

The ice winter frequency in the Wadden Sea is estimated as one every eight to ten years (Beukema 1979; Strübing 1996), probably related to the North Atlantic Oscillation (NAO) (Kröncke et al. 1998). This mean frequency seems to be constant. With respect to storms, the situation is less clear. The interpretation of data of the last decades is debatable. A time series of strong wind events (wind force > 6) during the last thirty years seems to indicate that within the study area of ELAWAT strong wind events have occurred with an unusually high frequency since the end of the 1980s. The slight decrease in the elevation of the sandflat "Gröninger Plate" could be an indicator for this change in the wind regime. However, the qualifier "unusual" depends on the temporal and spatial scale of observation considered for this statement.

The meteorological data set of ELAWAT suggests that there are no "normal" or "usual" years in the Wadden Sea. As in all systems with high variability, the average over longer periods of time only has statistical significance and actually occurs only by chance and momentarily. The mean thus does not represent "normal" years, but can only be used to assess constancy or trends in the abiotic regime. Moreover, even if the mean is constant, for example the frequency of ice winters, the time interval between ice winters may vary considerably.

Stability mechanisms

In the Wadden Sea, stability properties of the meteorological boundary conditions and of the disturbance regime are determined by climatic processes of the entire Northern Atlantic. They are therefore "boundary" conditions imposed from outside which define the abiotic boundaries for the organisms and the processes in the Wadden Sea. Long-term predictions about the Wadden Sea are thus necessarily limited by possible climate changes.

Extreme scenarios

A drastic increase or decrease in the frequency of storms or ice winters would have different consequences for the various species of the Wadden Sea (Beukema et al. 1978; Beukema 1979, 1985). It seems unlikely that benthic species of the Wadden Sea would be driven to extinction by sea level variations, since most of them have been around during the sea level rise of the past 10 000 years (Wolff 1987). The chance for invading species to become established in the Wadden Sea in the course of climate change cannot be predicted in general nor can the effects be foreseen, as these are species-specific.

At the community level, a severe decrease in the frequency of disturbance events like ice winters or storms would increase the relative importance of biotic interactions, because biogenic epipenthic structures (e.g., mussel beds, "tube lawns" of *Lanice conchilega*) could exist for longer periods of time and on larger areas (Nehls & Thiel 1993). This effect is also predicted by the model presented in Chap. 8. However, without ice winters, the reproduction of mussels might be impaired, which could threaten the existence of their biogenic structures in the long-term. On the other hand, a drastic increase of disturbance frequency would diminish the significance of biotic interactions (Reise 1985). More storms could also lead to an increased net export of sediments from the Wadden Sea.

9.4.2 Hydrography, Morphology and Sediments

Variables of interest

The most important hydrographical variables of interest are tides and the variation in levels of low and high tide, current flow and direction and swell. Morphology is characterized by elevation, i.e. topographic height. The main aspects of morphology in the Wadden Sea are the tidal flats and the tidal channels. Sediments are ultimately characterized by grain size spectra and by the distribution of these spectra in space (Chaps. 3.3 and 3.4).

Significance of the ingredients for the Wadden Sea

The tides are the decisive process in the Wadden Sea. They shape the characteristic morphology comprising the extended tidal flats and the tidal channels. The existence of the large tidal flats is the cardinal property of the Wadden Sea.

The tidal flats of the Wadden Sea are a unique habitat. On the one hand, the regular alternation between low and high tide requires broad physiological and behavioural flexibility on the part of the organisms living in the tidal flats (Costanza et al. 1993). On the other hand, the low tide increases primary production in the sediment because benthic algae gain more light than in the subtidal. Moreover, tidal currents and swell allow for an effective exchange of oxygen and nutrients, as well as organisms, between water column and sediment.

The local hydrographical regime plays, as was also shown by ELAWAT (Chaps. 5.3 and 8), a pivotal role for the spatial distribution of macrobenthic species. Settlement, for example, is strongly determined by the local properties of the tidal currents (Chaps. 5.3 and 5.4). Yet on a regional scale, too, currents are decisive for many species because they enable the transport of larvae over long distances.

In addition to the tidal currents, swell contributes to the kinetic energy of the water body and thus promotes exchange processes between water column and sediment. In addition, swell affects sedimentation, and probably also larval settlement and the local availability of food and nutrients. However, there are no detailed empirical records of these effects of swell.

Finally, properties of the sediment are a determining parameter for all organisms living in the sediment. Many benthic species of the Wadden Sea prefer certain types of sediment (mudflats, muddy sand, sandflats; Dörjes 1978). Thus the spatial distribution of sediment types determines the distribution of benthic species, especially macrobenthic species, to a large degree.

Stability properties: constancy and resilience

Within a tidal cycle, living conditions change so sharply that, according to Costanza et al. (1993), tidal flats are mainly characterized by their unpredictability, i.e. the high variation in abiotic conditions (see definition of "unpredictability" above).

However, this notion is incomplete because of the apparent paradox that the unpredictability driven by the tides is by far the most predictable aspect of the Wadden Sea. After all, whatever happens, the next tide is bound to come in! Thus the process that moulds the Wadden Sea, i.e. the tides, is extremely constant with respect to the occurrence of low and high tides.

The seeming contradiction between the two different notions of tidal flats as being characterized by unpredictability (Costanza et al. 1993) or by constancy does not really exist, because Costanza et al. refer to the unpredictability of *states*, for example abiotic conditions, whereas we refer to the constancy (or predictability) of the *dynamics* of the living conditions. "Constancy" in this context refers to the overall characteristics of the dynamics, not to particular values of the variables of interest. There is, of course, variation in all variables of interest which are determined by the tides, e.g., the size of the areas falling dry (Chap. 3.3) and the time span over which the tidal flats fall dry, which is also influenced by the lunar cycles and the current wind direction.

Likewise, the currents driven by the tides are highly variable if considered on small spatial and temporal scales (Chap. 3.3), but considered over longer periods of time (decades) tidal currents presumably show high constancy. This has two main reasons: the predictability of the tides and the tidal range, and the fact that the tidal currents are mainly determined by the geomorpology of the tidal flats which, in turn, are constant and resilient (see below). In fact, the stability properties of the tidal currents and the geomorphology are inseparably linked.

It is not known whether the hydrographic regime of the Wadden Sea follows a long-term trend. Global warming would certainly change the parameters of the tides in the Wadden Sea. In the case of the sea level rising, a natural Wadden Sea, i.e. without dikes, would – as happened several thousand years ago – shift landwards without changing its main characteristics (unless the topography of the flooded land shows major irregularities; Wolff 1993). This natural adaptation to changing hydrographical conditions is no longer possible today because the landward boundaries of the Wadden Sea are fixed by dikes (cf. Wolff 1987; Reise et al. 1998).

The morphology of the tidal flats was extremely variable on small spatial and temporal scales. Certain parts of the "Swinnplate", for example, changed due to major transport processes in winter, biodeposition of the mussel beds in summer, and seasonally occurring, ephemeral small tidal channels (Chap. 3.4). All in all, however, it seems as if the position, size, and morphology of the sandflats in the Spiekeroog backbarrier area only change extremely slowly. For example, the relief and position of the "Gröninger Plate" have remained essentially constant during the past 30 years; only its topographic height has decreased slightly (see 9.4.1).

A comparison of historical maps of the Wadden Sea (Chap. 3.4) reveals a slow but continuous change of morphology because of the rising sea level and because morphology has had to react to coastal engineering for several centuries. On the other hand, notwithstanding these changes, the essential characteristics of the morphology seem to have remained unchanged for hundreds or even thousands of years: extended tidal flats with tidal channels, sandflats, mudflats and muddy sands as well as – before coastal engineering started – extended salt marshes.

A good demonstration of the resilience and high elasticity (short return time) of the morphology was recorded at the "Swinnplate" after the ice winter in 1995/96 (Chap. 3.4): despite the partly extreme erosion during winter, the morphology reorganized itself within just six months. The morphology of the backbarrier areas is determined by the geometry of the overall tidal basin (see references cited in Chap. 3.4) and is even resilient to changes induced by events like storms or ice winters.

Sediment types in the backbarrier area (sandflat, mudflat, muddy sand) show a large-scale zonation pattern which is determined by the landward gradient of the decreasing kinetic energy of the water body. The closer to the coast, the more chance the finer particles have of settling (Chap. 3.4). The majority of the tidal flats in the Wadden Sea comprises sandflats or muddy sands, whereas mudflats are restricted to a small strip adjacent to the coast (and to some protected bays and biogenic mudbanks; Chaps. 3.4 and 3.5). Historical comparisons of the Wadden Sea show that the proportion of mudflats is decreasing – a trend which will probably continue in the future (Reise et al. 1998).

Flemming and Nyandwi (1994) interpret the small strips of mud near the dikes as natural mudflats which are cut off by the dikes. Without dikes, or if the dikes lay behind their current position, the mudflats would be much larger. They conclude that the energy of the water near the dikes is too high to allow for the sedimentation of finer particles. This results in a net export of fine material from the Wadden Sea, which in turn further decreases the mudflat areas.

The construction of dikes and the reclamation of land thus constrain the natural processes of sedimentation and morphodynamics responsible for the diversity of structure and habitat in the Wadden Sea. This means that from a sedimentological point of view, there is no constancy or reference dynamics beyond a time frame of, say, one or two decades. On the other hand, the loss of mudflats proceeds so slowly that it has neither the dynamics nor the drama sufficient to arouse public alarm and ultimately motivate decision-makers to develop and implement conservation plans. An additional problem in this context is that trends in the Wadden Sea may easily be masked by disturbance events (storm tides, ice winters), making it methodologically difficult to verify the trends.

Stability mechanisms

Hydrography and morphology mutually determine each other, and together they determine the conditions for sedimentation. On a time scale of several decades, there is a "morphodynamical equilibrium", i.e. morphology and, in turn, tidal currents change only marginally.

Equilibria, however, can only emerge if negative feedback between some components of the system occurs (Wissel 1989). This is indeed the case with the morphodynamics of the Wadden Sea: fast currents lead to erosion, whereas slow currents allow for the re-sedimentation of eroded material. Since the morphology determines domains of fast and slow currents, negative feedback exists between currents and morphology: very slow currents increase sedimentation, which in turn changes morphology such that, roughly speaking, according to Bernoulli's law, higher current velocities result, and vice versa.

This negative feedback and the resulting equilibrium are the decisive stability mechanisms which have allowed the landscape "Wadden Sea" to follow the rising sea level for thousands of years while maintaining its characteristic features (morphology with tidal channels, sand and mudflats, and salt marshes).

Strictly speaking, we should refer to "quasi-equilibrium" because "equilibrium" means by definition that the variables of interest considered do not change at all, whereas "quasi-equilibrium" means that these variables still change, albeit very slowly. Under natural conditions (i.e., without dikes), however, these changes are not relevant for the organisms living in the Wadden Sea, because they proceed much slower than the typical life cycles of the organisms, and because they do not alter the essential properties of the landscape.

Extreme scenarios

The most extreme scenario conceivable with respect to hydrography would be the construction of dikes connecting the barrier islands. Although this scenario is absurd, it helps highlight the "essence" of the Wadden Sea: the tides and the key processes driven by the tides. This essence is abiotic and will be very important for the following considerations in this chapter.

A less unrealistic scenario, which has indeed actually emerged in the North Frisian part of the Wadden Sea, is the construction of dams along the tidal watersheds. They would, as is known from the Sylt-Rømø-bight (Reise et al. 1998), have enormous consequences on hydrography, morphology and sedimentation. This is because a large proportion of the exchange of water between a tidal basin and other regions (basically the North Sea) takes place across the tidal watersheds (for the area behind Spiekeroog, about 30 %; Chap. 3.3). Additionally, reflection of the tides at the dams would strongly alter the morphology of the landscape. Damming of the Zuiderzee in the Netherlands in 1932 has caused major changes in the hydrography and geomorphology of the western Dutch Wadden Sea (Steyaert & Bakker 1994).

An extreme sedimentation scenario would be the complete loss of mudflats (which is indeed not unlikely). In this scenario, the diversity of benthic species and birds would decrease because some species depend on mudflats as their habitat or feeding grounds (for example, the amphipod *Corophium volutator*, the mudsnail *Hydrobia ulvae*, the avocet *Recurvirostra avosetta*, the redshank *Tringa totanus*, juveniles of flatfish and brown shrimps; Reise 1985, 1998).

Less clear are the consequences of a total disappearance of the mudflats for the primary production of microphytobenthos and plankton, as well as for species which prefer sandflats as adults but use mudflats as "breeding grounds", i.e. as a habitat for the juveniles (*Arenicola marina*). Living in breeding grounds during the juvenile stages avoids competition between juveniles and adults (Reise 1985). This mechanism, which seems to be part of the "success story" of *Arenicola marina* in the Wadden Sea, would work much less effectively without breeding grounds in the upper intertidal.

Biogenic mudbanks cannot substitute mudflats, because they differ with respect to position and other properties from mudflats, and because they are strongly affected by the disturbance regime imposed by storms and ice winters (Chaps. 3.6, 5.2 and 8).

9.4.3 Chemistry of the Sediments

Variables of interest

Numerous chemical and physical variables can be used to characterize the chemical properties of the sediment. Concerning the entire ecosystem of the Wadden Sea, however, one aspect is of crucial importance: the existence of a layer on the surface of the sediment which contains oxygen. The thickness of this oxic layer is the variable of interest which is considered with respect to resilience, whereas as far as persistence is concerned, merely the existence of this layer forms the (binary) variable of interest.

Significance of the ingredient for the Wadden Sea

Apart from bacteria with an obligate or optional anaerobic metabolism, all organisms depend on oxygen and cannot therefore survive in an anoxic medium. Moreover, anoxic sediment is often enriched with sulphide, which is toxic for higher organisms, although tolerances to different concentrations and exposure time vary among species (Chaps. 5.2.2 and 6).

A completely and permanently anoxic sediment would be almost exclusively populated by bacteria. For this reason, and because sulphide concentrations in the water column might become toxic, a Wadden Sea without an oxic layer at the sediment surface could no longer be a habitat for benthic organisms, juvenile fish, crabs or birds. Thus, the existence and resilience of the oxic layer is a decisive precondition for "higher" life in the Wadden Sea, i.e. for organisms beyond bacteria. "Higher" life includes the usage of the Wadden Sea by man.

Stability mechanisms: persistence and resilience

The oxic layer exists because of the diffusion of oxygen from the air and water into the sediment. This basic mechanism is supported by the turbation of the sediment by the tidal currents and swell, drainage of oxygen rich water during ebb through the sediment, the bioturbation by the infauna, and the oxygen produced by the microphytobenthos.

Since the physical mechanisms which contribute to the oxic layer are related to the tides, which are the most predictable aspect of the Wadden Sea, there can be no doubt about the persistence of the oxic layer under natural conditions. For the same reason, the oxic layer is also resilient. If the layer were to be destroyed, it would re-form within a few tides because anoxic sediment on the surface would be oxygenized by oxygen from the air and water very quickly (Chap. 6).

The thickness of the oxic layer depends on the physical and chemical properties of the sediment, season and biotic activities. The thickness decreases as the amount of organic matter in the sediment increases, because bacteria use oxygen when decomposing organic matter. This is the case when the organic matter of plankton blooms is decomposed and was particularly pronounced after the bloom of *Coscinodiscus concinnus* in June 1996 (Chap. 7).

The anoxic sediment horizon reaches the surface of the sediment ("black spots") when locally very high amounts of dead organic matter are processed by the bacteria, for example dead *Mya arenaria* or macroalgal mats, which are buried in the sediment (Höpner & Michaelis 1994). With layers of macroalgal mats, in addition to the organic matter decomposing, the diffusion of oxygen from water and air into the sediment is hampered.

The local input of large amounts of organic matter can be considered as a disturbance of the oxic layer. This disturbance is caused not by the organic matter itself, but by the decomposition of the material by the bacteria, which reduces the oxygen concentration. If the oxic layer re-forms after such a disturbance, this can be interpreted as resilience. This kind of resilience has so far been observed for all disturbances caused by organic input into the sediment. However, elasticity, i.e. the return time until the oxic layer rebuilds and reaches its original thickness, depends on the strength of the disturbance. It can take weeks or even months before the dead organic matter is decomposed in black spots and the activity of heterotrophic bacteria is in turn decreased to a level where the oxic layer may build up again.

Although black spots are detrimental for the higher benthic fauna, they are probably not "warning signals" (Lozán et al. 1994) *per se*. Instead, the increase in the frequency and size of black spots (which then become "black areas") may be a warning signal, i.e. an indicator of the consequences of the increased anthropogenic pollution of the Wadden Sea with nutrients (Höpner & Michaelis 1994).

After the ice winter in 1995/1996, in some backbarrier areas of the East Frisian Wadden Sea, large black areas emerged the size of soccer fields. They looked dramatic, because the sediment was covered with dead benthic fauna. It turned out, however, that these black areas could be explained as the result of a chain of rare but natural events, including an "oily" layer which was presumably caused by a bloom of the diatom *C. concinnus* (Chap. 7). Even for these large black areas, the oxic layer was re-established after the organic pollution had decomposed. It could, however, have taken much longer before the normal physical exchange processes between the sediment, air and water re-formed the oxic layer, because these processes were hampered by the lipid film on the sediment surface (Chap. 7). However, a storm which occurred a few weeks after the emergence of the black areas seems have to destroyed this lipid film and thus re-established the normal exchange processes between sediment, air and water.

Stability mechanisms

The decisive stability mechanisms for the persistence of the oxic layer are physical and chemical processes which are assisted by biotic processes (bioturbation, primary production of the microphytobenthos). The tides ensure that these processes can take place because they enable the diffusion of oxygen from the air during low tide, as well as turbation caused by tidal currents. Another mechanism which is relevant for the oxic layer seems to be storms.

Bacteria are mainly responsible for the resilience of the oxic layer to locally increased amounts of dead organic matter. If they were not able to switch to anaerobic metabolism, the decomposition of organic matter would be incomplete and the oxic layer would not be re-established. Thus, the resilience and, in turn, the persistence of the oxic layer rely not only on physical and chemical processes, but also on biotic activities.

Extreme scenarios

The scenario of an enduring anoxic intertidal with a rotten odour of hydrogen sulphide and where only bacteria can survive is – under natural conditions – very unlikely. The risk of collapsing to lasting anoxic conditions, which exists for more static waters like lakes or lagoons in warmer climates, does not threaten the Wadden Sea because of the tides which move the water, facilitating the diffusion of oxygen from the air during low tide, and ensuring the input of oxygen-rich water from the North Sea into the Wadden Sea. However, the massive anthropogenic input of nutrients or toxicants or even a major oil tanker disaster would considerably increase the regeneration time of the oxic layer (Kuiper et al. 1984).

9.4.4 Summary: Abiotic Ingredients of the Wadden Sea

The attempt to assess the stability properties of the abiotic ingredients of the Wadden Sea revealed in almost all cases the same pattern: on a short time scale (from a few days to a year) and in small areas – low constancy (or high unpredictability); on an intermediate time scale (years to several decades) and larger areas – constancy; and on a large time scale (several decades to centuries) and for the entire Wadden Sea – slow and continuous change which, however, under natural conditions does not alter the persistence of the Wadden Sea, i.e. the essential properties of a landscape shaped and determined in all respects by the tides.

This general pattern forms the background for all following assessments of the stability properties of the biotic ingredients of the Wadden Sea. In particular, reference dynamics can only be specified for intermediate temporal and spatial scales.

We tried to demonstrate that the tides do not only shape the landscape but are also responsible for the stability properties of three key ingredients of the Wadden Sea: the existence of extended tidal flats, which is the result of an interplay between tides, historical sea level rise, the flat morphology of the coast, and sedimentation; the characteristic morphology of the Wadden Sea; and the existence and resilience of an oxic layer on the sediment surface.

9.5 The Biotic Ingredients of the Wadden Sea

9.5.1 Bacteria

Variables of interest

Species diversity among the bacteria is presumably higher than that of all other species in the Wadden Sea taken together. With respect to the whole Wadden Sea, however, it is not the individual bacteria species which matter, but rather their overall performance, i.e. the remineralization of dead organic matter. The main variable of interest characterizing the activity of heterotrophic bacteria is thus the rate of remineralization. Another variable of interest is the overall biomass of bacteria, because they represent an important food resource for higher organisms. This variable has, however, not yet been sufficiently studied in quantitative terms.

Significance of the ingredient for the Wadden Sea

The high rate of primary production in the Wadden Sea (see 9.5.2) leads to high consumption rates of nutrients. Thus the high rate of primary production depends on a sufficient supply of nutrients. This supply does not only occur via exchange with water from the North Sea or input from rivers, but mainly via the remineralization of dead organic matter in the sediment by bacteria. The activity of the bacteria is thus crucial for the high primary production in the Wadden Sea.

Stability properties: persistence and resilience

The performance of bacteria is extremely persistent with respect to nutrient cycling in the Wadden Sea, and extremely resilient as far as local disturbances of higher inputs of organic matter are concerned (Chap. 5.2). This high persistence and resilience are very likely to apply as well to their role as food supply for the meio- and macrofauna (Reise 1985).

Stability mechanisms

The decisive mechanism responsible for these stability properties is the extremely high *functional diversity* of bacteria, which is a consequence not only of the high species diversity but also of the high *metabolic flexibility*. For any type of resources (e.g., fresh or refractory organic matter) and for almost any type of local environment (e.g., oxic, anoxic, different concentrations of sulphate, nitrate, etc.; Chap. 5.2) different species or different metabolic pathways exist which can very quickly start remineralization at the highest rate possible in that particular environment. This high adaptability of bacteria to even sudden and extreme changes of their environment was demonstrated by ELAWAT in laboratory experiments in which organic matter was added to sandy sediment (Chap. 5.2). Twelve hours after incubation with the organic substrate, only 5 % of the substrate had decomposed, this figure rising to 90 % after just 36 hours. Thus, the bacteria are able to produce the particular enzymes needed for the decomposition of particular organic matter very quickly, either by population growth of certain species, or by changes in metabolic pathways.

In this context much has been written in the ecological literature about possible relationships between (species) diversity and "stability". In most cases this relationship is not clear because the terms used are not unequivocal (Pimm 1984). However, with respect to the performance of bacteria in the Wadden Sea we can conclude that the stability properties persistence and resilience rely on the high functional diversity of the bacteria (cf. Reise et al. 1998).

One necessary requirement for the functional diversity of bacteria to become effective is a close coupling between the compartment where remineralization takes place and the entire system. This close coupling is ensured in the Wadden Sea by a water body which is dynamic because of tides and swell. Tides and swell lead to turbation on the surface of the sediment, which enables an effective exchange of material between sediment and water. The effects of turbation are enhanced by the effects of bioturbation caused by the infauna.

Extreme scenarios

Without bacteria, the high rate of primary production in the Wadden Sea would be impossible because all the nutrients available would have been consumed very quickly. Hence, all consumers, including fish, birds and seals would disappear from the Wadden Sea. But such a scenario, i.e. a Wadden Sea without bacteria, is extremely unlikely because of the high physiological flexibility and functional diversity of bacteria. Extreme scenarios regarding bacteria would thus not be scenarios without bacteria, but scenarios in which bacteria exist to the virtual exclusion of any higher organisms. Such a scenario would occur after long-lasting periods of anoxic condition on the surface of the sediment; but this, as has been pointed out above, is very unlikely because of the turbation caused by tides, swell and storms (see above, 9.4.3).

9.5.2 Microphytobenthos and Phytoplankton

Variables of interest

As with bacteria, it is the overall physiological performance of microalgae that matters with respect to the entire Wadden Sea, i.e. primary production. As far as benthic algae are concerned, spatial distribution is another important variable of interest. This distribution may be patchy because of local heterogeneities or small-scale disturbances. With phytoplankton, the body size of the different species is a relevant variable because the consumer of the planktonic algae may prefer particular size classes (Chap. 5.1). Species composition is thus a relevant variable of phytoplankton, which becomes even more relevant after rare extreme events (e.g., ice winter), when species which are not typical for the Wadden Sea become dominant for some time (Chaps. 5.1 and 7).

Significance of the ingredient for the Wadden Sea

The single-cell algae of microphytobenthos and phytoplankton are the most important primary producers of the Wadden Sea. They are a food source of meio- and macrobenthos. The high rate of primary production is a characteristic property of the Wadden Sea.

Another important role of the microphytobenthos is that they reduce the erodibility of the sediment surface (Holland et al. 1974). This "stabilization" of the sediment is presumably an important requirement for the larval settlement of many macrobenthic species (cf. Chap. 5.3).

Stability properties: persistence and resilience

The primary production of microphytobenthos and phytoplankton might be regarded as persistent if in the definition of persistence we allow for a wide range of variation of the rate of primary production. Primary production does not, however, seem to be constant because in some parts of the Wadden Sea a marked increase of primary production has been observed during the past decades, presumably due to an increased input of nutrients to the Wadden Sea (Asmus et al. 1998).

The overall abundance and species composition of phytoplankton within a certain year is as variable as the variables which mainly determine abundance and species composition, temperature and nutrient concentration (see Chap. 5.1). In addition, rare episodic events may occur which increase variability, e.g., the blooms of *Coscinodiscus concinnus* and *Phaeocystis pouchetii* after an ice winter (Chap. 5.1 and 7). Both of these species are adapted to temperatures lower than those typical of the Wadden Sea and thus had a temporary competitive advantage during the long-lasting low temperatures in spring 1996, which led to a bloom of these species in the East Frisian Wadden Sea – at the expense of "normal" species which did not bloom here as usual.

The presence of benthic diatoms in the sediment is resilient to disturbances (e.g., storms, drift ice, algal mats; Chaps. 6 and 7) and in turn extremely persistent. Nevertheless, the phenology of the abundance of benthic diatoms showed marked differences in different years, which are mainly due to variations of the abiotic conditions and of grazing pressure (Reise 1992).

Stability mechanisms

The mechanisms responsible for the stability properties of the microalgae in the Wadden Sea are similar to those of the bacteria: high potential population growth rates and high species diversity. In addition, the abiotic conditions which are required for high primary production (e.g., better light conditions in the intertidal during low tides, dynamic water body, availability of nutrients and remineralization) are persistent and resilient. However, benthic diatoms themselves contribute to the stability properties of the abiotic ingredient "oxic layer" by the production of oxygen.

Two mechanisms are responsible for the resilience of benthic algae to local disturbances: Firstly, disturbed areas are recolonized from undisturbed areas via the tidal currents (cf. Chap. 6). Grubb & Hopkins (1986) call this mechanism "resilience by migration", i.e. at the disturbed site all organisms are killed so that the re-establishment of the original species composition depends on colonization ("migration") from other, undisturbed areas. In this context it does not matter whether "migration" is active or passive ("migration" usually refers to active movements).

The second mechanism of resilience is called "*in-situ*" resilience (Grubb & Hopkins 1986). In this case not all organisms are killed at the disturbed site. When after the disturbance the abiotic conditions become similar to the original, pre-disturbance conditions, the regeneration of populations may occur "*in-situ*" and does not necessarily require the input of individuals from other sites. Examples of this kind of resilience include seed banks of plants, serotinous seeds (which only germinate after fires) or the resprouting of burned plants (Enright et al. 1996). Some species of benthic algae are tolerant of burial and many live heterotrophic (Admiraal et al. 1979). As was shown by the recolonization experiments during ELAWAT (Chap. 6), some diatoms exhibited "*in-situ* " resilience because they were even able to survive several weeks of anoxic conditions.

Extreme scenarios

A Wadden Sea without microphytobenthos would have lost an essential ingredient and would therefore no longer be the Wadden Sea we know. Primary production would be much lower. Many species which feed upon microphytobenthos would disappear, with far-reaching consequences for the food web of the Wadden Sea. In addition, sediment would be less "stable", which would increase the turbidity of the water and would also presumably affect larval settlement.

Such a scenario, however, is hard to envisage because the decisive abiotic ingredients which enable the high primary production of the microphytobenthos (i.e., sufficiently high input of light, carbon dioxide and nutrients) are very persistent and resilient. Whether, however, the actual rate of primary production remains more or less constant, or follows an increasing or decreasing trend, mainly depends on the anthropogenic input of nutrients.

With respect to phytoplankton, one extreme scenario would be such a high turbidity of the water body that photosynthesis is strongly reduced. Such strong turbidity could be caused by a much more dynamic water body containing much more suspended matter, as it occurs, for example, in the turbidity maximum of estuaries. Indeed, the kinetic energy of the water body is already higher than it would be under natural conditions because of the dikes.

9.5.3 Macrozoobenthos

Variables of interest

The assessment of stability properties is much more complex for higher, heterotrophic species than for autotrophs because there is no highly aggregating variable which would have the same significance as, for example, primary production or the rate of remineralization. There is no single variable that would allow to characterize the "performance" of the higher heterotrophic benthic animals (worms, mussels, crabs, etc.) in the Wadden Sea. Thus, the following considerations will be more complex and more detailed because we will have to change the level of description from whole functional groups (bacteria, microphytobenthos) to communities and, in particular, populations. On this level of description, the most important variable of interest is the abundance or the density of the species considered.

As a consequence of this new level of description and variables of interest, the spatial distribution of the organisms must be explicitly discussed in the following. In soft-bottom communities, distributions are often "patchy", where "patch" is defined as "an area which, on a specified scale of observation, has some level of consistency in the density of individuals" (Hall et al. 1994, p. 333; Chap. 8).

Besides single species, there is a long tradition of studying communities in the Wadden Sea (Reise 1980, 1990a). However, it is often impossible to relate the classification of communities used in early studies to each other and to the actual distribution of species (Michaelis & Böhme 1994). The types of "biofacies" defined by Hertweck (1995; Chap. 3.5) are more suitable because they consider both the dominant species and the type of sediment in a certain area.

Significance of the ingredient for the Wadden Sea

The extremely high abundance of macrobenthic species (polychaetes, molluscs, snails, crustacea) is a characteristic property of the Wadden Sea. This high abundance is exploited by fish and birds, as well as by humans (e.g., harvesting mussels). The quantitative aspects of the activity of the macrobenthos are well-known but still impressive: before their decline in the Lower Saxonian part of the Wadden Sea (Zens et al. 1997), mussels filtered all the water in their region within just a few days. The polychaetes *Heteromastus filiformis*, *Arenicola marina*, and the Baltic tellin *Macoma balthica* move enough sediment within one year to equal a layer of sediment 35 cm deep (Cadée 1990). The qualitative aspect, i.e. the ecological significance of the macrofauna, is a topic of research in the Wadden Sea. The questions studied most intensively are the causes and consequences of spatial and temporal distribution, interactions between species, rates of recovery after local disturbances, succession, and the concept of community (Reise 1985, Hall et al. 1994, Raffaelli & Hawkins 1996).

Stability properties

Constancy and variability – The evaluation of constancy, or the inverse property, variability, depend upon the spatial and temporal scale considered. At the local level, e.g., a square metre, the temporal variability of abundance is extremely high. As the ELAWAT studies show, abundance fluctuates on a scale of just a few days, as well as within a year and from year to year. At larger spatial scales variability decreases because small scale heterogeneities average out. However, how much abundances vary among years on the scale of the Spiekeroog backbarrier tidal basin cannot be assessed because of methodological problems and the short duration of ELAWAT. On a time scale from years to decades, the distributions and abundances of macrobenthic species are driven by episodic events (storms, ice winters; Chap. 8) that can cause considerable fluctuations. Besides this, there are records of historical changes of the distribution patterns and abundances due to human influence (Reise 1994).

Spatial variability (at a given point in time) also depends on scale. On the scale of a few centimetres to decimetres the spatial distribution of individuals is extremely variable, which is indicated by the wide range of abundances even in individual samples of a "multicorer" (Chap. 5.5). Depending on the species considered, site and sampling date, distribution patterns may be random but are mostly clumped or patchy (cf. Thrush et al. 1989).

At larger spatial scales, one finally arrives at the "zonation patterns" described in Chap. 8 by a model, i.e. typical assemblages of species that occur together in certain areas and are mostly dominated by a few characteristic species (Dörjes 1978; Reise 1985; Hertweck 1995). In contrast to rocky shores, where zonation patterns often follow a linear gradient and therefore the zones are linearly arranged, zonation patterns in the intertidal of the Wadden Sea are complex because of the complex morphology of the tidal flats and the marked disturbance regime.

Resilience – The large-scale zonation pattern is apparently resilient because it becomes re-established within a few years after major disturbance events. For example, the re-establishment of the typical distribution pattern of the sand mason (*Lanice conchilega*) took only three or four years (Hertweck 1995; cf. Chaps. 5.4 and 8). The ELAWAT studies on the Gröninger Plate show that after the ice winter in 1996/97, during which virtually all *L. conchilega* in the intertidal were killed, *L. conchilega* settled in the same areas where this species had lived before. Likewise, *Mytilus edulis* seems to prefer the same areas of a certain sandflat for settlement, as indicated by the relief casts of Hertweck (1995; Chap. 3.5) and a comparison of the distribution of *M. edulis* of the past 30 years (Michaelis et al. 1995).

Species composition and mean abundance are also resilient on smaller spatial scales of up to ten metres. On this scale, elasticity, i.e. the time needed to return to the pre-disturbance situation, depends on the size of the disturbed areas as well as on the type, duration and date of the disturbance. This is well-known for soft-bottom communities (Hall et al. 1994) and has been confirmed by ELAWAT's recolonization experiments (Chap. 6).

Persistence – In contrast to terrestrial ecosystems, where the risk of population extinction has become a central issue in ecology and conservation biology in the last decade, the extinction and persistence of soft-bottom species are usually not discussed. The reason for this is the high resilience of macrobenthic species which, due to several mechanisms (see below) guarantees persistence. In the Wadden Sea, so far two species which were able to build large epibenthic structures are recorded as having been extinguished by mankind: *Ostrea edulis* and *Sabellaria spinulosa*. The latter species built sandy reefs in the tidal channels. *Ostrea* was overfished, while *Sabellaria* reefs were presumably destroyed by bottom trawlers. In turn, several species exclusively associated with oyster banks or *Sabellaria* reefs have also disappeared (Reise 1982, 1991, 1994; Michaelis & Reise 1994).

Besides these historical extinctions, the diversity or, to be more precise, species richness of the macrobenthos is almost constant because of the persistence of the individual species. Changes of species richness and composition would only occur as a result of further anthropogenic influences or drastic alterations of the abiotic boundary conditions. Part of the anthropogenic influences are alien invasions, which so far have become established in the Wadden Sea without ousting any resident species (Reise 1991). Whether this will continue to be the case in the future cannot be predicted.

Stability mechanisms

Mobility – The following processes are responsible for the high local and temporal variability of the abundance of the macrobenthic species: the heterogeneous distribution of resources, the heterogeneous input of larvae and juveniles, the immigration and emigration of juveniles and adults, and local mortality.

The input of larvae firstly depends on the regional supply of larvae whose size and phenology are extremely variable for all species studied under ELAWAT (Chap. 5.1). This is partly due to abiotic factors, e.g., ice winters. In addition, meroplanktonic larvae are more or less coupled to the dynamics of their food resources, i.e. the phytoplankton. The actual local input of larvae which are ready to settle depends on local conditions. *Lanice conchilega* (Chap. 5.3) and *Mytilus edulis* need hard substrate for settlement; most other species settle in areas with low velocities of the near-bottom flow due to higher rates of sedimentation.

Another source of local variability in abundance is the secondary dispersal of postlarvae, i.e. young-of-the-year (Günther 1992). Most juveniles leave the areas of their initial settlement and search actively or passively for suitable areas (Chap. 5.3). The adults of most macrobenthic species are also mobile. Exchange rates of individuals between sediment and water column are extremely high, as has been demonstrated by the staining technique (Chap. 5.3). Thus, a complete turnover of the individuals at a certain location may occur within a few days or weeks. Local mortality, finally, depends on age or on predation, for example by epibenthic predators (e.g., *Carcinus maenas*, Chap. 5.6).

The reasons for the small-scale patchiness of the distribution patterns are the same as those which are responsible for the high temporal variability on this spatial scale: because of their mobility the organisms are able to react quickly to small-scale differences of food resources and to the presence or activity of other organisms.

Resilience and abiotic boundary conditions – The resilience of the large-scale "zonation pattern" is mainly due to the resilience of the abiotic boundary conditions, i.e. of hydrography, morphology and sedimentation. Even after extreme events like ice winters, the original morphology and current regime become quickly re-established.

Phenotypic plasticity – Organisms are not machines which function to a rigid programme. Instead, they can adjust growth rate, the age of reproduction, the allocation of assimilated energy to growth, egg production, or the production of shells etc. to varying environmental conditions. According to the theory of life history evolution (Stearns 1992), this "phenotypic plasticity" may be considered a genetically fixed mechanism to cope with a broad range of environmental conditions, e.g., to achieve the highest reproductive values possible for any environmental condition.

An impressive demonstration of the wide range phenotypic plasticity may entail is *Mytilus edulis* which shows, depending on habitat type (higher parts of the sandflats, lower parts at the edge of tidal channels, or in the subtidal), extremely different life histories (Ruth 1991). Essink et al. (1989) showed that *M. edulis* does adapt its feeding apparatus in response to high (Wadden Sea) or low (North Sea) concentrations of suspended matter. An example observed in ELAWAT is the spawning behaviour of *Macoma balthica* (Chap. 5.1) with a minor spawning in spring and a second, much stronger spawning in summer 1995 (Fig. 5.1.6). After the following ice winter, the larval abundance of *M. balthica* was extremely low. This spawning behaviour by *M. balthica* had not been recorded before (Günther et al. 1998); *M. balthica* usually shows a strong recruitment after ice winters.

Phenotypic plasticity is the mechanism at the individual level which leads at the population level to resistance or low sensitivity with respect to environmental changes. Another example of this mechanism recorded by ELAWAT are the "red" and "green" colour forms of *Carcinus maenas* (Wolf 1997, Chap. 5.6).

"Resilience by migration" and metapopulations – After the ice winter in 1995/96, another important stability mechanism of macrobenthic species was observed when the adults of *Lanice conchilega* were killed not only in the intertidal but also in the subtidal. Therefore, the larval supply in the following summer must have originated in remote regions. This "resilience by migration" is also characteristic of other macrobenthic species which are sensitive to certain types of disturbances (e.g., *Mytilus edulis*). This stability mechanism would break down if the last areas where a species occurs were so close to each other that a disturbance event could affect all of them. In this case, a species could disappear from the Wadden Sea, as has been reported for the oyster (*Ostrea edulis*; Reise 1982, 1991, 1994; Michaelis & Reise 1994).

Yet it is not clear whether the populations of macrobenthic species occurring in different regions (different parts of the Wadden Sea, North Sea, English Channel, French or British coast) should be interpreted as parts of a metapopulation (Hanski & Gilpin 1997; Reich & Grimm 1996). In metapopulations, local extinctions which may be due to disturbance events can be remedied by recolonizations from subpopulations that still exist. In this "classical" notion of metapopulations (Levins 1969) all local populations are prone to extinction, whereas on a regional scale the persistence of the whole metapopulation is possible if certain conditions are fulfilled.

Another notion of metapopulations is the "mainland-island" scenario, in which some populations are so safe (because of being very large or living on protected sites; e.g., populations of *Lanice conchilega* off the French coast) that they are the ultimate source of the temporary existence of populations which are smaller or which are more exposed to disturbances. Safe populations would constitute the "mainland" upon which the "island" population depends. Unfortunately, little is known about the origin of larvae in the Wadden Sea except that many of them will originate from the North Sea. Therefore it is not yet possible to fully discuss whether the concept of metapopulation applies to macrobenthic species on spatial scales extending beyond the Wadden Sea.

However, at the spatial scale of the Wadden Sea there is a stability mechanism which resembles the "mainland-island" scenario: environmental fluctuations and disturbance events like storms or drift ice mainly affect organisms in the intertidal. In the subtidal, living conditions for most macrobenthic species are much more favourable (less physiological stress, more predictable environmental conditions, less exposed to disturbances). Therefore, subtidal populations are likely to buffer negative influences on intertidal populations (cf. Chesson & Huntley 1988; Günther 1990). However, there are only a few studies of subtidal populations in the Wadden Sea (but see the ELAWAT study of *Carcinus maenas*, Chap. 5.6).

Finally, even at smaller spatial scales of a few metres or so (Chap. 6), resilience relies on "migration". In this case, in addition to mobile larvae, the mobility of juveniles and adults comes into play.

Functional diversity – As a consequence of the high variability of the abiotic boundary conditions in the Wadden Sea within a year and from year to year, the quality and quantity of food supply for macrobenthic species change all the time. For this reason, many macrobenthic species are generalist feeders, i.e. they can utilize a more or less wide spectrum of food resources. This may even mean that they can change their feeding mode. This generalist feeding behaviour is, like the above-mentioned phenotypic plasticity, another mechanism at the level of the individual which at the population level promotes resistance to environmental fluctuations.

Thus, the relatively low number of macrobenthic species, i.e. low diversity, may be misleading if stability properties are to be assessed, because low resilience is often inferred from low diversity. However, Costanza et al. (1993) point out that the wide food spectra of macrobenthic species living in tidal flats or estuaries mean that there is a high *functional* diversity. Instead of many different species, which would allow for alternative paths in a food web, the different "feeding functions" of just a few species enable food webs with certain stability properties. Only if there are alternative paths in a food web will the temporary loss of certain single paths have no fatal consequences for the entire web (MacArthur 1955).

Extreme scenarios

An extreme scenario involving macrobenthic species would be a Wadden Sea with a short circuit of nutrient cycles, i.e. the existence of just algal primary production and bacterial remineralization. There would be no macrobenthos at all. The consequences of this scenario for all the species which depend on macrobenthic species as the major food resource would, of course, be disastrous: birds and young fish would have no more food.

However, this scenario is not very likely for two reasons. Firstly, two preconditions for the existence and the high abundance of the macrobenthos possess marked stability properties themselves: the existence of an oxic sediment layer and high primary production. Secondly, the macrobenthic species of the Wadden Sea have so many very effective stability mechanisms that their total disappearance is very unlikely under natural conditions. Anthropogenic influences have, however, caused some changes in the species composition of the macrobenthos (Reise 1982, 1991, 1994).

9.5.4 Migratory Birds

Variables of interest

Variables of interest considered are the abundance, species composition and spatial distribution of migratory birds in the course of a year. But since the migratory birds only spend more or less a short part of the year in the Wadden Sea, no conclusions are possible from the abundances recorded in the Wadden Sea about the full population dynamics of the birds. Additional variables of interest studied by ELAWAT include the consumption of individual birds and the estimated total consumption of the entire avian community (Exo, Ketzenberg unpubl.).

Significance of the ingredient for the Wadden Sea

The significance of the Wadden Sea for migratory birds with arctic or subarctic breeding grounds is well-known (Rose & Scott 1994). On their way from or to their winter sites in the south, the migratory birds build or fill up their energy reserve necessary for their flight.

Less clear is the significance of the migratory birds for the Wadden Sea. In view of the extremely high number of birds utilizing the Wadden Sea year after year, one might well wonder about the extent to which total consumption by the avian community controls the macrozoobenthos. For example, it has been hypothesized that the decline of *Mytilus edulis* in the East Frisian part of the Wadden Sea may partly be a consequence of the increasing number of birds observed in recent years (cf. Michaelis et al. 1995).

Stability properties

The Wadden Sea is an important area for waders and waterbirds with about 10–12 million birds using the area over the year (Meltofte et al. 1994). The phenology of the abundance of the different species follows the same pattern each year. In the studied tidal flats dominated by *Lanice conchilega*, the relative consumption of the birds per month in autumn was estimated as 5–14 % of the biomass (Petersen & Exo 1999). For certain types of prey and locations, however, consumption may be higher. For example, it has been estimated for mussel beds in the North Frisian part of the Wadden Sea that birds consume 32 % of the biomass (Nehls et al. 1998). Feeding pressure exerted by birds seems to be relatively constant if considered on the time scale of years. In recent years, however, an increase in bird abundance has been observed. If this trend continues, consumption by birds may increase and consequently diminish the recruitment of macrobenthic species.

Observation after the ice winter in 1995/96 revealed that the birds are able to cope with extreme climatic conditions when the quantity and quality of their prey are considerably reduced or when the preferred prey species temporarily disappear completely. This resistance is due to the plasticity of individual behaviour including temporary migration to ice-free areas. Whether resilience also exists is not known because it could not be ascertained whether the ice winter actually had demographic consequences, for example by resulting in a decline of certain bird species.

Stability mechanisms

Little can be inferred from the studies under ELAWAT about the mechanisms responsible for the relative constancy of bird abundance because these mechanisms mainly function outside the Wadden Sea. It seems clear, however, that the constantly high abundance of food resources in the Wadden Sea is an important factor in the predictability of the overall abundance of the migratory birds. Wadden Sea "petrol station" is very reliable from the point of view of the migratory birds.

Resistance to ice winters relies on mechanisms at the individual level. Flexibility with respect to their prey properties or prey species enables the birds to even survive extreme weather conditions. Moreover, their migratory behaviour is flexible. Birds leave the Wadden Sea earlier than usual under bad weather conditions.

Extreme scenarios

In the view of birds which use the macrozoobenthos as food, evidently a Wadden Sea without macrozoobenthos would be disastrous. From the point of view of the other fauna, especially the macrozoobenthos, the total disappearance of the birds would probably have less dramatic consequences. Because of the rather small proportion of available biomass consumed by birds, it can be assumed that birds do not control the dynamics of the benthic species but merely influence them numerically. However, there is too little empirical evidence and theoretical insight into the mode of interaction between birds and their prey in the Wadden Sea to be sure about this point.

9.5.5 Summary: Biotic Ingredients of the Wadden Sea

The assessment of stability properties of the biotic ingredients of the Wadden Sea follows a similar pattern as for the abiotic ingredients (9.4.4): on a short time scale (from a few days to a year) and small areas – low constancy (or high unpredictability); on an intermediate time scale (a few years to several decades) and larger areas – constancy (or low variability), with resilience in the case of disturbances; and on a long time scale (many decades and longer) and the entire Wadden Sea – persistence. It must be emphasized, however, that in the long term abiotic boundary conditions may continue to change slowly but surely as a result of anthropogenic influences. Several species were exterminated in the Wadden Sea by overexploitation. Furthermore, the loss of species reported for the past was often linked to the loss of key habitats such as oyster reefs, *Sabellaria* reefs, or salt marshes (Reise 1992, 1991). The ongoing loss of mudflats may have similar consequences in the future.

As with meteorological conditions, there are no reference "states" for biotic components: there are no "normal" years. At best, reference dynamics can be identified, i.e. typical elements of the phenology, and typical ranges of abundances for certain species.

The resilience of the biotic ingredients is mostly considered in the case of spatially localized disturbances: resilience in the performance of bacteria and benthic algae in disturbed areas, as well as resilience in the distribution pattern of macrobenthic species and in the species composition of benthic communities.

The mechanisms which are responsible for the stability properties are similar to those of the abiotic ingredients. All these mechanisms enable resistance or resilience and thus persistence.

9.6 Conclusions

In this chapter, we have tried to assess the stability properties of the abiotic and biotic ingredients of the Wadden Sea which were studied by ELAWAT. Stability properties are emergent properties which can only be perceived on a certain level of description and have no direct meaning at lower levels of description. Our enumeration showed that the ingredients exhibit certain stability properties on certain spatial and temporal scales. However, it is impossible for practical and principle reasons to attribute stability properties for the entire Wadden Sea. The practical reasons for this are that, firstly, ELAWAT could only study a subset of ingredients over too short a time span (Chap. 10), and secondly, the significance of interactions between ingredients and species was not sufficiently studied. The principle reasons are, besides those listed by Grimm et al. (1992) and Grimm (1996), that the identification of emergent properties at the level of the ecosystem depends on the identification of the ecosystem itself (cf. Wiegleb & Bröring 1996; Jax et al. 1998) – a task which has not yet been fully performed for the Wadden Sea.

In the following we will summarize our findings concerning the scale dependence of stability properties and the conditions required to enable regeneration after disturbances.

Scale dependence of stability properties

For both abiotic and biotic ingredients, our enumeration showed that on small spatial and temporal scales the variables of interest show little constancy or high variability. Constancy, and in the case of disturbances, resilience only shows up at larger scales. On even larger scales (many decades and more) and for the entire Wadden Sea or large parts thereof, persistence would prevail under natural conditions. However, historical records and long-term studies seem to indicate slow changes among both abiotic ingredients (e.g., morphology) and of biotic ingredients (species composition, primary production, biomass, bird abundance).

For many ingredients it proved difficult to identify reference dynamics. It turned out that certain states, e.g. certain snapshots of the Wadden Sea, are unsuitable as a reference. Instead, the state of the Wadden Sea *is* its dynamics!

Survey of stability mechanisms

Stability mechanisms of abiotic ingredients are often coupled to supra-regional processes (e.g., climate) or are due to negative feedback loops within the Wadden

Sea (e.g., between tides, flow regime, sedimentation and morphology). For the biotic ingredients, the following mechanisms at the individual or population level are responsible for stability mechanisms:

- high diversity (taxonomic or functional)
- high reproduction
- high mobility
- phenotypic plasticity
- flexible feeding behaviour
- physiological tolerance

It was shown that organisms of the microfauna, meiofauna and macrofauna are able to react to disturbances by both "resilience *in-situ*" and "resilience by migration" (Chaps. 6 and 7).

Using the "4-box-model" (Holling's terminology, 1986), this list of stability mechanisms suggests that the Wadden Sea of today is permanently in the "exploitation phase". According to Holling, complex systems first have an exploitation phase and then proceed to a "conservation phase" which is characterized by the dominance of K-strategists, sessile species and biogenic structures, which effectively buffer environmental fluctuations. A typical example of this transition is the succession of tree species, which ends in a local climax community where the canopy trees provide the biogenic structure buffering environmental fluctuations. The original Wadden Sea of 1000 or more years ago may have been in a conservation phase, but long-lasting and permanent human impact (loss of habitats and large, long-living species) has changed the system and still continues to do so (Wolff 1993 and pers. comm.).

The mobility and flexibility in life cycles of the organisms of the Wadden Sea has also been reported for other soft-bottom communities and in general for ecosystems with a natural variability of driving forces and is regarded as the precondition for their resilience (Chap. 6; Costanza et al. 1993; DeAngelis & White 1994).

Conditions for regeneration after disturbances

Virtually all ingredients of the Wadden Sea have stability properties which are directly or indirectly linked to the tides or to the existence of the tidal flats. The tides determine morphology, flow regime and sedimentation which, in turn, determine almost all processes which are decisive for life in the Wadden Sea (e.g., settlement, dispersal). If anthropogenic impacts affect this tide-driven web of processes, important stability mechanisms may break down. For example, areas which may serve as sources of recolonization as well as the unrestricted transport of dispersal stages are crucial for the resilience of benthic organisms to disturbances (Chaps. 5.3 and 6). If anthropogenic impact leads to a loss of habitat heterogeneity and the flow regime, regeneration after disturbances may be impeded. In order for the natural stability mechanisms to fully unfurl their potential, the natural, tide-driven abiotic framework must continue to exist. To enable the stability mechanisms of the organisms to work, care must be taken that the abiotic processes which enable these mechanisms in the first place are a primary focus of conservation.

Acknowledgements

We thank all our colleagues who have contributed to the discussions about this chapter. A reviewer has given valuable comments on an earlier draft, contributed some substantial ideas to this chapter and provided literature relevant to the issues discussed here. The project was funded by the German Bundesministerium für Bildung, Wissenschaft, Forschung und Technologie (BMBF) under grant number 03F0112 A and B. The responsibility for the contents of the publication rests with the authors.

References

Admiraal W, Peletier H (1979) Sulphide tolerance of benthic diatoms in relation to their distribution in an estuary. Br Phycol J 14:185–196

Asmus R, Jensen MH, Murphy D, Doerffer R (1998) Energiefluß und trophischer Transfer im Sylt-Rømø Wattenmeer. In: Gätje C, Reise K (Eds) Ökosystem Wattenmeer. Austausch-, Transport- und Stoffumwandlungsprozesse. Springer, Berlin Heidelberg New York. pp 367–391

Beukema JJ (1979) Biomass & species richness of the macrobenthic animals living on a tidal flat area in the Dutch Wadden Sea: effects of a severe winter. Neth J Sea Res 13: 203–223

Beukema JJ (1985) Zoobenthos survival during severe winters on high and low tidal flats in the Dutch Wadden Sea. In: Gray JS, Christensen ME. Marine Biology of Polar Regions and Effects of Stress on Marine Organisms. John Wiley & Sons, Ltd. pp 351–361

Beukema JJ, de Bruin W, Janssen JJM (1978) Biomass and species richness of the macrobenthic animals living on the tidal flats of the Dutch Wadden Sea: long-term changes during a period with mild winters. Neth J Sea Res 12: 58–77

Cadée GC (1990) Feeding traces and bioturbation by birds on a tidal flat, Dutch Wadden Sea. Ichnos 1: 22–30

Chesson PL, Huntly N (1988) Community consequences of life-history traits in a variable environment. Ann Zool Fenn 25: 5–16

Cole J, Lovett G, Findlay S (Eds), (1991) Comparative analyses of ecosystems. Springer, New York

Connell JH, Sousa WP (1983) On the evidence needed to judge ecological stability or persistence. Am Nat 121: 789–824

Costanza R, Kemp WM, Boynton WR (1993) Predictability, scale, and biodiversity in coastal and estuarine ecosystems: implications for management. Ambio 22: 88–96

DeAngelis DL, White PS (1994) Ecosystems as products of spatially and temporally varying driving forces, ecological processes, and landscapes: A theoretical perspective. In: Davies SM, Ogden JC (Eds) Everglades. The Ecosystem and its Restoration. Boca Raton, USA, St. Lucie Press, p 9–27

Dörjes J (1978) Das Watt als Lebensraum. In: Reineck H-E (Ed) Das Watt. Ablagerungs- und Lebensraum. Waldemar Kramer, Frankfurt, pp 107–143

Enright NJ, Lamont BB, Marsula R (1996) Canopy seed bank dynamics and optimum fire regime for the highly serotinous shrub, *Banksia hookeriana*. J Ecol 84: 9–17

Essink K, Tydeman P, de Koning F, Kleef HL (1989) On the adaptation of the mussel *Mytilus edulis* L. to different environmental suspended matter concentration. In: Klekoswksi RZ et al. (Eds) Proceedings of the 21st EMBS. Polish Academy of Sciences – Institute of Oceanology, Ossolineumn, Warsaw, p 41–51

Fenchel T (1992) What can ecologists learn from microbes: life beneath a square centimetre of sediment surface. Funct Ecol 6: 499–507

Flemming BW, Nyandwi N (1994) Land reclamation as a cause of fine-grained sediment depletion in backbarrier tidal flats (Southern North Sea). Neth J Aquat Ecol 28: 299–307

Gigon A, Grimm V (1997) Stabilitätskonzepte in der Ökologie: Typologie und Checkliste für die Anwendung. In: Fränzle O, Müller F, Schröder W (Eds) Handbuch der Umweltwissenschaften. Ecomed, Landsberg, pp III-2.3/1–20

Grimm V (1996) A down-to-earth assessment of stability concepts in ecology: dreams, demands, and the real problems. Senckenbergiana marit 27: 215–226

Grimm V (1998) To be, or to be essentially the same: the 'self-identity of ecological units'. Trends Ecol Evol 13: 298–299

Grimm V, Schmidt E, Wissel C (1992) On the application of stability concepts in ecology. Ecol Model 63: 143–161

Grimm V, Wissel C (1997) Babel, or the ecological stability discussions: An inventory and analysis of terminology and a guide for avoiding confusion. Oecologia 109: 323–334.

Grubb PJ, Hopkins AJM (1986) Resilience at the level of the plant community. In: Dell B, Hopkins AJM, Lamont BB (Eds) Resilience in Mediterranean-type Ecosystems. Junk, Dordrecht, pp 21–38

Günther C-P (1990) Zur Ökologie der Muschelbrut im Wattenmeer. Dissertation, Universität Bremen

Günther C-P (1992) Dispersal of intertidal invertebrates: a strategy to react to disturbances of different scales? Neth J Sea Res 30: 45–56

Günther C-P, Boysen-Ennen E, Niesel V, Hasemann C, Heuers J, Bittkai A, Fetzer I, Nacken M, Schlüter M & Jaklin S (1998) Observations of a mass occurrence of *Macoma balthica* larvae in midsummer. J Sea Res 40: 347–351

Hall CAS, DeAngelis DL (1985) Models in ecology: paradigms found or paradigms lost? Bull Ecol Soc Am 66: 339–346

Hall SJ, Raffaelli D, Thrush SF (1994) Patchiness and disturbance in shallow water benthic assemblages. In: Giller PS, Hildrew AG, Raffaelli DG (Eds) Aquatic Ecology. Scale, Pattern and Process. Blackwell Science, Oxford, pp 333–375

Hanski IK, Gilpin ME (1997) Metapopulation Biology: Ecology, Genetics, and Evolution. Academic Press, San Diego

Hertweck G (1995) Verteilung charakteristischer Sedimentkörper und Benthossiedlungen im Rückseitenwatt der Insel Spiekeroog, südliche Nordsee. I. Ergebnisse der Wattkartierung 1988–92. Senckenbergiana marit 26: 81–94

Holland AF, Zingmark RG, Dean JM (1974) Quantitative evidence concerning the stabilization of sediment by marine benthic diatoms. Mar Biol 27: 191–196

Holling CS (1986) The resilience of terrestrial ecosystems: local surprise and global change. In: Clark WC, Munn RE (Eds) Sustainable Development of the Biosphere. Cambridge University Press, Cambridge, pp 292–317

Höpner T, Michaelis H (1994) Sogenannte "schwarze Flecken" - ein Eutrophierungssymptom des Wattenmeeres. In: Lozán JL, Rachor E, Reise K, von Westernhagen H, Lenz W (Eds) Warnsignale aus dem Wattenmeer. Blackwell Wissenschafts-Verlag, Berlin, pp 153–159

Jax K, Jones CG, Pickett STA (1998) The self-identity of ecological units. Oikos 82: 253–264

Jax K, Vareschi E, Zauke G-P (1993) Entwicklung eines theoretischen Konzepts zur Ökosystemforschung Wattenmeer. Umweltbundesamt Berlin, Texte 47/93

Jones CG, Lawton JH, Shachak M (1994) Organisms as ecosystem engineers. Oikos 69: 373–386

Kröncke I, Dippner JW, Heyen H, Zeiss B (1998) Long-term changes in macrofaunal communities off Norderney (East Frisia, Germany) in relation to climate variability. Mar Ecol Prog Ser 167:25-36

Kuiper J, de Wilde P, Wolff WJ (1984) Effects of an oil spill in outdoor model tidal flat ecosystems. Marine Pollution Bulletin 15:102–106

Levins R (1969) Some demographic and genetic consequences of environmental heterogeneity for biological control. Bull Entom Soc Am 15: 237-240

Lozán JL, Rachor E, Reise K, von Westernhagen H & Lenz W (Eds) (1994) Warnsignale aus dem Wattenmeer. Blackwell Wissenschafts-Verlag, Berlin

MacArthur RH (1955) Fluctuations of animal populations and a measure of community stability. Ecology 36: 533–536

Meltofte H, Blew J, Frikke J, Rösner H-U & Smit CJ (1994) Numbers and distribution of waterbirds in the Wadden Sea. Results and evaluation of 36 simultaneous counts in the Dutch-German-Danish Wadden Sea 1980–1991. IWRB Publication 34/Wader Study Group Bull 74, Special issue

Michaelis H, Obert B, Schultenkötter, Böcker L (1995) Die Miesmuschelbestände der niedersächsischen Watten, 1989–1991. Ber Forsch-Stelle Küste 40: 55–70

Michaelis H, Böhme B (1994) Benthosforschung im Ostfriesischen Wattenmeer. Umweltbundesamt, Berlin. Texte 24/94

Michaelis H, Reise K (1994) Langfristige Veränderungen des Zoobenthos im Wattenmeer. In: Lozán JL, Rachor E, Reise K, von Westernhagen H & Lenz W (Eds) Warnsignale aus dem Wattenmeer. Blackwell Wissenschaftsverlag, Berlin, pp 106–117

Müller F, Bredemeier M, Breckling B, Grimm V, Malchow H, Nielsen SN, Reiche EW (1997) Emergente Ökosystemeigenschaften. In: Fränzle O, Müller F, Schröder W (Eds) Handbuch der Umweltwissenschaften. Ecomed, Landsberg, pp III-2.5/1–20

Nehls G, Hertzler I, Ketzenberg C, Scheiffahrt G (1998) Die Nutzung stabiler Miesmuschelbänke durch Vögel. In: Gätje C, Reise K (Eds) Ökosystem Wattenmeer. Austausch-, Transport- und Stoffumwandlungsprozesse. Springer, Berlin Heidelberg New York, pp 421–435

Nehls G, Thiel M (1993) Large-scale distribution patterns of the mussel *Mytilus edulis* in the Wadden Sea of Schleswig-Holstein: do storms structure the ecosystem? Neth J Sea Res 31: 181–187

Petersen B & Exo KM (1999) Predation of waders and gulls on *Lanice conchilega* tidal flats in the Wadden Sea. Mar Ecol Prog Ser 178: 229–240

Pimm SL (1984) The complexity and stability of ecosystems. Nature 307: 321–326

Pimm SL (1991) The Balance of Nature? University of Chicago Press, Chicago

Raffaelli D, Hawkins S (1996) Intertidal Ecology. Chapman & Hall, London

Reich M, Grimm V (1996) Das Metapopulationskonzept in Ökologie und Naturschutz: Eine kritische Bestandsaufnahme. Z Ökologie u Naturschutz 5: 23–139

Reise K (1980) Hundert Jahre Biozönose. Die Evolution eines ökologischen Begriffes. Naturw Rundschau 33: 328–335

Reise K (1982) Long-term changes in the macrobenthic invertebrate fauna of the Wadden Sea: are polychaetes about to take over? Neth J Sea Res 16: 29–36

Reise K (1985) Tidal flat ecology. Springer, Berlin Heidelberg New York

Reise K (1990a) Karl Möbius: dredging the first community concept from the bottom of the sea. Dt Hydrog Z Erg -H B 22: 149–152

Reise K (1990b) Grundgedanken zur ökologischen Wattforschung. Umweltbundesamt Berlin, Texte 7/90: 138–146

Reise K (1991) Dauerbeobachtungen und historische Vergleiche zu Veränderungen in der Bodenfauna des Wattenmeeres. Laufener Seminarbeiträge 91: 55–60

Reise K (1992) Grazing on sediment shores. In: John DM, Hawkins SJ, Price JH (Eds) Plant-Animal Interactions in the Marine Benthos. The Systematics Association, Clarendon Press, Oxford. Spec Vol 46: 133–145

Reise K (1994) Changing life under the tides of the Wadden Sea during the 20th century. Ophelia Suppl 6: 117–125

Reise K (1998) Coastal change in a tidal backbarrier basin of the northern Wadden Sea: are tidal flats fading away? Senckenbergiana marit 29:121–127

Reise K, Köster R, Armonies W, Asmus H, Asmus R, Hickel W, Riethmüller R (1998) Austauschprozesse im Sylt-Rømø Wattenmeer: Zusammenschau und Ausblick. In: Gätje C, Reise K (Eds) Ökosystem Wattenmeer. Austausch-, Transport- und Stoffumwandlungsprozesse. Springer, Berlin Heidelberg New York, pp 538–558

Rose PM, Scott DA (Eds) (1994) Waterfowl Population Estimations. IWRB Special Publ 29

Ruth M (1991) Miesmuschelfischerei im Nationalpark "Schleswig-Holsteinisches Wattenmeer" – Ein Fischereizweig im Interessenkonflikt zwischen Ökonomie- und Naturschutzinteresse. Arbeiten des deutschen Fischerei-Verbandes 52: 137–68

Schwegler H (1985) Ökologische Stabilität. Verh Ges Ökol 13: 263–270

Shrader-Frechette KS, McCoy ED (1993) Method in ecology – strategies for conservation. Cambridge University Press, Cambridge

Sousa WP (1980) The responses of a community to disturbance : the importance of successional age and species life histories. Oecologia 45: 72–81

Stearns SC (1992) The Evolution of Life Histories. Oxford University Press, New York

Steyaert FHI, Bakker JF (1994) Folgen des Abschlußdeiches für die frühere Zuidersee und das Wattenmeer. In: Lozán JL, Rachor E, Reise K, von Westernhagen H, Lenz W (Eds) Warnsignale aus dem Wattenmeer. Blackwell Wissenschafts-Verlag, Berlin, pp 175–178

Strübing K (1996) The ice winter of 1995/96 on the German coasts between Ems and Oder, with a survey of the entire Baltic area. Dt Hydrogr Z 48: 73-87

Thrush SF, Hewitt JE, Pridmore RD (1989) Patterns in the spatial arrangements of polychaetes and bivalves in intertidal sandflats. Mar Biol 102: 529–535

White PS, Pickett STA (1985) Natural disturbance and patch dynamics: an introduction. In: Pickett STA, White PS (Eds) The Ecology of Natural Disturbance and Patch Dynamics. Academic Press, New York. pp 3–13

Wiegleb G, Bröring U (1996) The position of epistemological emergentism in ecology. Senckenbergiana marit 27: 179–193

Williamson M (1987) Are communities ever stable? Br Ecol Soc Symp 26: 353–371

Wissel C (1989) Theoretische Ökologie - Eine Einführung. Springer, Berlin Heidelberg New York

Wolf F 1997. Untersuchungen zum Auftreten roter und grüner Farbvarianten bei der Strandkrabbe, *Carcinus maenas* (Linneaus 1758) (Crustacea: Decapoda: Brachyura). Dissertation, Universität Frankfurt

Wolff WJ (1987) Ecological effects of a rapid relative increase of sea level. In: Wind HG (Ed) Impact of sea level rise on society. A.A. Balkema, Rotterdam, pp 153–155

Wolff WJ (1993) Netherlands – Wetlands. Hydrobiologia 265: 1–14

Zens M, Michaelis H, Herlyn M, Reetz M (1997) Die Miesmuschelbestände der niedersächsischen Watten im Frühjahr 1994. Bericht Forsch-Stelle Küste 41: 141–155

10 Assessing Stability Properties: How Suitable is this Approach for Ecosystem Research in the Wadden Sea?

Sabine Dittmann

Abstract: With stability properties as the conceptual framework to study processes in the Wadden Sea ecosystem, attention was focused on qualitative aspects. The relationships between the approach of ELAWAT and historical developments of ecosystem research and ecological concepts are briefly described. While the realization of this approach was restricted by the dynamic nature of the system under study and the limited duration of the project, a potential for future research is seen in the consideration of the single criteria of the ecological checklist used. Furthermore, quantitative and qualitative approaches should be combined in the future to allow an integrated analysis of dynamic ecosystems like the Wadden Sea.

Introduction

The approach of the ecosystem research project ELAWAT taking stability properties as the conceptual framework for an ecosystem analysis was the first of its kind (Chap. 2). Therefore we discuss the usefulness of this approach for ecosystem analysis here at the end of the project. From the pilot studies (Jax et al. 1993) over the main phase (Dittmann et al. 1997) into the synthesis, this approach and the underlying concepts and theories were discussed among the participants. This chapter gives an overview on the (continuing) debate.

With its conceptual framework, ELAWAT tried to achieve an understanding of ecological relations in the Wadden Sea ecosystem by analysing the reaction to disturbances and by the specification of stability properties. The most important step thereby was the identification of stability mechanisms, as the analysis of processes relevant for regeneration should result in recommendations for a protection of processes (Chaps. 2, 8, 9 and 12).

The analysis of stability properties meant to direct the attention to qualitative properties of the Wadden Sea ecosystem and followed in the tradition of qualitative research according to Reise (1995). The approach of ELAWAT was thus a deviation from ecosystem approaches focussing on tropho-dynamics (Lindemann 1942), productivity and energy flow (Odum 1971) or exchange processes (Gätje & Reise 1998). That is also why our definition for the Wadden Sea "ecosystem" was based on the initial ecosystem definition by Tansley (1935) in the version of Likens (1992) (see Chap. 2). Unlike ecosystem approaches studying e.g. energy flow and defining an ecosystem by functional relations, ELAWAT studied the properties of the system in a spatially defined area, a landscape unit.

Relation to ecological concepts and other ecosystem approaches

Throughout the history of ecology, several attempts were made to understand and investigate the complexity of ecological systems (Bradbury et al. 1996). O'Neill et al. (1986) distinguish two approaches, a "population-community" and a "process-functional" approach. The former focuses on organisms and interactions of biota while the latter targets rates of processes, energy flow and productivity. O'Neill et al. (1986) suggest a hierarchical approach to ecosystem research, but also the approach taken in ELAWAT is suitable to combine the "population-community" and a "process-functional" approaches (Leuschner & Scherer 1989). Therefore the relevance of processes causing stability properties on population or community levels has to be assessed for higher hierarchy levels. However, missing long-term and large-scale studies make the realization of this assessment more difficult (Chap. 9).

The species with their specific life histories are the active players for stability properties on the population and community level and for the recolonization after disturbances. To take this into account, ELAWAT turned away from (super-) organismic approaches (Richardson 1980) and was more related to the individualistic concept of Gleason (1939) (Dittmann et al. 1997). This dispute also fuelled a workshop on concepts of ecosystem (see articles in Dittmann et al. 1996). Starting off with a given stability property in the title of ELAWAT, a lively debate developed in the course of the project about the (non-)existence of emergent properties at the ecosystem level and whether the Wadden Sea can have emergent properties at all.

The analysis of stability properties is closely related with several ecological concepts addressing the relevance of disturbances in ecosystems and succession. These are the patch-dynamic (Pickett & White 1985, Paine & Levin 1981) and the mosaic-cylce theory (Remmert 1991). However, it was disputed whether these concepts can be applied to the ecology of the soft-sediment realm of the Wadden Sea (see Chap. 8).

The discussion on "stability" can be followed throughout the ecological literature of the past decades (MacArthur 1955; May 1971; Holling 1973; Orians 1975; Gray 1977; Pimm 1984; Grimm et al. 1992). The most often debated relation between diversity and complexity with stability properties was not a topic in ELAWAT. To apply stability concepts for the analysis of an ecosystem, the "checklist" by Grimm et al. (1992) was useful for focusing on clearly defined ecological situations. "Stability" encompasses a number of properties (Grimm & Wissel 1997) and those specified in ELAWAT are defined in Chap. 9.

Realization of the approach

The realization of the approach of ELAWAT which focussed on qualitative properties required a proper way to handle the concept (Bradbury et al. 1986), a knowledge of theoretical backgrounds by all scientists involved and a spatial and temporal frame of the project adjusted to the topic.

The methods used in ELAWAT to analyse stability properties and mechanisms encompassed the techniques developed by mathematicians to analyse and compare stages of systems (Chap. 4) and ecological modelling coming along with a theoretical consulting (see Chap. 11).

Restrictions for the study of stability properties were set both by the dynamic nature of the system and by the restricted time frame of the project. A stability property such as resilience can only be identified when a return to a reference state or a reference dynamic has occurred. This implies a sufficiently long time frame for observations and a knowledge of the reference dynamic. The restricted duration of the project and the ice winter 1995/96 interfered with the assessment of a return to a reference state (see Chaps. 3.5, 6 and 7). The time frame to assess the regeneration has to be adjusted e.g. to the life cycles of the organisms studied (Connell & Sousa 1983; Grimm et al. 1992). Thus the potential of ELAWAT to analyse stability properties was limited by the duration of the project.

A further difficulty was the assessment of reference states or reference dynamics. The studies in ELAWAT showed a high variability on the scale of tidal cycles, days and years that made it impossible to define a reference situation in most cases (see e.g. Chap. 5.1). As said in Chap. 9, the "state" of the Wadden Sea is its dynamics, obstructing the definition of reference situations. Furthermore, given the dependence on boundary conditions such as climatic variations, the results would have been different if the studies would have taken place in another set of years (see Chap. 3.2). This demonstrates that long-term studies are an essential prerequisite for studies of stability properties.

The ecological checklist (see above, Chap. 9) used for the analysis of stability properties ensured a complete characterisation of the ecological situation for respective stability properties, covering the variable of interest, level of description, reference dynamic, type of disturbance and temporal and spatial scales. This method (Grimm 1996; Grimm et al. 1992) helped to give a precise enumeration of stability properties for defined ecological situations.

Evaluation of the approach

The approach taken in ELAWAT focused on processes that are essential for a persistence of components of the Wadden Sea ecosystem. The analysis of stability properties enabled an assessment of the qualitative relevance of components and processes and was a suitable approach for an ecosystem research of the Wadden Sea. As Chap. 9 shows, stability properties could be identified for many components of the Wadden Sea, but were dependent on temporal and spatial scales of observation. It was also shown that the stability mechanisms of many biotic components in the Wadden Sea can only operate in an essentially unchanged setting of abiotic boundary conditions. This analysis was also essential for the recommendations given in Chap. 12 for the protection of processes in the Wadden Sea.

As the Wadden Sea faces a variety of environmental threats, a variety of research approaches is required (Reise 1995). The two approaches realized in the basic research projects SWAP and ELAWAT within the "Ecosystem Research Wadden Sea" (see Chap. 2), focussing more on quantitative or qualitative aspects respectively, complement each other and allow an integrated view on the Wadden Sea. For this open ecosystem, knowledge about exchange processes is essential, but since variations in species compositions, abundances and distribution patterns affect the budgets of nutrients and sediment, knowledge about the resilience of biotic components is required for an assessment of future developments (Reise 1995; Dittmann 1996).

Outlook

The further development of qualitative ecosystem approaches in marine ecology was already demanded by Bradbury et al. (1986). The approach taken in ELAWAT offers the possibility to do justice to the qualitative nature of ecological systems. For the Wadden Sea, characterized by a tidal dynamic, a consideration of qualitative aspects is rather vital for any research. However, every approach in ecology and ecosystem research has a different view of a problem and can contribute meaningful insights. Combining a variety of approaches will be a profitable path for future studies of complex systems.

The analysis of stability properties directs the attention to processes and mechanisms relevant for a persistence of the system and its components. The single criteria of the checklist for the analysis of stability properties have their own relevance in ecology (Grimm 1996) and offer a diagnostic tool for assessing the relevance of variables of interests, disturbances, scales and hierarchy levels and for defining reference dynamics. This is a perspective for the future application of this approach in ecosystem research.

Acknowledgements

I thank Volker Grimm, Carmen-Pia Günther and two reviewers for comments and discussions on this chapter. The project was funded by the German Bundesministerium für Bildung, Wissenschaft, Forschung und Technologie (BMBF) under grant number 03F0112 A and B. The responsibility for the contents of the publication rests with the author.

References

Bradbury RH, Green DG, Reichelt RE (1986) Qualitative patterns and processes in marine ecology: A conceptual programme. Mar Ecol Prog Ser 29: 299–304

Bradbury RH, van der Laan JD, Green DG (1996) The idea of complexity in ecology. Senckenbergiana marit 27: 89–96

Connell JH, Sousa W (1983) On the evidence needed to judge ecological stability or persistence. Am Nat 121: 798–824

Dittmann S (1996) Resilience and exchange processes: integrated view on the ecosystem of the Wadden Sea. Wadden Sea News Letter 1996-3: 8–9

Dittmann S, Kröncke I, Albers B, Liebezeit G (Eds), (1996) The concept of ecosystems. Senckenbergiana marit 27: 81–255

Dittmann S, Marencic H, Roy M (1997) Ökosystemforschung im Niedersächsischen Wattenmeer. In: Fränzle O, Müller F, Schröder W (Eds) Handbuch der Umweltwissenschaften. Ecomed. Landsberg, V-4.1.2

Gätje C, Reise R (1998) Ökosystem Wattenmeer. Austausch-, Transport- und Stoffumwandlungsprozesse. Springer, Berlin Heidelberg New York

Gleason HA (1939) The individualistic concept of the plant association. Amer Midl Nat 21: 92–110

Gray JS (1977) The stability of benthic ecosystems. Helgoländer Meeresunters 30: 427–444

Grimm V (1996) A down-to-earth assessment of stability concepts in ecology: dreams, demands, and the real problems. Senckenbergiana marit 27: 215–226

Grimm V, Schmidt E, Wissel C (1992) On the application of stability concepts in ecology. Ecological Modelling 63: 143–161

Grimm V, Wissel C (1997) Babel, or the ecological stability discussions: An inventory and analysis of terminology and a guide for avoiding confusion. Oecologia 109: 323–334

Holling CS (1973) Resilience and stability of ecological systems. Annual Reviews of Ecology and Systematics, 4: 1-23

Jax K, Vareschi E, Zauke G-P (1993) Entwicklung eines theoretischen Konzeptes zur Ökosystemforschung Wattenmeer. Umweltbundesamt Berlin. Texte 47/93

Leuschner C, Scherer B (1989) Fundamentals of an applied ecosystem research project in the Wadden Sea of Schleswig-Holstein. Helgoländer Meeresunters 43: 565–574

Likens GE (1992) The Ecosystem Approach: Its Use and Abuse. Excellence in Ecology 3. Kinne O (Ed) Ecology Institute, Oldendorf/Luhe

Lindemann RL (1942) The trophic-dynamic aspect of ecology. Ecology 23: 399–418

MacArthur RH (1955) Fluctuations of animal populations and a measure of community stability. Ecology 36: 533–536

May RM (1971) Stability in multi-species community models. Mathematical Biosciences 12: 59–79

O'Neill RV, DeAngelis DL, Waide JB, Allen TFH (1986) A Hierarchical Concept of Ecosystems. Monographs in Population Biology 23. Princeton University Press, Princeton

Odum EP (1971) Fundamentals of Ecology (3rd edition). Saunders, Philadelphia

Orians GH (1975) Diversity, stability and maturity in natural ecosystems. In: van Dobben WH, Lowe-McConnel RH (Eds) Unifying Concepts in Ecology. Junk, The Hague, pp 139–149

Paine R-T, Levin SA (1981. Intertidal landscapes: disturbance and the dynamics of pattern. Ecol Monographs 51: 145–178

Pickett STA, White PS (1985) The Ecology of Natural Disturbance and Patch Dynamics. Academic Press, Orlando

Pimm SL (1984) The complexity and stability of ecosystems. Nature 307: 321–326

Reise K (1995) Predictive ecosystem research in the Wadden Sea. Helgoländer Meeresunters 49: 495–505

Remmert H (1991) The Mosaic Cycle Concept of Ecosystems. Springer, Berlin Heidelberg New York

Richardson JL (1980). The organismic community: resilience of an embattled ecological concept. BioScience 30: 465–471

Tansley AG (1935) The use and abuse of vegetational concepts and terms. Ecology 16 (3): 284–30

11 Joint Research Projects: Experiences and Recommendations

Sabine Dittmann, Andreas Hild, Volker Grimm, Carmen-Pia Günther, Verena Niesel, Marlies Villbrandt, Hauke Bietz & Ulrike Schleier

Abstract: ELAWAT ("Resilience of the Wadden Sea ecosystem") was one of the larger joint projects ever undertaken in German marine sciences. As this project is ending, we have compiled our experiences with that joint research approach in this chapter, as this could be helpful in respect to future projects and research policies. In order to learn from this experience, we try to be frank in stating deficits, mistakes and matters that were overlooked. These were mainly caused by the severe structural, financial and personnel problems found in the German University system at the moment, so that we comment on that situation as well. This chapter contains only those experiences and recommendations drawn from ELAWAT; in other parts of the "Ecosystem Research Wadden Sea" there were different experiences. In addition, we have listed several research questions which remained either unanswered or which arose during the "Ecosystem Research Wadden Sea", and we also recommend areas for future research emphasis.

11.1 Realization of ELAWAT

Temporal realization

The basic research within the context of the "Ecosystem Research Wadden Sea" was initiated in the state of Lower Saxony in 1990 and by the end of 1991/early 1992, 15 sub-projects commenced their work for the pilot studies of ELAWAT. The tasks of these pilot studies were mainly to assess methodological aspects and determine appropriate temporal and spatial scales for the investigations of the different disciplines involved. The pilot studies ended in February 1993 and their results were compiled in a two-volume report.

Due to delays in the grant-application for the main study phase there was no continuity between these two stages of the project. Funds for the main study phase were granted in autumn 1993 and only a few institutions had been able to continue their research throughout the summer of that year. Furthermore, not all of the sub-projects involved in the pilot studies were continued and some new sub-projects included in the main study phase. A better combination of these two stages and a continuity of sub-projects would have been more effective. A further hindrance in this early stage was that no steering group was in place for 1992.

The main study phase for ELAWAT started in October 1993, initially with a duration of 39 months. Eleven sub-projects of 8 different institutions were involved. When the opportunity arose to study the effects of the ice winter 1995/96, ELAWAT was extended for 3 months. In the course of the project, demand grew for hydrographic and ecological modelling to help with the analysis of temporal and spatial distribution patterns, and both efforts were finally added to the project in 1996. ELAWAT was further prolonged for a synthesis of the results. Altogether the project thus ran until December 1997. The quick and flexible reactions of the funding agency to the requirements and research needs were very favourable for the overall success of ELAWAT.

In spite of the extension, a time constraint remained which demanded a concentration on the major goal and did not allow a flexible reaction to some specific events which arose during the course of the research. For example, it was not possible to investigate the anoxic areas in the backbarrier tidal flat of the island of Baltrum, which developed in summer 1996, although that would have been an ideal large-scale natural defaunation experiment (Chap. 6; 7). Thus, for the topic of ELAWAT, an exceptional situation was missed. This shows that an investigation of sudden events is not possible if funds have to be applied for first. Only established institutions with a sufficient budget and experienced scientific staff can react to such situations.

The sub-projects had half a year to analyse their data and write their reports, which proved to be too short a time, especially for combining results from several sub-projects, as was the idea of a joint project. The writing of publications based on the collaboration of several sub-projects was postponed beyond the end of the project.

During ELAWAT, a normal, a mild and an ice winter occurred, and the investigations covered a wide range of possible boundary conditions. If the project would have taken place with a time shift of just a year or two, these boundary conditions would have been different again and our results probably would have been as well. Disturbance effects could also only be temporarily assessed, but not with their full consequences. For example, for studying the effect on a population the entire life span of the respective organism has to be dealt with (Grimm et al. 1992, Chap. 10), and this exceeded the project time for most benthic organisms in the Wadden Sea. The limited project time was also an obstacle for the assessment of the reference dynamic, which was essential for the analysis of stability properties. Thus long-term data from other parts of the Wadden Sea had to be used as well for the analysis given in Chap. 9. An ecosystem approach targeting stability properties demands a longer-termed project.

Spatial realization

The concentration on a main study area for all sub-projects was rational to facilitate interdisciplinary collaboration. The field work for ELAWAT was carried out in the main study area of the "Ecosystem Research Wadden Sea" in Lower Saxony, the backbarrier system of the island of Spiekeroog (Chap. 3.1). The choice fell on this backbarrier system because of logistic reasons and of prior research that had been carried out there by some of the institutions involved in the ecosystem research. The ecosystem research concentrated mainly on the intertidal; both supratidal (salt marshes and dunes) and subtidal areas (with the exception of investigations on the shore crab and in the water column) were not part of the research. This deficit resulted from research priorities and limited funding.

Looking back, the choice of the main study area and the two specifically studied tidal flats in there (Swinnplate and Gröninger Plate, Chap. 3.1) was not a wise decision. The studied tidal flats were suitable for the investigation of specific topics, e.g. the biodeposit gradient on the Swinnplate, but unsuitable in other respects (e.g. the size and shifting location of the mussel bed there) or idiosyncratic, so that a generalization of the results is hampered. Also, the benthic fauna on the Gröninger Plate was relatively poor in species and individuals compared to other parts of the Wadden Sea. In the sense of an encompassing, spatially extensive ecosystem research, the investigations should have been carried out over a much wider spatial scale. However, to work on a wider scale requires more personnel and funding, which has to be considered during the application of a project.

An initial idea in ELAWAT was to switch the study to another benthic community each year. This was given up for the benefit of a more detailed investigation of the associations around *Lanice conchilega*. Thus seasonal and year to year variations could be determined and more detailed experimental investigations on the settlement behaviour of this species and its associated fauna could be carried out. As a consequence, ELAWAT provides data on selected benthic communities only, but these are based on thorough investigations, which proved to be essential for the analysis of stability properties and mechanisms (Chap. 9).

Organisation and coordination

Joint projects require an organisation and coordination of the sub-projects involved. For the management of the "Ecosystem Research Wadden Sea" experiences from the "Ecosystem Research Berchtesgaden" were used and a steering committee called into place (Leuschner 1989). In Lower Saxony, the project leadership consisted of a scientific board and the actual steering group (Dittmann et al. 1997). This structure ensured a connection of the applied and basic research parts of the "Ecosystem Research Wadden Sea". The tasks of the steering group encompassed the coordination and scientific supervision of the projects, however, the hierarchy in the project leadership was often a hindrance for an efficient performance. The steering committee was funded through the applied part of the and employed by the National Park administration. When the applied part ended, the coordination of ELAWAT was funded through this project and carried out from the Research Centre Terramare, which facilitated day to day work.

The range of tasks necessary for coordination developed in the course of the project. They included revising the concept of the project, applications, some administration, cross-checks and summarisation of reports, as well as organisation of seminars, hearings and international workshops. Association and communication between the sub-projects were promoted and some sub-projects were partly supervised. The coordination of the project would have been facilitated by a clear definition of the duties and rights for this position from the very beginning with all people involved in the project, as there were moments of confusion and also a lack of acceptance by some sub-project leaders. Both the responsibilities of the project leadership as well as those of the leaders of single sub-projects should be specified and agreed upon at the start of a joint research project. Three spokespersons were initially nominated for ELAWAT, which proved to be a poor idea as the result was at times an uneven presentation of the project to the public.

The Research Centre Terramare was the main applicant for this joint project and arranged the entire logistics, organisation of boat times, employment of student assistants and the financial administration. This proved to be an efficient way of administering and organising a joint project.

The single sub-projects of ELAWAT were structured with a sub-project leader at the head, scientific (mostly PhD students) and technical staff. Mainly the scientific staff of the various sub-projects collaborated well and shouldered the load of the field work, thus filling this joint project with life and substance. Some of the sub-project leaders were not much engaged in the project. It would have been beneficial when sub-project leaders had participated more in the project and would have had the time required to do so. In a joint project the demand for scientific supervision is high, as the concerns of the joint research venture with an overall topic and questions encompassing several sub-projects dominate over the scientific interests of single sub-projects. Sub-project leaders have to be aware of this challenge. However, the effort to fulfil this task certainly reaches the limits of their work capacity. And since, nowadays, tenure track positions for academic staff are rare in German universities, this deficit could not be offset in any way.

To enhance the understanding between the different disciplines and thus promote their collaboration, a number of common activities were started right from the beginning of ELAWAT. These included seminars where colleagues reported about their studies and the respective scientific background, methods and objectives of their work. Further seminars dealt with the ecological theories and concepts relevant to the project. The statistical sub-project carried out seminars to teach methods of data analysis and data handling.

A further level of interdisciplinary collaboration was realized by working groups, set up at the start of the project according to its major scientific topics. In the course of the project, some new working groups were formed and others merged. At these meetings, reports were given on the progress of work, results and literature reviews were presented and arrangements were made for the next combined field work or later on for the joint analysis of data. The speakers of the working groups took part at the meetings of other working groups, thus allowing for cross-relations. The working groups were also a good forum to get to know each other, which fostered the good and friendly atmosphere of collaboration realized in ELAWAT. Working groups are thus a suitable means to facilitate the organisation of a joint project.

All these activities succeeded in promoting collaboration and increasing knowledge. Still, although everyone involved in the project welcomed the seminars and working groups, they were overshadowed by the time they demanded. As the institutions collaborating in ELAWAT were scattered over all of northern Germany, a lot of time and travel was required to take part in these meetings. This time was missing elsewhere, a matter which is of concern for joint projects. It was beneficial for the organisation and corporate identity of ELAWAT when the new building of the Research Centre Terramare, opened in 1994, could host the seminars and meetings.

In spite of all these efforts, the required collaboration did not work out in all cases. Sometimes there were problems with the thematic delimitation required for PhD theses. This is unavoidable as long as projects rely on PhD students as their sole or main scientific manpower.

Next to the interdisciplinary collaboration in a joint project, the involvement in the individual disciplines must still be maintained. Therefore, scientific partners at the institutions or universities have to ensure a continuous dialogue within the respective discipline. This requires also regular participation in conferences to stay up-to-date with specific topics.

A stronger feedback with experienced scientists on the progress of work would have been welcomed by ELAWAT. This does not only ask for the supervision by sub-project leaders, but e.g. also for written recommendations by the reviewers at the annual hearings. Such a feedback could also raise the motivation.

International workshops

Since the topic and approach of ELAWAT touched on aspects of marine and ecosystem sciences that are under debate in the international scientific community, we looked for discussions on these issues. This was realized in two workshops with the participation of experts from all over the world. One workshop targeted conceptual problems with the ecosystem approach (see articles in Dittmann et al. 1996), the other the relevance of disturbances in coastal marine ecosystems. At both workshops the presentations enjoyed lively and inspiring discussions. For ELAWAT, these workshops gave the opportunity to talk about theoretical aspects and reflect on our approach, which resulted in the advancement of the concept in the course of the project. We could also put our results into an international perspective. Constructive recommendations and the dialogue between the participants at the workshops helped to promote work in ELAWAT. Furthermore, our project thus became known both nationally and internationally. Workshops on specific topics with a small number (50–100) of international scientists have thus proven to be very favourable for the advancement of this joint ecosystem research.

Statistical consulting

A further successful approach in ELAWAT was to have a sub-project solely designated to statistical questions. This enabled consultations in statistical questions which were essential, as the complex ecological questions could not be analysed with ordinary procedures, but required the advancement or new developments of statistical tests (see also Chap. 4).

The statistical consulting offered by the sub-project was in high demand throughout the entire project. This included the planning of experiments and sampling designs, the choice of suitable statistical procedures for data analysis, the choice for and practical help with statistical software, the interpretation of statistical results, the (further) development of statistical models and procedures according to the demands of the sub-projects as well as the development of specific software.

The experiences showed that for a sound scientific use of statistical procedures a mutual understanding between the disciplines is essential. The statisticians have to be able to understand ecological questions in order to transfer them into appropriate models. The ecologists have to be prepared to accept statistical thinking in order to assess the validity of ecological inferences based on statistical results. The collaboration between this sub-project and all other sub-projects also resulted in an overview on the entire project, which was favourable for the collaboration and synthesis.

Data management

The task of the data management was to develop a unified system for the data handling in all sub-projects and to provide the essential tools therefore, with the aim to facilitate the exchange of data between sub-projects as well as the import of data into statistical software and various computer programmes. This was realized with a PC-based data handling (Ortleb 1997). With this approach ELAWAT was exempted from contributing to the centralized data base of the "Ecosystem Research Wadden Sea". Consultations for data handling were given during the entire project and in most cases suitable structures could be found in the form of relational data bases. However, the complete inventory of all data in the approved structure took time beyond the duration of the sub-projects. Furthermore, some projects had concerns about the copy rights on the data and hesitated to supply their data for the data base of ELAWAT. Thus not all of the data were approved at the end of the project. This hampered the synthesis of the results, as in some cases a further analysis required tedious restructuring of the original data. In the end, almost all data were compiled on CD-ROM.

A successful data management also requires the mutual understanding of the disciplines. The data base manager must be able to consider ecological questions and their required analysis for the choice of an appropriate data structure. The ecologists have to learn scientific criteria for the establishment of data tables to assist in the development of data structures for the respective ecological questions.

In general, a decentralized data handling is preferred, where data are available and can be accessed from anywhere in the project. This structure allows a close and unhindered collaboration between the sub-projects, the overall project coordinators, and the statisticians and modellers. But, beforehand, any legal concerns on data security and rights have to be clearly settled to the satisfaction of all concerned.

Theoretical consulting and ecological modelling

The approach in ELAWAT, which for the first time directed the focus of an ecosystem analysis on stability properties, required an intensive and continuous theoretical and conceptual discourse about this theme (Chap. 2). Joint projects can only reach their goal with an encompassing conceptual framework. Often enough the participants of a project are overloaded or not trained for the development of a conceptual framework. In ELAWAT, the integration of an ecological modelling also provided a kind of theoretical consulting, to the profit of the entire project and the synthesis. The joint development of the models (Chaps. 5.4 and 8) supported goal-oriented scientific studies. It would have been helpful to have had the theoretical consulting available right from the start of the project.

Synthesis

To synthesize the results of ELAWAT, nine months were available, which proved to be much too short. Some sub-projects were running late in submitting their reports and some results of the sub-projects have still not been published. The missing joint analysis of the results of the collaboration between sub-projects could not be compensated by the synthesis. In some cases, data still had to be analysed, which was further held up as some data were submitted rather late. This time was missing for the critical evaluation of results.

The synthesis was carried out by eight scientists from different disciplines, which proved to be a necessary combination to fulfil the job lying ahead in the given time. A further advantage was that we could all dedicate our time solely to the synthesis and could collaborate as a team in one building.

11.2 Problems and Potentials of Joint Projects

Joint projects result when there is a necessity for interdisciplinary collaboration. This is based on the idea that it is possible to gain more knowledge in a joint integrated project than can be achieved through several individual and only vaguely-linked projects. For ecosystem research, the need for interdisciplinary collaboration is self-evident. The "Ecosystem Research Wadden Sea" was realized by four major joint projects. We have outlined in Sect. 11.1. how the joint ELAWAT project was realized, and in Sect. 11.4. we will present recommendations from the projects´ perspective. In this section, also from the experience of ELAWAT, we discuss particular aspects of joint projects.

The decision as to whether a research project should be realized as a joint project or as a single one has to be based on the objectives of the project. If the central question requires an interdisciplinary approach, there is no alternative to a joint project (see also Reise 1997; Haber 1998). But joint projects pose special demands in multiple ways, regarding their organisation and coordination and the nature of scientific collaboration.

Aspects of subject

The more precise an interdisciplinary question is, the stronger is the advantage of a joint research approach, resulting in a possible gain of quality. In this case, the results of a joint project are more than the sum of results of a number of single projects. But this calls for an intensive contact and scientific collaboration already during the planning and application phase. It is required that the connections between sub-projects are known right from the beginning. It is further advisable that all sub-projects and collaborators enter the project with the same level of knowledge.

In a joint project the sub-projects are closely linked and dependent on each other for the success of the project. Thus the joint project sets a rather rigid frame allowing little flexibility to react to eventual requirements for modifications in single sub-projects. Such changes would always effect other sub-projects as well and thus have to be discussed by the entire joint project.

With the background of a poor financial situation of the universities these days, one of the main motives to apply for funds is to obtain PhD positions. The scientific interest in the respective topic often comes second. This is especially fatal for joint projects. The PhD students have to finish their thesis in the first place and can be less concerned about the success of the whole enterprise. Often enough they have not yet got the scientific experience to react to research results in the course of a project. Thus a particularly intensive supervision is required if a joint project relies almost exclusively on PhD students. The limitation on PhD students has a further disadvantage for science policy, namely that experienced scientists are usually released into unemployment nowadays, while a new generation starts from zero again. This seems to be an uneconomic waste of resources and leaves scientific opportunities unused.

The scientific efficiency of a research project, measured in the quality and number of publications, depends to a large degree on the commitment and experience of the scientists involved. This also requires an intensive supervision, especially since interdisciplinary publications should result from a joint project. The requirements for interim reports should be reduced to leave more time for publishing.

Furthermore, joint projects can provide training in a variety of disciplines. This is important given an increasingly stringent time frame and demand to specialize at universities today. Scientists and PhD students involved in a joint project can acquire a wide range of knowledge and are thus well equipped for interdisciplinary endeavours in the future.

To summarize, with the potential for higher quality of scientific results, joint projects are scientifically more demanding than single projects and pose higher requirements to scientific work. This is not restricted to the time of field work, but valid for the entire period from the initial stages to a final synthesis.

Aspects of organisation

The organisational efforts of joint projects are considerable. They cover not only the scientific coordination, but financial and technical organisation as well. For all of these aspects additional staff is needed. On the other hand, the combined accomplishment of organisational and technical duties avoids double work and sub-projects are financially relieved. The incorporation of theoretical consulting, statistical consulting and data management are usually unaffordable for single projects, but can be provided for by joint projects. With the research being carried out in the sub-projects, a joint research project still requires an additional position for scientific coordination. This is an organisational expression of the higher scientific needs for the overall result of a joint project.

For a joint project, a number of meetings are necessary to advance the collaboration (see Sect. 11.1). If, as in ELAWAT, the sub-projects are carried out by institutions from several places, the requirements for travel time and costs grow in addition.

With all efforts for the organisation of a joint project, it will never be possible to reach all scientists involved to the same degree. The insight into the needs of interdisciplinary collaboration does not exist automatically and the motivation for co-operation can well be different. A joint project cannot enforce collaboration, but provide a framework for a facilitated collaboration. Also for this reason, the initial stages of planning and conceptualising are significant. Since all-embracing criteria are essential in a joint project, the sub-project leaders cannot have the final say for all aspects. This requires decisions during the planning stage as well.

Thus a joint project is also more demanding in organisational terms than single projects. Extra time is required for the necessary contact and communication between the disciplines involved for any scientific or organisational matters.

11.3 Gaps of Knowledge and Research Recommendations

ELAWAT and the other parts of the "Ecosystem Research Wadden Sea" have been a decisive step forward in the investigation of the Wadden Sea. Without this research scheme, our understanding of the system would not have advanced as much in the same time. Still, the queries never stop with what has been achieved and research always leaves new questions. In the previous chapters suggestions for future research have been given here and there, which are compiled below.

We regard an outlook on future research in the Wadden Sea as an integral part of the synthesis we have provided here. In spite of the considerable extent of ELAWAT in terms of finances and personnel involved, many aspects could only be touched upon. There will always be restrictions to science, so that it is essential to focus the available budgets on meaningful questions. Therefore we recommend topics based on the experiences and understanding we have gained now. They are combined into three major aspects.

Can small-scale processes be extrapolated?

The search for possible emergent properties of the Wadden Sea ecosystem, as e.g. stability properties, required an extrapolation of results, which has been carefully attempted in Chap. 9. But the extrapolation of research results derived from investigations on small temporal and spatial scales to phenomena or properties on larger scales is, if at all, only possible with great care (Hall et al. 1994; Flach 1996; Thrush et al. 1997). The results presented in the other chapters of this book show great variations in abiotic and biotic parameters over and over again, so that a high temporal and spatial variability characterizes the studied ecosystem. Thus, how representative are results from small-scale studies? Whether results can be extrapolated is especially crucial for studies on interactions, as these are often methodologically constrained to a small-scale.

The studies and synthesis of ELAWAT have shown that an evaluation of the results, e.g. on abundances and distributions of benthic organisms, required a comparison with long-term records (Chap. 9). But it became obvious that relatively few data are available on spatial and temporal distributions and abundances for many species of the Wadden Sea. Thus, long-term quantitative surveys of species- and individual densities should be carried out at several locations to assess a reference dynamic and to be able to diagnose and analyse possible changes.

All existing long-term data series of biological parameters and historical comparisons (Michaelis & Reise 1994) are incredibly precious. The relevance of long-term data series grows the longer the record lasts and the more events can be identified over the course of time (Beukema et al. 1998). Existing long-term data series should definitely be continued and be in a continuous feedback with topical research efforts. Wherever possible, historical archives such as sediment profiles, biofacies and literature records should be used in addition (cf. Reise 1995).

Are sedimentation and turnover in the Wadden Sea determined by extreme events?

Weather and security reasons usually restrict research in the Wadden Sea to summertime. In ELAWAT, some investigations were carried out in winter, e.g. during the ice winter 1995/96, but these results could not be related to other winter situations due to a lack of data. Exchange rates and the turnover of matter and energy can be fairly easily quantified during calm weather conditions. But what do their values mean in relation to extreme events, like e.g. storms? Extreme events can exert a strong influence on the Wadden Sea (Nehls & Thiel 1993) and should thus be studied further. Yet, extreme events are often unpredictable and require a spontaneous research action. This can only be realized by existing institutions having staff and technical resources at stand-by (see also Sects. 11.1; 11.4).

The results of ELAWAT pose the question for a budget of fine grained sediments in the Wadden Sea. Given the debate about the loss of mudflats and sedimentation areas in the Wadden Sea (Chap. 3.4, Flemming & Nyandwi 1994), the exchange rates of particulate and dissolved matter should be assessed during storm events, as these can transfer large amounts of matter. So far methodological difficulties have prevented such investigations. Should these be overcome, the quality of the matter should be determined simultaneously.

What are the connections between sub- and intertidal areas in the Wadden Sea?

Many benthic species in the Wadden Sea have sub-populations both in the inter- and subtidal areas. The investigations of ELAWAT indicate that subtidal populations might have a considerable importance for the recruitment and recolonization of intertidal areas. Over the last decades, research efforts have mainly concentrated on the intertidal of the Wadden Sea which is easily accessible. But the distribution and composition of the benthos in the subtidal realms of the Wadden Sea is almost unknown. This gap has to be filled with quantitative surveys on species compositions, population dynamics and abundances of subtidal species and sub-populations. Also, the transfer and exchange of organisms between the sub- and intertidal has to be quantified.

There is a growing awareness that benthic organisms are much more mobile than previously assumed. Secondary dispersal and drift have been documented for many species. But it is still unclear for a great many species, whether they are only mobile as juveniles or as adults as well and what the reasons and effects of specific drift behaviours are. Here lies an immediate research demand, as mobility affects recordings of benthic distribution patterns and has to be considered for sampling schemes. Mobility is important for recolonization and the ecological function of benthos in these soft sediments. Relations between drift and current regimes, the vertical distribution in the water column, transport distances and seasonal variations in drift patterns should all be investigated. The staining technique developed in ELAWAT is a promising tool to study the exchange rates of benthic organisms between the sediment and the water column (Chap. 5.4).

The recolonization after the ice winter indicated that the recruitment and recolonization could originate from source populations in "mainland"-populations of more southerly areas (e.g. English Channel) or the deeper North Sea. Genetic or enzymatic methods should be used to clarify the population size and exchange between sub-populations.

11.4 Recommendations for the Realization of Joint Projects

Duration

- The duration of a project has to be adjusted to the requirements of the topic. Alternatively, approaches compensating for restrictions in time have to be included (e.g. literature surveys, ecological modelling).
- Time-limited projects do not allow flexible reactions to episodic events. Event-oriented research can only be carried out when experienced scientific staff and finances are available straight away, without the need to apply for external funds.
- Long-term research is absolutely necessary and cannot be fulfilled by simple monitoring. Long-term research should be carried out by established scientific institutions with experienced scientific staff (i.e. with permanent employment) and sufficient budgets.

Joint meetings

- Meetings are essential to promote collaboration in joint projects. The time required for such seminars and working groups has to be taken into consideration during the planning stages.
- For participation at these joint meetings, it is necessary to have sufficient dedicated travel allowances identified in the overall project budget.
- The integration of all scientist involved in a joint project into an ecosystem research centre can reduce the cost and time required for joint meetings. Furthermore, this spatial federation would enhance interdisciplinary exchange.

Organisation of joint projects

- Joint projects require a central coordination, administration and logistics.
- Further central tasks are statistical consulting, data management, modelling, theoretical consulting and public relations. The general view obtained by these tasks benefits the entire project. Statistical and theoretical consulting should contribute to the planning stage already, so that experiments are designed in view of the statistical analysis of the data. Data management should be provided from the start of a project to ensure a sound handling of data. Modelling and theoretical consulting are helpful for joint projects and should also be included from the beginning. This way the understanding of relations and interactions is continuously challenged and both overriding comprehension or deficits in knowledge are recognized much faster.
- Joint projects can only reach their goals efficiently with an interdisciplinary conceptual framework. To think in models, which facilitates a focused and problem-oriented scientific work, requires a training in theoretical ecology and fundamentals of modelling. Ecological models can help to integrate the results from various sub-projects and to extrapolate their findings beyond the temporal and spatial frame set by the project. The formulation and analysis of ecological models requires a lot of time and can only be accomplished by skilled personnel.
- For the planning of organisational structures in a joint project, experiences made in other areas, e.g. in economy, should be considered as they could give suggestions for an efficient and economic structure of projects. It should be further considered to get a professional training for co-ordinators and project leaders.
- The organisation and coordination of research projects should be realized through scientific institutions. In the "Ecosystem Research Wadden Sea", the steering group responsible for the coordination was employed at the National Park Administration to allow for a quick transfer of scientific results into the administrative work. However, for the coordination of a research project it was not appropriate to be lodged within an administrative body. Furthermore, to give recommendations for an integrated coastal zone management, a scientific project should be independent and not work for any public authority. An open dialogue with all stakeholders is only possible if the research project stands free from the interests of administrative bodies and NGO´s.
- Joint projects should identify a designated spokesperson. The co-ordinator would be suitable for this task because of his/her overview of the project. Sub-project leaders should be spokesperson for their respective sub-project.

Communication

- Working groups are a suitable means for the promotion of interdisciplinary work in a joint project. They should be structured already during the planning stage and sub-projects should be allocated to working groups. The working groups should be linked by the mutual participation of their speakers on the

meetings, which would enhance the connections in the joint project. Working groups should have precise goals and focus on common publications.
- For the flow of information between the sub-projects, a central "info-exchange" is suggested, which can be realized nowadays via e-mail or web-sites (inter-/intranet) for all participants. Thanks to these new means of communication a fast, free and cheap exchange of information is possible, which does not need to be mediated through the co-ordinator. Yet, qualified personnel is required for the maintenance of this service.

Synthesis

- The synthesis of the results is an absolutely essential step for the analysis of a joint project.
- A synthesis group has to be sufficiently staffed and consist of scientists from various disciplines.
- The synthesis group should be lodged at one institution to work efficiently in a team.
- A synthesis builds up on reports and publications of a project. To avoid delays, a synthesis should either follow straight after the end of the main study phase or overlap with it towards the end. The latter way would allow for a collective analysis of the results in the direction of the synthesis with the colleagues from the sub-projects.

Efficiency

- Based on our experiences (Sect. 11.1), our recommendations for the realization of future research (Sect. 11.3) can be condensed to one major prerequisite: the strengthening of scientific personnel and finances of universities and research institutions. This would allow for a continuity in the study of specific research subjects again, which would immediately improve the education of young scientists and the thorough analysis of certain topics. The poor supervision, insufficient literature knowledge and -supply and a declining quality of scientific results can only be counteracted with the establishment of scientific "schools".
- For the supervision of research projects and young scientists, a sufficient tenured academic staff is indispensable. The universities have to be equipped with respective permanent positions. This would also take financial burdens off joint projects (cf. Haedrich 1995).
- A continuity is also required for technical personnel. Since technical staff is increasingly externally funded through projects as well, deficits occur in the maintenance of (often enough expensive!) equipment and routine measurements.

A better basic supply of universities and institutions does not at all imply to turn away from interdisciplinary collaboration, but is rather a precondition for it. Interdisciplinarity requires intensive contact between scientists, the interest for combined research and the ability for dialogue. Joint projects can always just set the stage for this and are just one possible way to create and foster interdisciplinary studies.

Acknowledgements

Several colleagues and a reviewer have contributed fruitful comments and discussions to this chapter. The project was funded by the German Bundesministerium für Bildung, Wissenschaft, Forschung und Technologie (BMBF) under grant number 03F0112 A and B. The responsibility for the contents of the publication rests with the authors.

References

Beukema JJ, Cadée GC, Dekker R (1998) How two large-scale "experiments" illustrate the importance of enrichment and fishery for the functioning of the Wadden Sea ecosystem. Senckenbergiana marit 29: 37–44

Dittmann S, Marencic H, Roy M (1997) Ökosystemforschung im Niedersächsischen Wattenmeer. In: Fränzle O, Müller F, Schröder W (Eds) Handbuch der Umweltwissenschaften. Ecomed, Landsberg, V-4.1.2

Dittmann S, Kröncke I, Albers B, Liebezeit G, (Eds) (1996) The Concept of Ecosystems. Senckenbergiana marit 27: 81–255

Flach EC (1996) Distribution of *Corophium* at different scales. Senckenbergiana marit 27: 119–127

Flemming BW, Nyandwi N (1994) Land reclamation as a cause of fine grained sediment depletion in backbarrier tidal flats (Southern North Sea). Neth J Aquat Ecol 28: 299–307

Grimm V, Schmidt E, Wissel C (1992) On the application of stability concepts in ecology. Ecol Model 63: 143–161

Haber W (1998) Ökosystemforschung und Fachwissenschaft. Oldenburger Universitätsreden Nr. 99. Bibliotheks- und Informationssystem der Universität Oldenburg, pp 7–22

Haedrich RL (1995) What does marine biological research cost? A case history of 25 years at a university research station. Helgoländer Meeresunters 49: 423–428

Hall SJ, Raffaelli D, Thrush SF (1994) Patchiness and disturbance in shallow water benthic assemblages. In: Giller PS, Hildrew AG, Raffaelli D (Eds) Aquatic Ecology. Scale, Pattern and Process. Blackwell Science, Cork. pp 333–375

Leuschner C (1989) Ökosystemforschung Wattenmeer. Hauptphase Teil 1. Erarbeitung der Konzeption sowie der Organisation des Gesamtvorhabens (Forschungsverbund). Umweltbundesamt Berlin. Texte 10/98

Michaelis, H. & Reise, K., (1994) Langfristige Veränderungen des Zoobenthos im Wattenmeer. In: Warnsignale aus dem Wattenmeer. J.L. Lozán, E. Rachor, K. Reise, H. v. Westernhagen & W. Lenz (Eds.). Blackwell, Berlin. pp 106–117

Nehls G, Thiel M (1993) Large-scale distribution patterns of the mussel *Mytilus edulis* in the Wadden Sea of Schleswig-Holstein - Do storms structure the ecosystem? Neth J Sea Res 31 (2): 181–187

Ortleb H (1997) Datenmanagement im Projekt ELAWAT. DGM-Mitteilungen 1-2/97, pp 41–44

Reise K (1995) Predictive ecosystem research in the Wadden Sea. Helgoländer Meeresunters 49: 495–505

Reise K (1997) Reflexion zur interdisziplinären Küstenforschung - eine SWAPliche Nachlese. DGM-Mitteilungen 1–2/1997: 5–8

Thrush SF, Schneider DC, Legendre P, Whitlatch RB, Dayton PK, Hewitt JE, Hines AH, Cummings VJ, Lawrie SM, Grant J, Pridmore RD, Turner SJ & McArdle BH (1997) Scaling-up from experiments to complex ecological systems: Where to next? J Exp Mar Biol Ecol 216: 243–254

12 Protection of Processes in the Wadden Sea

Marlies Villbrandt, Carmen-Pia Günther & Sabine Dittmann

Abstract: The protection of processes has been proclaimed both on a national level in the National Park declarations for the Wadden Sea and on an international level by the 6th Trilateral Governmental Conference. This approach means that all natural processes and dynamic developments are allowed to happen in an ecosystem in an unrestricted way. However, the realization in the Wadden Sea National Parks is hampered as the current zonation concept does not protect sufficiently large and coherent areas, and because the integration of human activities in a sustainable way (as required by the "Man and the Biosphere" programme) has not yet been achieved. Furthermore, the public acceptance for the protection of processes has to grow. Based on the results of ELAWAT it is recommended to take the relevance of source populations, dispersal and habitat heterogeneity into account for coastal zone management, as they are essential for regeneration processes after disturbances. The protection of entire tidal basins could ensure the capacity for resilience in this ecosystem.

12.1 What is Protection of Processes?

The protection of processes means to refrain from any human intervention in an ecosystem. This is based on the philosophy "to let nature be nature" (Bibelriether 1993 a, b) and implies to allow dynamic developments and all natural processes in an unrestricted way (Knapp 1998). This can only be realized in sufficiently large areas (e.g. National Parks), which contain all important faunal and floral elements (key-species and ecosystem engineers with all their development stages), typical successional stages and a variety of habitats. As a further logical consequence of the protection of processes, natural disturbances (e.g. floods, fires) should not be obstructed (Remmert 1988).

Unlike the more conservational approaches of protecting species and habitats, the protection of processes profits from dynamic characteristics of an ecological system (Finck et al. 1998). The classic conservation approaches imply the danger that certain successional stages are artificially prolonged. Given a sufficiently large area, the protection of processes can be seen as a superior conservation strategy, embracing the protection of species and habitats. Since the large areas required for the protection of processes are often only provided by National Parks, the protection of species and habitats has to continue in respective conservation areas.

The protection of processes has a scientific base in the patch-dynamics concept of Pickett & White (1985). This implies that change and disturbance are natural in many systems and have important ecological functions. A further base for the protection of processes is the mosaic-cycle concept (Remmert 1991) which assumes local successional cycles following disturbance events, leading to a climax community. Unlike the mosaic-cycle theory, the patch-dynamic concept is independent from assumptions about equilibrium (Jax 1994; Böhmer 1997).

Often enough, National Parks are established in areas which cannot be cultivated because of harsh environmental conditions and are thus incidentally left in an almost pristine state. A comparatively small part of about 2 % of the total area of Germany are National Parks and only 0.4 % are zero-use areas suitable for the protection of processes (Stock et al. 1996). Thus only few areas remain where the protection of processes can be realized without creating serious conflicts with human activities, in particular with economic interests. According to the guidelines of the German National Parks, at least 75 % of the area within a National Park should fall under this protection category. These areas should be designated as strictly protected nature zones without management measures on a long-term basis (Bibelriether 1997; Knapp 1998). However, such a high share of strictly protected areas is not intended by the German Wadden Sea National Parks.

12.2 Protection of Processes in the Wadden Sea

12.2.1 Historical Background and Prerequisites for Nature Conservation in the Wadden Sea

The esteem for the Wadden Sea as a valuable wetland grew in the 20th century and efforts to achieve a conservation have undergone an evolution of approaches from functional to a natural development approach (Wolff 1997, Dankers & de Vlas 1994). In the Wadden Sea of Lower Saxony the first conservation areas were established in the late 19th century to protect species or habitats. This included e.g. breeding- or roost sites of birds or sensitive habitats such as salt marshes or dunes which should be sheltered from damaging human activities (Dittmann et al. 1996). In 1976, the Wadden Sea of Lower Saxony was certified as a wetland of international importance according to the RAMSAR-convention.

With the establishment of three Wadden Sea National Parks in Schleswig-Holstein (1985), Lower Saxony (1986) and Hamburg (1990), the protection of processes was officially proclaimed in the National Park regulations by stating that the natural processes in these habitats with their species rich plant- and animal stock should continue to exist.

In 1993, the Wadden Sea National Park of Lower Saxony also received the status of a biosphere reserve according to the UNESCO-programme "Man and the Biosphere" (MAB). The other two Wadden Sea National Parks are biosphere reserves as well, where human activities and nature conservation have to be harmonized (Erdmann & Nauber 1995).

The Wadden Sea is endangered by indirect and direct effects of human activities (e.g. coastal engineering) (see Lozán et al. 1994). Thus the species stock of the Wadden Sea has changed in the course of this century and in many cases the loss of species meant a loss of associated fauna and flora (Reise 1994, Reise et al. 1989). Also the reduction of habitats such as mud flats, salt marshes and seagrass beds caused declines in fauna and flora (de Jong et al. 1993, Michaelis & Reise 1994, Reise et al. 1994, Nordheim et al. 1996, Schories et al. 1997, Reise 1998).

Still, as a landscape, the Wadden Sea is relatively unaffected by humans and therefore of high ecological relevance. As one of the last coherent large-scale landscapes of Central Europe left in a relatively pristine state (apart from the alpine regions of the Alps), the Wadden Sea is a valuable area in various regards (Reise 1992; Broggi 1996). The total size of the three Wadden Sea National Parks in Lower Saxony, Hamburg and Schleswig-Holstein with approx. 524 700 hectares under protection would be sufficiently large for the protection of processes. The declaration of the National Parks with the clear commitment for the protection of processes is a legal prerequisite for the continuity of the natural dynamic in this ecosystem.

The protection of processes is also a base for international Wadden Sea policies. At the 6th Trilateral Governmental Conference in Esbjerg in 1991, the ministers for the environment of Germany, Denmark and The Netherlands agreed on a Guiding Principle for the trilateral Wadden Sea policy and management, which is "to achieve, as far as possible, a natural and sustainable ecosystem in which natural processes proceed in an undisturbed way."

However, protection of processes in the Wadden Sea cannot prevent anthropogenic impacts. Their solution requires other political decisions. Natural processes are e.g. also affected by pollution from the land, rivers, the North Sea and its adjacent seas, and by climate change. Thus, protection of the Wadden Sea can never stand alone, but requires an improvement of boundary conditions as well. A reduction of pollution and other anthropogenic impacts is the goal of international conservation endeavours.

12.2.2
Realization of the Approach and Problems of Acceptance

To realize the protection of processes in the Wadden Sea National Park of Lower Saxony which has still got a high share of human activities (e.g. tourism, fishing), the area was divided into three zones of different protection level. The highest protection level is realized in 54 % of the area (zone I), followed by areas for restricted use (zone II) and areas for recreation (zone III). But even in zone I human activities are not totally banned, a fact further complicated by varying legislation depending on high and low tide. Another disadvantage for the efficiency of the protection in zone I is their patch-work like arrangement in space without regard to ecological requirements.

To improve the protection, the results of the applied part of the "Ecosystem Research Wadden Sea" in Schleswig-Holstein formed the scientific base for recommendations for a new concept of the National Park there (Stock et al. 1996). It is suggested therein to protect entire tidal basins as the smallest ecological "units" in the Wadden Sea ecosystem. Tidal basins contain all habitats and connect them by the hydrographic regime. Thus, certain entire tidal basins should be designated as zero-use nature reserves (Reise 1992, 1993a), where all human activities are restricted despite for guided tours. This recommendation is a compromise already, suggesting to realize the protection of processes in certain areas of the Wadden Sea while allowing a sustainable use in others.

When the recommendations for a new concept for the Wadden Sea National Park in Schleswig-Holstein were published, they encountered violent resistance from stakeholders. However, unlike sometimes stated in the media, no unrealistic demands (e.g. to tear down the dikes) were proclaimed, but suggestions were given to consider the reestablishment of sedimentation areas wherever it is possible, e.g. by opening a summer polder (Reise 1993b, Reise et al. 1998).

The protest against the reestablishment of natural processes has occurred in all National Parks where the protection of processes was proclaimed. It is difficult for the public to understand the idea of this conservation concept and to accept that this approach is of benefit to nature although it is often not foreseeable where the development goes. Scientists or conservation groups do not profit from the realization of the protection of processes. For a society where economy has a high esteem, education measures have to accompany this conservation concept to increase the appreciation for a pristine nature and the acceptance of the idea to "let nature be nature".

12.3 Arguments from ELAWAT for the Protection of Processes

The project ELAWAT focussed on the reaction of the Wadden Sea ecosystem to disturbances and the identification of stability properties. The results of the synthesis were expected to contribute to the discussion about the protection of processes to ensure a natural development in the Wadden Sea, e.g. by recommendations to sustain processes essential for the properties of the system and regeneration following disturbances (Dittmann et al. 1997).

The stability properties enumerated in Chap. 9 indicate resilience for many parameters of the Wadden Sea ecosystem on a certain temporal and spatial scale. To ensure resilience in the future, the abiotic frame and especially all processes driven by the tides, have to be undisturbed. As an example, the processes and components of the system necessary for recolonization by benthic organisms are listed here:

- **Source population**: recolonization can only take place, when parts of a population are unaffected by a disturbance and can function as a source for recolonization. The probability for the existence of undisturbed parts of a population grows with the size of the protected area and the variety of habitats. This includes also subtidal environments.
- **Dispersal**: both the annual settlement of benthic organisms and the recolonization depend on an undisturbed transport of larvae, juvenile and adult organisms in the water column. Most species have mobile life stages where they are subject to dispersal and migration processes (Günther 1992). The high mobility of

the organisms in the Wadden Sea is essential for the resilience of their populations to disturbances (see Chap. 9).

- **Diversity of habitats**: many benthic organisms in the Wadden Sea need multiple settlement sites before they reach a favourable habitat (Beukema 1993; Reise 1985; Günther 1991). If one of these habitats is missing, e.g. the mudflat as settlement sites of young *Arenicola marina*, the entire population of this species in the Wadden Sea is affected. For the successful settlement and an undisturbed dispersal, a coherent and natural variety of habitats is required. These conditions could be provided best by the protection of entire tidal basins in the Wadden Sea (Reise 1993a, Reise et al. 1998).

The results from ELAWAT showed that many species and other parameters in the Wadden Sea could react with resilience to brief, small-scale disturbances which are part of the natural dynamic of this ecosystem (Chap. 9). Important mechanisms for this behaviour are e.g. the mobility of many species, a functional diversity of the micro-organisms and a phenotypic plasticity of some species. Disturbances caused by human activities can modify boundary conditions and thus prohibit resilient reactions. Wolff (pers. comm.) argues that either the continuous human impact on the Wadden Sea or the slow speed of recovery processes prevent a return of this ecosystem to the conservation phase *sensu* Holling (1986) (see Chap. 9). Compared with an original Wadden Sea, many habitats and species have been lost over the past decades. For the most recent case, the decline of mudflats, a compensation by other muddy habitats (fringes of tidal channels, biodeposit layers in mussel beds) is not possible, because of different biotic and abiotic conditions (Flemming & Delafontaine 1994, Kröncke 1996). The loss of sedimentation areas following coastal engineering has been postulated by Flemming & Nyandwi (1994) and Reise (1998). To counteract this development, a partial opening of summer polders should be discussed with all stakeholders on the coast of Lower Saxony.

It is not necessary for a protection of processes to give more weight to certain processes than to others. Thus, we do not need to nominate a "hitlist" of "relevant" processes based on the results of ELAWAT. For the coastal zone management of the Wadden Sea, we recommend to give priority to the processes and components mentioned above to ensure an ecosystem where natural processes can proceed in an undisturbed way, as stated in the Esbjerg Declaration 1991. Altogether, the protection of processes could be much better realized if entire tidal basins fall under one protection category, instead of the present patchwork of protection zones.

12.4 Outlook

ELAWAT was expected to contribute ideas to the management of the Wadden Sea and our results could support the need for a natural development and protection of processes. But as a basic research project, ELAWAT cannot advise on human use or strategies to achieve sustainable development in this ecosystem. Most of the Wadden Sea in Germany is both National Park and biosphere reserve, posing a conflict for conservation. As a National Park, the protection of all natural processes has to be guaranteed, and as a biosphere reserve sustainable development should be achieved. While the former gives priority to natural development undisturbed by any human impacts (zero-use), the latter always implies the use of natural resources by man.

Existing regulations and the recommendations for a new National Park legislation in Schleswig-Holstein (Stock et al. 1996) try to solve this conflict with the classification of zones, where either one of the two approaches is realized. However, this compromise can only comply with the protection of processes as required by national and trilateral governmental declarations if:

- the area for protection of processes is large enough and coherent (such as tidal basins) and
- if a true sustainable use is accomplished in the remaining areas.

Even if entire tidal basins were protected, it has to be ensured that e.g. fishery or tourism in the other zones follow the guidelines of sustainable use. This is currently counteracted e.g. by increasing sizes of fishing vessels with a higher catch efficiency, a development criticised by German fisherman themselves. Thus the protection of the Wadden Sea ecosystem still requires many efforts, both on the conceptual level and in the debate with stakeholders. Future research in the Wadden Sea will face the challenge to contribute to a coastal zone management that ensures the integration of both a natural dynamic and sustainable development in this ecosystem.

Acknowledgements

Many colleagues were involved in the discussion about this chapter and we have appreciated all of their comments. We thank especially Vera Knoke for stimulating comments. The project was funded by the German Bundesministerium für Bildung, Wissenschaft, Forschung und Technologie (BMBF) under grant number 03F0112 A and B. The responsibility for the contents of the publication rests with the authors.

References

Beukema JJ (1993) Successive changes in distribution patterns as an adaptive strategy in the bivalve *Macoma balthica* (L.) in the Wadden Sea. Helgoländer Meeresunters 47: 287–304

Bibelriether H (1993a) Wälder in Naturschutzgebieten – pflegen oder wachsen lassen? Bericht über die FINNPE- Generalversammlung in Helsinki, Finnland

Bibelriether H (1993b) Klare Ziele für Europäische Kulturlandschaften. Nationalpark 78: 37–39

Bibelriether H (1997) Studie über bestehende und potentielle Nationalparke in Deutschland. Angewandte Landschaftsökologie 10: 31–35

Böhmer H-J (1997) Zur Problematik des Mosaik-Zyklus-Begriffes. Natur und Landschaft 7/8: 333–338

Broggi MF (1996) Die Bedeutung von Nationalparken für die Bewahrung des Naturerbes in Europa. Ber Int Nationalparksymp. pp 16–21

Dankers N, de Vlas J (1994) Ecological targets in the Wadden Sea. Ophelia Suppl 6: 69–77

Dittmann S, Enemark J, Farke H, Marencic H (1996) Zusammenarbeit zwischen Niedersachsen und den Niederlanden: Schutz und Erforschung des Wattenmeeres. In: Wittheit zu Bremen (Ed) Bremen und die Niederlande. Jahrbuch 1995/1996. Verlag Hauschild GmbH, Bremen. pp 255–265

Dittmann S, Marencic H, Roy M (1997) Ökosystemforschung im Niedersächsischen Wattenmeer. In: Fränzle O, Müller F, Schröder W (Eds) Handbuch der Umweltwissenschaften. Ecomed, Landsberg, V-4.1.2

Erdmann K-H, Nauber J (1995) Der deutsche Beitrag zum UNESCO-Programm "Der Mensch und die Biosphäre" (MAB) im Zeitraum Juli 1992 bis Juni 1994: mit einer englischen Zusammenfassung. Deutsches Nationalkomitee für das UNESCO-Programm "Der Mensch und die Biosphäre" (MAB), Bonn BMU

Finck P, Klein M, Riecken U & Schröder E (1998) Wege zur Förderung dynamischer Prozesse in der Landschaft. Schr-R f Landschaftspfl u Natursch. BfN, Bonn-Bad Godesberg, H 56: 413–424

Flemming BW, Delafontaine MT (1994) Biodeposition in a juvenile mussel bed of the East Frisian Wadden Sea (southern North Sea). Neth J Aquat Ecol 28: 289–297

Flemming BW, Nyandwi N (1994) Land reclamation as a cause of fine-grained sediment depletion in backbarrier tidal flats (southern North Sea). Neth J Aquat Ecol 28: 299–307

Günther C-P (1991) Settlement of *Macoma balthica* on an intertidal sandflat in the Wadden Sea. Mar Ecol Prog Ser 76: 73–79

Günther C-P (1992) Dispersal of intertidal invertebrates: a strategy to react to disturbances on different scales? Neth J Sea Res 30: 45–56

Holling CS (1986) The resilience of terrestrial ecosystems: local surprise and global change. In: Clarke WC, Munn RE (Eds). Sustainable Development of the Biosphere. Cambridge University Press, Cambridge, pp 292–317

Jax K (1994) Mosaik-Zyklus und Patch-dynamics: Synchrone oder verschiedene Konzepte? Eine Einladung zur Diskussion. Zeitschrift für Ökologie und Naturschutz 3: 107–112

Jong de F, Bakker J F, Dahl K, Dankers N, Farke H, Jäppelt W, Koßmagk-Stephan, Madsen, P B (1993) Quality Status Report of the North Sea. Subregion 10: The Wadden Sea. Common Wadden Sea Secreteriat, Wilhelmshaven

Knapp HD (1998) Freiraum für natürliche Dynamik – "Prozeßschutz" als Naturschutzziel. Schr-R f Landschaftspfl u Natursch. BfN, Bonn-Bad Godesberg, H 56: 401–412

Kröncke I (1996) Impact of biodeposition on macrofaunal communities in intertidal sandflats. Mar Ecol P.S.Z.N.I. 17: 159–174.

Lozán JL, Rachor E, Reise K, v. Westernhagen H, Lenz W (1994) Warnsignale aus dem Wattenmeer. Blackwell, Berlin

Michaelis H & Reise K (1994) Langfristige Veränderungen des Zoobenthos im Wattenmeer. In: Lozán JL, Rachor E, Reise K, von Westernhagen H, Lenz W (Eds) Warnsignale aus dem Wattenmeer. Blackwell, Berlin, pp 106–117

Nordheim von H, Andersen OL, Thissen J (1996) Red Lists of Habitats, Flora and Fauna of the Trilateral Wadden Sea Area, 1995. Helgoländer Meeresunters 50 (Suppl)

Pickett STA, White PS (1985) The Ecology of Natural Disturbance and Patch Dynamics. Academic Press, Orlando

Remmert H (1988) Gleichgewicht durch Katastrophen: Aus Forschung und Medizin 1: 7–17

Remmert H (1991) The Mosaic-Cycle Concept of Ecosystems. Springer, Berlin Heidelberg New York

Reise K (1985) Tidal Flat Ecology. An experimental approach to species interactions. Springer, Berlin Heidelberg New York

Reise K (1992) The Wadden Sea as a pristine nature reserve. Neth J Sea Res 20: 49–53

Reise K (1993a) Welchen Naturschutz braucht das Wattenmeer? Wattenmeer International 4: 6–8

Reise K (1993b) Die verschwommene Zukunft der Nordseewatten. In: Klimaänderung und Küste, Schellnhuber H-J & Sterr (Eds), Springer, Berlin Heidelberg New York, pp 223–229

Reise K (1994) Changing Life under the tides of the Wadden Sea during the 20th Century. Ophelia 6: 117–125

Reise K (1998) Coastal change in a tidal backbarrier basin of the northern Wadden Sea: are tidal flats fading away? Senckenbergiana marit 29: 121–127

Reise K, Herre E, Sturm M (1989) Historical changes in the benthos of the Wadden Sea around the island of Sylt in the North Sea. Helgoländer Meeresunters 43: 417–433.

Reise K, Kolbe K, de Jonge V (1994) Makroalgen und Seegrasbestände im Wattenmeer. In: Lozán JL, Rachor E, Reise K, von Westernhagen H, Lenz W (Eds) Warnsignale aus dem Wattenmeer. Blackwell, Berlin. pp 90–100

Reise K, Köster R, Müller A, Armonies W, Asmus H, Asmus R, Hickel W Riethmüller R (1998) Austauschprozesse im Sylt-Rømø Wattenmeer: Zusammenschau und Ausblick. In: Gätje C, Reise K (Eds) Ökosystem Wattenmeer. Austausch-, Transport- und Stoffumwandlungsprozesse. Springer, Berlin Heidelberg New York. pp 529–558

Schories D, Albrecht A, Lotze H (1997) Historical changes and inventory of macroalgae from Königshafen Bay in the northern Wadden Sea. Helgoländer Meeresunters 51: 321–341

Stock M et al (1996) Ökosystemforschung Wattenmeer – Synthesebericht: Grundlagen für einen Nationalparkplan. Schriftenreihe des Nationalparks Schleswig-Holsteinisches Wattenmeer, Heft 8

Wolff WJ (1997) Development of the conservation of Dutch coastal waters. Aquatic Conservation: Marine and Freshwater Ecosystems 7: 165–177

13 Summary

What are the processes and mechanisms relevant for stability properties in the Wadden Sea ecosystem? As part of the "Ecosystem Research Wadden Sea", the interdisciplinary project ELAWAT ("Resilience of the Wadden Sea Ecosystem") was carried out from 1993 to 1997 to increase the understanding of processes essential for regeneration and ecological relations within this ecosystem. Temporal and spatial distribution patterns of abiotic and biotic components were investigated as well as the regeneration following natural and experimentally induced disturbances. The main study area was the backbarrier system of the island of Spiekeroog in the East Frisian Wadden Sea and two tidal flat areas therein, the Gröninger Plate and Swinnplate, were studied in detail. This book covers a synthesis of the results of this project and provides a case study on the ecology of an intertidal backbarrier system.

The approach taken in ELAWAT put stability properties into the centre of the conceptual framework for this ecosystem analysis. Special attention was given to resilience, i.e. the ability to return to a reference state or reference dynamic following a temporary disturbance. In most cases the results revealed a high variability which made it difficult to specify reference states for this ecosystem. Instead, it became apparent that the state of the Wadden Sea is its dynamic.

Based on the results of the investigations and experiments, stability properties were enumerated for several abiotic and biotic components. The analysis of relevant processes for regeneration after disturbance events also lead to recommendations for the protection of processes in the Wadden Sea. In the following, some results of this research project are summarized.

- The backbarrier system of Spiekeroog covers an area of 73.5 km^2 with more than 80 % tidal flats. The water volume related to mean sea level amounts to 110 Mill m^3, whereof 70 % are exchanged at each low tide through the tidal inlet of the Otzumer Balje between the islands of Spiekeroog and Langeoog. Mesotidal conditions prevail in the backbarrier system of Spiekeroog with a tidal range of 2.7–2.9 m. The current velocities were higher on the Swinnplate than on the Gröninger Plate and while the first area was dominated by ebb currents, flood currents prevailed on the latter.
- The majority of the tidal flats in the backbarrier system of Spiekeroog were muddy sand or sandflats, mudflats were subordinate. The deficit of fine particles is seen as a consequence of coastal engineering. The two tidal flats studied in detail showed little variation in their grain size distribution, but biodeposition by blue mussels contributed to > 50 % of the mud content over the summer months on the Swinnplate. In the winter months, the mud content was as low as 10 %, as the mud fraction was resuspended or had been exported in suspension.
- The meteorological situation during the three study years was characterized by a high variability. While 1994 and 1995 were warmer than average, 1996 was

1 °C colder than the long term mean. The winter of 1995/1996 classified as a strong winter with 49 days of ice cover, which was the longest period of ice cover in 33 years.

- The Swinnplate was subject to seasonal morphological changes, characterized by sediment dislodgements over the winter months and increased biodeposition by blue mussels over the summer. After the ice winter it took 6 months until general morphological features were re-established. The relief and location of the Gröninger Plate remained unchanged over a longer time frame.
- Eight distinct types of biofacies were distinguished in the backbarrier area of Spiekeroog. In the central parts of the Swinnplate, sandy tidal flats prevailed which were colonized by the lugworm *Arenicola marina*, the tube worm *Lanice conchilega* and the blue mussel *Mytilus edulis*. It was estimated that blue mussels have occurred on the Swinnplate for the past 50 years. However, the mussel beds were subject to annual change in their location and density. Sandflats with *A. marina* and *L. conchilega* in varying densities occurred in the central parts of the Gröninger Plate.
- With the seasonal input of biodeposits, the mud- and POC-content as well as other biogeochemical properties of the sediment changed in the direction of the ebb current away from the mussel bed on the Swinnplate. The chemical composition of the sediment was closely coupled to the grain size distribution and not affected by the activity of the mussels. The input and subsequent decay of phytoplankton blooms and macroalgal cover could be followed with specific biomarkers. The quality of organic matter was decisive for further biotic activities. Organic components of high quality (photosynthetic pigments, amino acids, lipids) decreased in concentration with increasing distance from the mussel bed. Microbial biomass and activity was tightly linked to the availability of fresh organic matter and decreased further away from the mussel bed. The distribution of macrobenthic assemblages also differed along a gradient of decreasing biodeposits. In the mussel bed, small opportunistic polychaetes dominated the assemblage. Species and individual numbers as well as biomass of benthic organisms decreased from the fringe of the mussel bed to the sandflat.
- The carrying capacity of the sediment chemistry was studied on the Gröninger Plate. Monthly samples of pore water contents revealed a high variability, both along vertical depth profiles and with time in dependence of the activity of microbial remineralization. Experimental additions of organic matter caused the appearance of anoxic spots on the sediment surface which took more than one year to regenerate, depending on the amount of organic matter added.
- In the water column, high temporal and spatial variabilities were recorded for concentrations of dissolved nutrients as well as for abundances and species compositions of phyto- and zooplankton on the scale of tidal cycles and years. Not all phytoplankton species were able to reproduce in the Wadden Sea. Mainly benthic microalgal species had their habitat here. Out of 66 genera (162 species) of diatoms, 44 % lived exclusively benthic, 14 % exclusively in the water column and 42 % occurred both in the water column and sediment. The zooplankton recordings revealed an export out of the Wadden Sea over the summer months of larvae of the mud snail *Hydrobia ulvae*, spionid polychaetes and cirripedia. The timing and intensity of phytoplankton blooms affected the occurrence of meroplanktonic larvae.

- The distribution patterns of the polychaetes *Lanice conchilega*, *Heteromastus filiformis* and *Scoloplos armiger* were mainly determined by their initial settlement while those of the molluscs *Macoma balthica*, *Ensis americanus*, *Cerastoderma edule* and *Mya arenaria* followed from postlarval dispersion. However, postlarval dispersion can also be decisive for polychaetes, e.g. for the recolonization after ice winters.
- The polychaete *Lanice conchilega* requires hard substrate for settlement. No correlation was found between the larval density and number of settling larvae. Results of a model indicated that the density of a developing *Lanice*-patch is determined by the near-bottom current velocity.
- The occurrence of *L. conchilega* on the tidal flats was subject to high spatial and temporal variations. Abundances and species numbers of benthic organisms were higher within natural *Lanice*-patches and artificial tube fields than in ambient sediments without *L. conchilega*. Juvenile mussels attached to the protruding tube structures. Among the infauna, predatory polychaetes and spionid polychaetes were more abundant in the *Lanice*-patches. Small-scale associations around the *Lanice* tubes were detected for nematode- and copepod species.
- The mobility of the benthic organisms in the Wadden Sea contributes to the high variability of densities. A staining technique revealed a high turn-over. For example, all stained *Pygospio elegans* were exchanged on a test field within 6 days, although the low numbers of worms caught in drift nets and sediment traps had indicated a low capability for dispersal.
- A survey of the shore crab *Carcinus maenas* showed that crabs occurring in the intertidal are not a representative part of the entire population. The intertidal is especially important for early life stages which occur in mussel beds. Furthermore, medium-sized crabs migrate into the tidal flats, while large crabs remain in the subtidal. Shore crabs can occur in different colour forms which go along with physiological and behavioural differences (phenotypic plasticity). Green crabs were more abundant intertidally. The colour forms are discussed as life stage strategies beneficial for living in a constantly changing environment like the Wadden Sea.
- Experimentally disturbed sandflat sites were recolonized within a few days or weeks by diatoms and nematodes and within several months by smaller macrofauna. Recolonization was faster in spring than in autumn. Varying the intensity of disturbance gave faster recolonization rates the smaller the disturbance. The course of the recolonization depended mainly on the availability of larvae and postlarvae. Juvenile polychaetes were more abundant on disturbed than control sites. Among the first species to colonize the experimentally disturbed sites were the polychaete *Pygospio elegans*, ostracods and amphipods. They were also frequent after the ice winter. The results gave indications for both resilience *in-situ* and resilience by migration.
- The ice winter 1995/1996 had different effects on the organisms in the Wadden Sea. Benthic diatoms were only slightly affected. Three cold-adapted species appeared which had not been recorded in preceding years. The bloom of *Coscinodiscus concinnus* in the East Frisian Wadden Sea and nearshore North Sea is seen as the main course for the development of large anoxic sediment areas in several backbarrier systems. Abundances of nematodes and many macrobenthic species were reduced right after the ice winter, but recovered within a year. Re-

covery was slower for cold-sensitive macrobenthic species, especially suspension feeders (blue mussels, cockles and *Lanice conchilega*) and predators such as *Carcinus maenas*. Compared to preceding years the settlement of benthic organisms occurred later in the year, starting in June/July, and molluscs had strong spatfalls.

- Statistical research and applied statistics contributed to the description of patterns and processes, the analysis of relations as well as distinctions between states of a system. A variety of exploratory statistical techniques were applied and a non-parametric multivariate test procedure was developed.
- Grid-based models were applied to analyse spatio-temporal distributions of benthic organisms in respect of patch-formation and succession following disturbance events. An outlook is given on the future use of this modelling approach in intertidal soft-bottom systems. It is concluded that the mosaic-cycle theory is not applicable for the Wadden Sea in its original, full meaning.
- Stability properties and mechanisms are analysed for a variety of abiotic and biotic ingredients in the Wadden Sea. Most ingredients exhibited high variability at small spatial and temporal scales, constancy and resilience (in the case of disturbances) at intermediate scales and persistence seems possible under natural conditions over a longer term. Among the important stability mechanisms of the organisms are high mobility, functional diversity, plasticity as well as flexibility, and high potential population growth rates. The stability mechanisms of the biotic ingredients can only unfold in a natural, tide-driven abiotic framework.

In the Wadden Sea National Parks, all natural processes should continue to exist. Based on the results in ELAWAT on recolonization by benthic organisms, it is recommended to secure undisturbed source populations, unrestricted dispersal and mobility of the organisms as well as diversity of habitats to accomodate all life stages. The protection of entire tidal basins could ensure the capacity for resilience in the Wadden Sea ecosystem.

Appendix

Sub-projects of ELAWAT

The titles for the sub-projects given here are abbreviations of their long German titles to indicate the focus of their respective work.

RESEARCH CENTRE TERRAMARE E.V.

Central Logistic

Sub-project leader:
Priv. Doz. Dr. Gerd Liebezeit
Forschungszentrum Terramare
Schleusenstr. 1
26382 Wilhelmshaven

Co-worker/Collaborator:
Dr. Sabine Dittmann
Dr. Sibet Riexinger
Dr. Volker Grimm
Heike Scheele
Petra Thesing
Christa Dohn
Martin Steffens

UNIVERSITY OF OLDENBURG

Applied statistics

Sub-project leader:
Prof. Dr. Dietmar Pfeifer
Institut für mathematische Stochastik
Bundesstr. 55
20146 Hamburg

Co- worker/Collaborator:
Dr. Hans-Peter Bäumer
HRZ - Angewandte Statistik
Uhlhornsweg 19-45
26015 Oldenburg

Dr. Ulrike Schleier
Heidrun Ortleb

Dynamics of microalgae

Sub-project leader:
Prof. Dr. Ulrich Sommer
Institut für Meereskunde
Düsternbrooker Weg 20
24105 Kiel

Co-worker/Collaborator:
Verena Niesel
Margrit Kanje

Endobenthic remineralization

Sub-project leader:
Prof. Dr. Thomas Höpner
ICBM
Postfach 25 03
26111 Oldenburg

Co-worker/Collaborator:
Ingo Langner
Bärbel Oelschläger
Dr. Carlos Neira
Regine Kaiser
Marianne Kleiner

Microbial and biogechemical processes

Sub-project leader:
Prof. Dr. Wolfgang-E. Krumbein
ICBM
Postfach 25 03
26111 Oldenburg

Co-worker/Collaborator:
Prof. Dr. Jürgen Rullkötter
Prof. Dr. Hans-J. Brumsack
Priv. Doz. Dr. Gerd Liebezeit
Dr. Bert Albers
Dr. Marlies Villbrandt
Brigitte Behrends
Andreas Hild
Thomas Leu
Kim Knauth-Köhler
Peter Brocks
Doris Rohjans
Dr. Michael Böttchner
Dr. Gisela Gerdes
Dr. Thomas Klenke
Dr. Bernhard Schnetger
Mechthild Tepe

Forschungsinstitut Senckenberg

Biosedimentary system

Sub-project leader:
Prof. Dr. Burghard-W. Flemming
Forschungsinstitut Senckenberg
Schleusenstr. 39 a
26382 Wilhelmshaven

Co-worker/Collaborator:
Dr. Günther Hertweck
Dr. Ingrid Kröncke
Dr. Monique Delafontaine
Dr. Alexander Bartholomä
Dr. Michael Türkay
Dr. E. Boysen-Ennen
Frank Wolf
Kerstin Adolph

University of Bremen

Settlement and disperal of macrobenthos

Sub-project leader:
Prof. Dr. W.E. Arntz
Alfred-Wegener-Institut

Co-worker/Collaborator:
Dr. Carmen-Pia Günther
Sandra Jaklin

Columbusstr.
27568 Bremerhaven

Jens Heuers

NIEDERSÄCHSISCHES LANDESAMT FÜR ÖKOLOGIE - FSK -

Hydrodynamics and macrobenthos

Sub-project leader:
Dipl.-Ing. Hanz Dieter Niemeyer
Forschungsstelle Küste
An der Mühle 5
26548 Norderney

Co-worker/Collaborator:
Dr. Hermann Michaelis
Jürgen Steuwer
Ralf Kaiser

GKSS-FORSCHUNGSZENTRUM GEESTHACHT

Small-scale distributions of infauna

Sub-project leader:
K.-H. van Bernem
GKSS
Max-Planck-Str.
21502 Geesthacht

Co-worker/Collaborator:
Agmar Müller
Dr. Dietrich Blome
Jörn Reichert
Ruth Zühlke
Georg Ramm
Roswitha Keuker-Rüdiger
Günther Sach
Michael Grotjahn

UNIVERSITY OF MÜNSTER

Recolonization experiments

Sub-project leader:
Dr. Reinhardt Wilhelm
Westf. Wilhelms-Universität Münster
Institut für Zoophysiologie
Hindenburgplatz 55
48143 Münster

Co-worker/Collaborator:
Anita Tecklenborg

INSTITUT FÜR VOGELFORSCHUNG

Distribution patterns and consumption of migratory birds

Sub-project leader:
Dr. Klaus-Michael Exo
Institut für Vogelforschung
An der Vogelwarte 21
26386 Wilhelmshaven

Co-worker/Collaborator:
Christiane Ketzenberg

UNIVERSITY OF HAMBURG

Hydrographic modelling

Sub-project leader:
Prof. Dr. Jan O. Backhaus
Institut für Meereskunde
Troplowitzstr. 7
22529 Hamburg

Co-worker/Collaborator:
Udo Hübner
Dr. Susanne Rolinski

Subject Index

SPRINGER NATURE

GPSR Compliance

The European Union's (EU) General Product Safety Regulation (GPSR) is a set of rules that requires consumer products to be safe and our obligations to ensure this.

If you have any concerns about our products, you can contact us on ProductSafety@springernature.com

In case Publisher is established outside the EU, the EU authorized representative is:

Springer Nature Customer Service Center GmbH
Europaplatz 3
69115 Heidelberg, Germany

The manufacturer's authorised representative in the EU is Springer Nature Customer Service Centre GmbH, Europaplatz 3, 69115 Heidelberg, Germany. If you have any concerns regarding our products, please contact ProductSafety@springernature.com

Printed and bound by CPI Group (UK) Ltd, Croydon, CR0 4YY
30/09/2026
02225601-0013